CAMBRIDGE STUDIES IN ADVANCED MATHEMATICS 216

Editorial Board
J. BERTOIN, B. BOLLOBÁS, W. FULTON, B. KRA, I. MOERDIJK,
C. PRAEGER, P. SARNAK, B. SIMON, B. TOTARO

ORIENTED MATROIDS

Oriented matroids appear throughout discrete geometry, with applications in algebra, topology, physics, and data analysis. This introduction to oriented matroids is intended for graduate students, scientists wanting to apply oriented matroids, and researchers in pure mathematics. The presentation is geometrically motivated and largely self-contained, and no knowledge of matroid theory is assumed. Beginning with geometric motivation grounded in linear algebra, the first chapters prove the major cryptomorphisms and the topological representation theorem. From there the book uses basic topology to go directly from geometric intuition to rigorous discussion, avoiding the need for wider background knowledge. Topics include strong and weak maps, localizations and extensions, the Euclidean property and non-Euclidean properties, the universality theorem, convex polytopes, and triangulations. Themes that run throughout include the interplay between combinatorics, geometry, and topology and the idea of oriented matroids as analogs to vector spaces over the real numbers and how this analogy plays out topologically.

Laura Anderson is an associate professor in the Department of Mathematics and Statistics at Binghamton University. Her research focuses on interactions between combinatorics and topology, particularly those involving oriented matroids, convex polytopes, and other concepts from discrete geometry.

CAMBRIDGE STUDIES IN ADVANCED MATHEMATICS

Editorial Board
J. Bertoin, B. Bollobás, W. Fulton, B. Kra, I. Moerdijk, C. Praeger, P. Sarnak, B. Simon, B. Totaro

All the titles listed below can be obtained from good booksellers or from Cambridge University Press. For a complete series listing, visit www.cambridge.org/mathematics.

Already Published
176 P. Fleig, H. P. A. Gustafsson, A. Kleinschmidt & D. Persson *Eisenstein Series and Automorphic Representations*
177 E. Peterson *Formal Geometry and Bordism Operators*
178 A. Ogus *Lectures on Logarithmic Algebraic Geometry*
179 N. Nikolski *Hardy Spaces*
180 D.-C. Cisinski *Higher Categories and Homotopical Algebra*
181 A. Agrachev, D. Barilari & U. Boscain *A Comprehensive Introduction to Sub-Riemannian Geometry*
182 N. Nikolski *Toeplitz Matrices and Operators*
183 A. Yekutieli *Derived Categories*
184 C. Demeter *Fourier Restriction, Decoupling and Applications*
185 D. Barnes & C. Roitzheim *Foundations of Stable Homotopy Theory*
186 V. Vasyunin & A. Volberg *The Bellman Function Technique in Harmonic Analysis*
187 M. Geck & G. Malle *The Character Theory of Finite Groups of Lie Type*
188 B. Richter *Category Theory for Homotopy Theory*
189 R. Willett & G. Yu *Higher Index Theory*
190 A. Bobrowski *Generators of Markov Chains*
191 D. Cao, S. Peng & S. Yan *Singularly Perturbed Methods for Nonlinear Elliptic Problems*
192 E. Kowalski *An Introduction to Probabilistic Number Theory*
193 V. Gorin *Lectures on Random Lozenge Tilings*
194 E. Riehl & D. Verity *Elements of ∞-Category Theory*
195 H. Krause *Homological Theory of Representations*
196 F. Durand & D. Perrin *Dimension Groups and Dynamical Systems*
197 A. Sheffer *Polynomial Methods and Incidence Theory*
198 T. Dobson, A. Malnič & D. Marušič *Symmetry in Graphs*
199 K. S. Kedlaya *p-adic Differential Equations*
200 R. L. Frank, A. Laptev & T. Weidl *Schrödinger Operators: Eigenvalues and Lieb–Thirring Inequalities*
201 J. van Neerven *Functional Analysis*
202 A. Schmeding *An Introduction to Infinite-Dimensional Differential Geometry*
203 F. Cabello Sánchez & J. M. F. Castillo *Homological Methods in Banach Space Theory*
204 G. P. Paternain, M. Salo & G. Uhlmann *Geometric Inverse Problems*
205 V. Platonov, A. Rapinchuk & I. Rapinchuk *Algebraic Groups and Number Theory, I (2nd Edition)*
206 D. Huybrechts *The Geometry of Cubic Hypersurfaces*
207 F. Maggi *Optimal Mass Transport on Euclidean Spaces*
208 R. P. Stanley *Enumerative Combinatorics, II (2nd Edition)*
209 M. Kawakita *Complex Algebraic Threefolds*
210 D. Anderson & W. Fulton *Equivariant Cohomology in Algebraic Geometry*
211 G. Pineda Villavicencio *Polytopes and Graphs*
212 R. Pemantle, M. C. Wilson & S. Melczer *Analytic Combinatorics in Several Variables (2nd Edition)*
213 A. Yadin *Harmonic Functions and Random Walks on Groups*
214 Y. Kawamata *Algebraic Varieties: Minimal Models and Finite Generation*
215 J. Gillespie *Abelian Model Category Theory*

Oriented Matroids

LAURA ANDERSON
Binghamton University

CAMBRIDGE
UNIVERSITY PRESS

Shaftesbury Road, Cambridge CB2 8EA, United Kingdom

One Liberty Plaza, 20th Floor, New York, NY 10006, USA

477 Williamstown Road, Port Melbourne, VIC 3207, Australia

314–321, 3rd Floor, Plot 3, Splendor Forum, Jasola District Centre, New Delhi – 110025, India

103 Penang Road, #05–06/07, Visioncrest Commercial, Singapore 238467

Cambridge University Press is part of Cambridge University Press & Assessment,
a department of the University of Cambridge.

We share the University's mission to contribute to society through the pursuit of
education, learning and research at the highest international levels of excellence.

www.cambridge.org
Information on this title: www.cambridge.org/9781009494113
DOI: 10.1017/9781009494076

© Laura Anderson 2025

This publication is in copyright. Subject to statutory exception and to the provisions
of relevant collective licensing agreements, no reproduction of any part may take place
without the written permission of Cambridge University Press & Assessment.

When citing this work, please include a reference to the DOI 10.1017/9781009494076

First published 2025

A catalogue record for this publication is available from the British Library

Library of Congress Cataloging-in-Publication Data
Names: Anderson, Laura (Associate Professor of Mathematics), author.
Title: Oriented matroids / Laura Anderson, State University of New York, Binghamton.
Description: Cambridge, United Kingdom ; New York, NY: Cambridge University
Press, 2025. | Series: Cambridge studies in advanced mathematics ; 216 |
Includes bibliographical references and index.
Identifiers: LCCN 2024045997 (print) | LCCN 2024045998 (ebook) |
ISBN 9781009494113 (hardback) | ISBN 9781009494076 (epub)
Subjects: LCSH: Oriented matroids.
Classification: LCC QA166.6 .A53 2025 (print) | LCC QA166.6 (ebook) |
DDC 511/.6–dc23/eng/20241203
LC record available at https://lccn.loc.gov/2024045997
LC ebook record available at https://lccn.loc.gov/2024045998

ISBN 9-781-009-49411-3 Hardback

Cambridge University Press & Assessment has no responsibility for the persistence
or accuracy of URLs for external or third-party internet websites referred to in this
publication and does not guarantee that any content on such websites is, or will
remain, accurate or appropriate.

For EU product safety concerns, contact us at Calle de José Abascal, 56, 1°, 28003 Madrid,
Spain, or email eugpsr@cambridge.org

To Tom Zaslavsky, whose dedication to good teaching and good writing never ceases to inspire.

Contents

Preface		*page* xi
1	**Realizable Oriented Matroids**	1
1.1	Some Notation	3
1.2	Discrete Models for Matrices	4
1.3	Arrangements	8
1.4	Subspaces	16
1.5	Cryptomorphisms for Realizable Oriented Matroids	24
1.6	Convex Polytopes	34
1.7	Deletion and Contraction	37
1.8	A Few Words about Unoriented Matroids	39
Exercises		40
2	**Oriented Matroids**	43
2.1	Scalar Multiplication and Hypersums of Sign Vectors	43
2.2	Vectors and Signed Circuits	44
2.3	Duality	52
2.4	Chirotopes	60
2.5	Summary on Cryptomorphism	77
2.6	Other Axiomatizations	78
2.7	Realizable vs. Nonrealizable Oriented Matroids	82
Exercises		90
3	**Elementary Operations and Properties**	92
3.1	Isomorphism	92
3.2	Reorientation and Cyclic/Acyclic Oriented Matroids	93
3.3	Parallel, Simple, Uniform	94
3.4	Direct Sum	95
3.5	Extensions	96

3.6	Poset Properties of $\mathcal{V}^*(\mathcal{M})$	97
3.7	Flats	99
Exercise		102

4 The Topological Representation Theorem   103
4.1	Shellings and Recursive Coatom Orderings	104
4.2	Convexity	113
4.3	Intervals in the Face Lattice	114
4.4	Tope Posets	116
4.5	A Recursive Coatom Ordering on the Face Poset of a Covector	118
4.6	Recursive Coatom Orderings on the Face Lattice and on Convex Sets	122
4.7	Covector Sets Give Pseudosphere Arrangements	124
4.8	Topological Interpretations of Oriented Matroid Concepts	126
4.9	Some Homotopy Results, and an Application	130
4.10	Historical Note	132
4.11	Final Notes	133
Exercises		134

5 Strong Maps and Weak Maps   136
5.1	Strong Maps	136
5.2	Weak Maps	142
5.3	Topological Representations of Maps	148
5.4	Some Additional Results on Strong Maps	153
Exercises		156

6 Single-Element Extensions   158
6.1	Preliminaries	158
6.2	Localizations	160
6.3	Proof of the Las Vergnas Characterization	163
6.4	Lexicographic Extensions	175
6.5	Adjoints	178
6.6	Intersection Properties	185
Exercises		211

7 The Universality Theorem   214
7.1	Realization Spaces	214
7.2	Some Elementary Results	218
7.3	An Oriented Matroid with Disconnected Realization Space	221
7.4	Semialgebraic Sets	225
7.5	Proof of the Universality Theorem	234
7.6	History	237

7.7	Universality and the Plücker Embedding	238
7.8	Applications	239
Exercises		241

8	**Oriented Matroid Polytopes**	**242**
8.1	Convex Polytopes	243
8.2	Face Lattices and Oriented Matroid Polytopes	245
8.3	Polarity and Adjoints	247
8.4	The Lawrence Construction	250
Exercises		257

9	**Subdivisions and Triangulations**	**258**
9.1	Some PL Topology	260
9.2	Geometric Fans	263
9.3	Oriented Matroid Fans	266
9.4	Stellar Subdivision in an Oriented Matroid Fan	271
9.5	Lifting Subdivisions	273
9.6	Superposition	277
9.7	Triangulations of Euclidean Oriented Matroids	286
9.8	Final Remarks	293
Exercises		294

10	**Spaces of Oriented Matroids**	**296**
10.1	Comparing Oriented Matroids and Subspaces of $\mathbb{R}^n$	296
10.2	Extension Spaces	298
10.3	$G(1, \mathcal{M})$ and $G(r-1, \mathcal{M})$	299
10.4	McCord's Theorem and Triangulations	301
10.5	Positive Results on Combinatorial Grassmannians	304

11	**Hints on Selected Exercises**	**308**
	References	310
	Index	317

Preface

One of the goofier potential titles for this book was *Oriented Matroids: A Matroid-Free Approach*. Despite their name, the origins of oriented matroids lie as much in the study of convexity as in matroid theory, and, thanks primarily to Lawrence's topological representation theorem, their study has long been topological as much as combinatorial. At the same time, important elements of oriented matroid theory were built on matroid theory and on other relatively specialized mathematical areas. The mathematician with geometric motivations to learn oriented matroids may struggle with the diverse background assumed in various presentations of the subject.

This book is an introduction to oriented matroids from a unified geometric perspective, grounded in linear algebra over the real numbers. I have endeavored to keep the presentation self-contained, geometrically motivated, and accessible to a broad mathematical audience. In service to this goal, some classic results get proofs quite different from their original ones – notably, Las Vergnas's characterization of single-element extensions and Fukuda's proof of duality of the Euclidean property.

Some of the most tantalizing topological questions in oriented matroid theory arise from viewing oriented matroids as analogous to subspaces of $\mathbb{R}^n$. This perspective is touched on throughout, and Chapter 10 surveys many of these questions and a few answers. This chapter and a few sections elsewhere require more than basic knowledge of topology, but the less topologically inclined reader can safely skip them.

With a few exceptions, problems appearing within a chapter are invitations to pause and think a small issue through, while exercises at the end of a chapter are more substantial.

In developing oriented matroids from a consistent perspective of geometry/topology, this book neglects many other valuable takes on the subject. The classic "Red Book" (Björner et al. 1999) is an invaluable reference surveying a much wider range of topics. Other books with different approaches include (Bachem and Kern 1992), which approaches oriented matroids from the linear programming perspective, and (Bokowski 2006), which focuses on computational aspects.

Many thanks to Michael Dobbins for some charming problems, and to Chris Eppolito for astute observations. Thanks to Nikolai Mnëv and Arnaldo Mandel for historical background. This book would be riddled with (even more?) errors were it not for many eagle-eyed students, among them Richard Behr, Amanda Ruiz, Nick Lacasse, Ting Su, Will Jones, Tara Koskulitz, Stefan Viola, Thomas Galvin, Han Lim Jang, Alireza Salahshoori, and Levi Axelrod.

1

Realizable Oriented Matroids

To adapt a phrase from Whitney (1935), the study of oriented matroids is the study of "the abstract properties of linear dependence over ordered fields." (For our purposes, the field in question may be assumed to be $\mathbb{R}$.) Briefly put, from a finite subset S of $\mathbb{R}^n$, one can extract a certain combinatorial structure, and from that structure one can reconstruct various dependency relationships in S. From properties of $\mathbb{R}^n$ we find properties satisfied by these structures, and we call any structure with these properties – whether it arises from a subset of $\mathbb{R}^n$ or not – an oriented matroid. Thus, oriented matroids are "combinatorial abstractions of finite vector arrangements in $\mathbb{R}^n$."

As we will see in this chapter, oriented matroids could as well be described as combinatorial abstractions of linear subspaces of $\mathbb{R}^n$, or of hyperplane arrangements in $\mathbb{R}^n$. Each of these interpretations has led to beautiful interplay between combinatorics, geometry, and topology. There are various other well-known mathematical objects that can be abstracted to oriented matroids (most notably, directed graphs), but we'll focus on direct connections from oriented matroids to linear algebra.

The first steps in learning about oriented matroids can be annoying, because every really honest introduction puts off the definition of oriented matroid until at least the second chapter. To justify this annoyance, here's a short preview. In Chapter 2 we will introduce several different axiom systems, writing preliminary definitions of the form:

Definition 1.1 An object $\mathcal{A}$ is a **Type 1 expression of an oriented matroid** if it satisfies the following axioms...

Definition 1.2 An object $\mathcal{B}$ is a **Type 2 expression of an oriented matroid** if it satisfies the following axioms...

Et cetera. We will then show that these different types of expressions are *cryptomorphic*, that is, there are nice bijections between the expressions of different types, so that each expression of one type determines a unique expression of each other type. Thus, each type encodes the same data in a different form. We will then define an oriented matroid to be this data, however expressed.

Why not stick with one type of expression? One answer is pragmatic – different types of expressions are easier to work with in different settings. A more interesting answer is that the different expressions reflect different aspects of the relationship between oriented matroids and linear algebra. Radon partitions, Grassmann–Plücker relations, orthogonality of vector spaces, and the combinatorics of hyperplane arrangements are some of the geometric notions that will have elegant combinatorial analogs in one or the other of the oriented matroid axiom systems. This wealth of different routes from $\mathbb{R}^n$ to oriented matroids is one indication that oriented matroids are the "right" combinatorial model for this kind of geometry. (Two other strong indications are the Topological Representation Theorem (Chapter 4) and the results in Section 10.5 on the *MacPhersonian*).

In this chapter we'll introduce our "Type X expressions" for oriented matroids via their concrete manifestations in $\mathbb{R}^n$. We'll look at different ways to extract combinatorial structures from a finite subset S of $\mathbb{R}^n$, and we'll see that these structures encode geometric data interesting in several different contexts. Finally, we'll see that these different structures each encode the same data about S. We call that data, however encoded, the *oriented matroid corresponding to S*.

The combinatorial structures arising from finite subsets S of $\mathbb{R}^n$ are called **realizable oriented matroids**. (S is then called a **realization** of that oriented matroid.) They're the examples that motivate the theory, but they're not the whole picture. In general, oriented matroids are defined in purely combinatorial terms, with no reference to $\mathbb{R}^n$, and not all oriented matroids are realizable. The relationship between oriented matroids and their realizations is a fraught topic, as Chapter 7 will show.

Despite the existence of nonrealizable oriented matroids, and the scandalously non-realizable behavior in which oriented matroids occasionally indulge, for the most part oriented matroids model real vector sets admirably. In working with the abstract combinatorics of general oriented matroids, it's almost always a good idea to be guided by one's intuition from $\mathbb{R}^n$. The point of this introductory chapter is to develop that intuition. The concepts and proofs of this chapter are mostly rather simple. In Chapter 2 they'll all be combinatorialized into more abstract notions, which will be less daunting if you keep the geometric inspirations in mind.

1.1 Some Notation

- Mat(r,n) will denote the set of all $r \times n$ real matrices of rank r.
- row(M) denotes the row space of the matrix M, and null(M) denotes the nullspace.
- A matrix will be viewed as a list of column vectors. Thus an element of Mat(r,n) will often be written as $(\mathbf{v}_1, \ldots, \mathbf{v}_n)$.
- If $M = (\mathbf{v}_1, \ldots, \mathbf{v}_n) \in \text{Mat}(r,n)$ then $M_{i_1,\ldots,i_r}$ denotes the square matrix $(\mathbf{v}_{i_1}, \ldots, \mathbf{v}_{i_r})$.
- For sets P and E, P^E denotes the set of all functions $E \to P$. An element of P^E will sometimes be written as $(p_e : e \in E)$.
- For a positive integer n, $[n]$ denotes $\{1, 2, \ldots, n\}$. For a set P and natural number n, $P^{[n]}$ will be abbreviated P^n.
- Depending on what's convenient, we will write elements of P^n as functions or as n-component vectors with entries in P. In particular, we will refer to the support of $X \in \{0, +, -\}^n$, which will mean the support as a function (that is, $\{i \in [n] : X(i) \neq 0\}$).
- For a sign vector X, we will sometimes denote $X^{-1}(+)$ by X^+, $X^{-1}(-)$ by X^-, and $X^{-1}(0)$ by X^0. If $A = X^+$, $B = X^-$, and $C = X^0$, we will sometimes denote X by A^+B^- or $A^+B^-C^0$. If $A = \emptyset$, then we may denote X by B^-, and if $B = \emptyset$, then we may denote X by A^+. (Aside: Sign vectors written in the A^+B^- notation are elsewhere sometimes called *signed sets*.)
- When the support of a sign vector has just a few elements, we may write it as a string of symbols $e^{X(e)}$, with e in the support. For instance, the sign vector $\{a,c\}^+\{b\}^-$ may be denoted $a^+b^-c^+$.
- For $\mathbf{x} = (x_1, \ldots, x_n) \in \mathbb{R}^n$, sign$(\mathbf{x})$ denotes the sign vector

$$(\text{sign}(x_1), \ldots, \text{sign}(x_n)) \in \{0, +, -\}^n.$$

- **0** denotes a vector all of whose components are 0. Context will tell us the number of components and whether the vector is a row or column vector.
- $\mathcal{P}(S)$ denotes the power set of S.

The set $\{0, +, -\}$ will be partially ordered by $+ > 0, - > 0, + \not> -, - \not> +$. The set $\{0, +, -\}^E$ will be ordered componentwise: If $X, Y \in \{0, +, -\}^E$, then $X \geq Y$ if and only if $X(e) \geq Y(e)$ for every $e \in E$.

For $X \in \{0, +, -\}^E$, the **orthant** in $\mathbb{R}^E$ corresponding to X is $\{\mathbf{x} \in \mathbb{R}^E : \text{sign}(\mathbf{x}) = X\}$, and the **closed orthant** corresponding to X is $\{\mathbf{x} \in \mathbb{R}^E : \text{sign}(\mathbf{x}) \leq X\}$. Thus our partial order on sign vectors corresponds to the partial order on closed orthants by inclusion.

1.2 Discrete Models for Matrices

It will be convenient to consider an ordered list of vectors spanning $\mathbb{R}^r$ as the set of columns of a matrix $M \in \text{Mat}(r,n)$. Such a list is called a *vector arrangement*.

We begin by considering some "discrete models for matrices." That is, for every $r, n \in \mathbb{N}$, we will consider some finite set $\mathcal{O}$ and function

$$\text{Mat}(r,n) \to \mathcal{O}$$

that seem somehow natural from the point of view of linear algebra.

The five models we'll look at are:

1. $\mathcal{V}: \text{Mat}(r,n) \to \mathcal{P}(\{0, +, -\}^n)$ defined by

$$\mathcal{V}(M) = \{\text{sign}(\mathbf{x}) : \mathbf{x} \in \text{null}(M)\}.$$

   $\mathcal{V}(M)$ is called the **set of vectors**[1] corresponding to M. It has a partial order as a subposet of $\{0, +, -\}^n$.

2. $\mathcal{V}^*: \text{Mat}(r,n) \to \mathcal{P}(\{0, +, -\}^n)$ defined by

$$\mathcal{V}^*(M) = \{\text{sign}(\mathbf{x}) : \mathbf{x} \in \text{row}(M)\} = \{\text{sign}(\mathbf{y}M) : \mathbf{y} \in \mathbb{R}^r\}.$$

   $\mathcal{V}^*(M)$ is called the **set of covectors** corresponding to M. Again, this is partially ordered as a subposet of $\{0, +, -\}^n$.

3. $\mathcal{C}: \text{Mat}(r,n) \to \mathcal{P}(\{0, +, -\}^n)$ takes each M to the set of minimal elements of $\mathcal{V}(M) \setminus \{\mathbf{0}\}$. Here, as always, minimality is with respect to the partial order on sign vectors described in Section 1.1. $\mathcal{C}(M)$ is called the **set of signed circuits** corresponding to M.

4. $\mathcal{C}^*: \text{Mat}(r,n) \to \mathcal{P}(\{0, +, -\}^n)$ takes each M to the set of minimal elements of $\mathcal{V}^*(M) \setminus \{\mathbf{0}\}$. $\mathcal{C}^*(M)$ is called the **set of signed cocircuits** corresponding to M.

5. $\chi: \text{Mat}(r,n) \to \{0, +, -\}^{[n]^r}$ defined by: If $M \in \text{Mat}(r,n)$, then $\chi(M): [n]^r \to \{0, +, -\}$ is the function taking each $(i_1, i_2, \ldots, i_r)$ to the sign of the determinant $|M_{i_1, \ldots, i_r}|$. $\chi(M)$ is called the **chirotope** corresponding to M.

Problem 1.3 Let

$$M = \begin{pmatrix} 1 & 0 & 2 & 0 & 0 & 0 \\ 0 & 1 & 0 & -1 & 0 & 0 \\ 0 & 0 & 0 & 0 & 1 & 0 \end{pmatrix}.$$

Determine $\mathcal{V}(M), \mathcal{V}^*(M), \mathcal{C}(M), \mathcal{C}^*(M)$, and $\chi(M)$.

---

[1] The terminology is terrible but firmly established.

1.2 Discrete Models for Matrices

Problem 1.4 For each of the sets $\mathcal{C}(N)$, $\mathcal{C}^*(N)$, and $\chi(N)$ associated to a matrix $N = (\mathbf{v}_1, \mathbf{v}_2, \ldots, \mathbf{v}_n)$, describe how to determine each of the following:

1. whether $\mathbf{v}_i = 0$,
2. whether $\mathbf{v}_i$ is a positive multiple of $\mathbf{v}_j$,
3. whether $\mathbf{v}_i$ is a negative multiple of $\mathbf{v}_j$,
4. whether a set of columns $\{\mathbf{v}_i : i \in I\}$ is independent, and
5. whether $\mathbf{v}_i$ is in the span of the remaining columns.

1.2.1 Invariance under Change of Coordinates

Problem 1.5 1. Prove that the set of vectors, set of covectors, set of signed circuits, and set of signed cocircuits of a vector arrangement in $\mathbb{R}^r$ are invariant under change of coordinates. That is, prove for each $A \in GL_r$ that $\mathcal{V}(AM) = \mathcal{V}(M)$, et cetera.
2. Prove that χ is invariant under orientation-preserving change of coordinates (that is, under left multiplication by $A \in GL_r^+$).

This shows one reason these models might be more interesting models for vector arrangements than, say, $\{\text{sign}(\mathbf{x}) : \vec{x} \text{ a column of } M\}$.

Problem 1.6 Prove that the set of vectors, set of covectors, set of signed circuits, set of signed cocircuits, and chirotope of a vector arrangement in $\mathbb{R}^r$ are invariant under scaling of columns by positive scalars. That is, prove for each diagonal $n \times n$ matrix D with all diagonal entries positive that $\mathcal{V}(MD) = \mathcal{V}(M)$, et cetera.

So far we have only considered vector arrangements that are expressed as the columns of $r \times n$ matrices of rank r. Thus we have assumed that our vector space is $\mathbb{R}^r$ and that our arrangement spans the space. But Exercise 1.5 points out one way to associate vector sets, covector sets, and so on, to a finite arrangement $(\mathbf{v}_i : i \in S)$ in any real vector space. One can simply fix a vector space isomorphism from the span $\langle \mathbf{v}_i : i \in S \rangle$ to $\mathbb{R}^r$ and then define $\mathcal{V}$, $\mathcal{V}^*$, and so on in terms of the corresponding arrangement in $\mathbb{R}^r$. This is actually the cheesy way to do it: A better way is to keep reading this chapter and see coordinate-free descriptions of each of our combinatorial structures. Either way it makes sense to talk about the oriented matroid of a finite arrangement of vectors in an arbitrary vector space over $\mathbb{R}$.

We will see in Section 1.5 that $\mathcal{V}(M)$, $\mathcal{V}^*(M)$, $\mathcal{C}(M)$, and $\mathcal{C}^*(M)$ encode the same data about M. ($\chi(M)$ encodes a bit more.) We will call this data, however encoded, the *oriented matroid* corresponding to M. The oriented matroids arising in this way are called *realizable*. The definition of general oriented matroids will come in Chapter 2.

1.2.2 Support-Minimality and Reduced Row-Echelon Form

Definition 1.7 For every $X, Y \in \{0, +, -\}^E$, their **separation set** is

$$S(X, Y) = \{e : \{X(e), Y(e)\} = \{+, -\}\}.$$

Remark 1.8 Observe that $X \geq Y$ if and only if $\text{supp}(X) \supseteq \text{supp}(Y)$ and $S(X, Y) = \emptyset$.

Proposition 1.9 *Let V be a linear subspace of $\mathbb{R}^n$, and let $\mathcal{F} = \{\text{sign}(\mathbf{x}) : \mathbf{x} \in V \setminus \{0\}\}$. Then $\min(\mathcal{F})$ is exactly the set of elements of $\mathcal{F}$ of minimal support.*

By applying Proposition 1.9 to the null space and to the row space of a matrix, we get the following.

Corollary 1.10 *$\mathcal{C}(M)$ consists exactly of the elements of $\mathcal{V}(M) \setminus \{0\}$ of minimal support, and $\mathcal{C}^*(M)$ consists exactly of the elements of $\mathcal{V}^*(M) \setminus \{0\}$ of minimal support.*

Proof of Proposition 1.9: Let $X \in \mathcal{F}$. Clearly if $\text{supp}(X)$ is minimal then $X \in \min(\mathcal{F})$.

If $\text{supp}(X)$ is not minimal then let $Y \in \mathcal{F}$ such that $\text{supp}(Y) \subset \text{supp}(X)$. We show that X is not minimal in $\mathcal{F}$ by induction on $|S(X, Y)|$. If $S(X, Y) = \emptyset$ then X is not minimal because $Y < X$. Otherwise, we will find a $Y' \in \mathcal{F}$ such that $\text{supp}(Y') \subset \text{supp}(X)$ and $S(X, Y') \subset S(X, Y)$.

Let $\mathbf{x}, \mathbf{y} \in V$ such that $X = \text{sign}(\mathbf{x})$ and $Y = \text{sign}(\mathbf{y})$, and consider the set $C = \{a\mathbf{x} + b\mathbf{y} : a, b > 0\}$ of all positive linear combinations of $\{\mathbf{x}, \mathbf{y}\}$. Notice that $C \subset V$, that $\text{supp}(\mathbf{v}) \subseteq \text{supp}(X)$ for each $\mathbf{v} \in C$, and that $S(X, \text{sign}(\mathbf{v})) \subseteq S(X, Y)$ for each $\mathbf{v} \in C$. Let $e \in S(X, Y)$. Then $\text{sign}(x_e) = -\text{sign}(y_e)$, and so $|y_e|x_e + |x_e|y_e = 0$. Thus we see an element $\mathbf{z} = |y_e|\mathbf{x} + |x_e|\mathbf{y}$ of C with $z_e = 0$. Let $Y' = \text{sign}(\mathbf{z})$. Then $\text{supp}(Y') \subseteq \text{supp}(X) \setminus \{e\}$ and $S(X, Y') \subseteq S(X, Y) \setminus \{e\}$. □

We can see the idea of the preceding proof by way of an example we can draw: This is also an opportunity to introduce a type of picture we'll be revisiting often. Consider a two-dimensional subspace V of $\mathbb{R}^4$. While we can't draw $\mathbb{R}^4$, we can draw V by itself, and in our drawing we can include the intersection of V with each coordinate hyperplane. Unless V is very special, each such intersection will be a line $L_i = \{\mathbf{v} \in V : v_i = 0\}$, and the half-space on one side of this line will be $\{\mathbf{v} \in V : v_i > 0\}$. In our picture we'll label L_i by the number i, and we'll draw a small arrow starting at L_i and pointing into the half-space $\{\mathbf{v} \in V : v_i > 0\}$. Figure 1.1 depicts such a V. This drawing is enough to see the poset $\mathcal{F}$ arising from V. For each W in the associated set

1.2 Discrete Models for Matrices

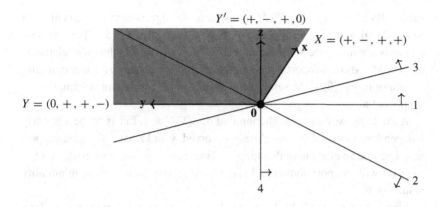

Figure 1.1 Illustration for the proof of Proposition 1.9.

$\mathcal{F}$, $\{\mathbf{v} \in V : \text{sign}(\mathbf{v}) = W\}$ is a cone: In Figure 1.1 some of these cones are labeled with the corresponding elements of $\mathcal{F}$. The figure also shows the cone C and the vector $\mathbf{z}$ from the proof of Proposition 1.9, for a particular choice of X, Y, $\mathbf{x}$, and $\mathbf{y}$, and with $e = 4 \in S(X, Y)$.

Lemma 1.11 *Let $M = (\mathbf{v}_1, \ldots, \mathbf{v}_n) \in \text{Mat}(r, n)$ and $S \subseteq [n]$. The space $V_S := \{\mathbf{x} \in \text{row}(M) : \text{supp}(\mathbf{x}) \cap S = \emptyset\}$ has dimension $r - \text{rank}\{\mathbf{v}_i : i \in S\}$.*

Proof: Since the rows of M are linearly independent, the map from $\mathbb{R}^r$ to row(M) taking each $\mathbf{w}$ to $\mathbf{w}M = (\mathbf{w}\mathbf{v}_1, \ldots, \mathbf{w}\mathbf{v}_n)$ has kernel $\{\mathbf{0}\}$. The preimage of V_S is $\{\mathbf{v}_i : i \in S\}^\perp$, and so the dimension of V_S is $r - \text{rank}\{\mathbf{v}_i : i \in S\}$. □

Remark 1.12 Lemma 1.11 is our first example of a useful, quirky aspect of oriented matroid theory: using information about the columns of M to derive information about the row space, or vice versa. We'll see the vice versa aspect in Section 1.3.3, when we give an interpretation of covectors for vector arrangements.

Definition 1.13 Let $M = (\mathbf{v}_1, \ldots, \mathbf{v}_n) \in \text{Mat}(r, n)$ and $B = \{i_1 < \cdots < i_r\} \subseteq [n]$. If $M_{i_1, \ldots, i_r}$ is rank r, then the **reduced row-echelon form of M with respect to B** is $(M_{i_1, \ldots, i_r})^{-1} M$.

This is familiar from linear algebra as the unique matrix N obtainable from M by elementary row operations such that $N_{i_1, \ldots, i_r} = I$.

Proposition 1.14 *Let $M = (\mathbf{v}_1, \ldots, \mathbf{v}_n) \in \text{Mat}(r, n)$. $\mathcal{C}^*(M)$ consists of all $\pm\text{sign}(\mathbf{x})$ such that $\mathbf{x}$ is a row in some reduced row-echelon form for M.*

Equivalently, $\mathcal{C}^*(M)$ consists of all $X \in \mathcal{V}^*(M)$ such that $\{\mathbf{v}_i : i \in X^0\}$ has rank $r - 1$.

Proof: By Corollary 1.10 $\mathcal{C}^*(M)$ consists of the sign vectors of elements **x** of row$(M)\backslash\{\mathbf{0}\}$ of minimal support. So consider such an **x**. Let $S = [n]\backslash\text{supp}(\mathbf{x})$, and consider the space V_S of Lemma 1.11. Notice that V_S is the set of elements of row(M) whose support is contained in supp(x), and so by our minimality assumption, $V_S\backslash\{\mathbf{0}\}$ is the set of elements of row(M) with support equal to the support of **x**.

Assume by way of contradiction that dim$(V_S) > 1$. Let $\{\mathbf{y}, \mathbf{z}\}$ be a linearly independent subset of V_S. As already observed, **y** and **z** have the same support as **x**. Let j be an element of this support. Then $z_j\mathbf{y} - y_j\mathbf{z}$ is a nonzero element of row(M) with support contained in supp$(\mathbf{x})\backslash\{j\}$, contradicting our minimality assumption.

Since dim$(V_S) = 1$, by Lemma 1.11, $\{\mathbf{v}_i : i \in S\}$ has rank $r - 1$. Take a maximal independent subset $\{\mathbf{v}_{i_1}, \ldots, \mathbf{v}_{i_{r-1}}\}$ of this set, and choose $a \in [n]$ so that $B = \{\mathbf{v}_{i_1}, \ldots, \mathbf{v}_{i_{r-1}}, \mathbf{v}_a\}$ is a basis of $\mathbb{R}^r$. Let M' be the reduced row-echelon form of M with respect to B. Since $x_{i_1} = \cdots = x_{i_{r-1}} = 0$ and **x** is a linear combination of the rows of M', we see that **x** is a scalar multiple of the row corresponding to the column indexed by a. Thus sign(**x**) is indeed a multiple of the sign of this row. □

1.3 Arrangements

In this section we'll look at two types of geometric objects:

- *vector arrangements* – finite ordered lists of vectors in $\mathbb{R}^r$. These vectors will always be viewed as columns of a matrix.
- *signed hyperplane arrangements* – finite ordered lists of signed hyperplanes in $\mathbb{R}^r$.

Actually an arrangement is not necessarily a list (a collection indexed by $[n]$) – see the definition of arrangement in Section 1.3.1.

We'll note how the discrete models of Section 1.2 describe fundamental notions associated to each of these objects.

1.3.1 Geometry Glossary

1. For us, a **hyperplane** is always a linear hyperplane in a vector space V over $\mathbb{R}^r$. We allow the **degenerate hyperplane** consisting of V itself. (In terms of an inner product on V, a hyperplane is the normal $\{\mathbf{x} : \mathbf{x} \cdot \mathbf{v} = 0\}$ to some vector **v**, and the degenerate hyperplane is the normal to **0**.)

1.3 Arrangements

2. A **signed hyperplane** in V is a triple $\mathcal{H} = (H^0, H^+, H^-)$, where H^0 is a hyperplane in V and H^+ and H^- are the two open half-spaces bounded by H^0. By convention, we call $(V, \emptyset, \emptyset)$ the **degenerate signed hyperplane**. H^+ is the **positive open half-space** of $\mathcal{H}$, and $H^+ \cup H^0$ is the **positive closed half-space** of $\mathcal{H}$. Likewise, H^- and $H^- \cup H^0$ are the **negative open half-space** and **negative closed half-space** of $\mathcal{H}$.
3. For a vector $\mathbf{v}$ in $\mathbb{R}^r$, we'll use $\mathbf{v}^\perp$ to denote the signed hyperplane with

$$(\mathbf{v}^\perp)^0 = \{\mathbf{x} \in \mathbb{R}^r : \mathbf{x} \cdot \mathbf{v} = 0\},$$

$$(\mathbf{v}^\perp)^+ = \{\mathbf{x} \in \mathbb{R}^r : \mathbf{x} \cdot \mathbf{v} > 0\},$$

$$(\mathbf{v}^\perp)^- = \{\mathbf{x} \in \mathbb{R}^r : \mathbf{x} \cdot \mathbf{v} < 0\}.$$

The degenerate signed hyperplane is $\mathbf{0}^\perp$.

4. For a signed hyperplane $\mathcal{H} = (H^0, H^+, H^-)$ in V and a subset W of V, let $\mathcal{H} \cap W$ denote the triple $(H^0 \cap W, H^+ \cap W, H^- \cap W)$. This will most commonly arise with W a linear subspace of V, in which case $\mathcal{H} \cap W$ is a signed hyperplane in W.
5. An **affine hyperplane in** V is a set $\mathbf{w} + W$, where W is a hyperplane in V and $\mathbf{w} \neq \mathbf{0}$. An **affine space in** $\mathbb{R}^r$ is an intersection of affine hyperplanes. We will denote a d-dimensional affine space by $\mathbb{A}^d$, or simply $\mathbb{A}$. When the particular ambient space V is not important, we will call elements of $\mathbb{A}$ **points**.
6. The **affine span** of $S \subseteq \mathbb{A}$ is the intersection of $\mathbb{A}$ with the linear span of S. The **relative interior** of $S \subseteq \mathbb{A}$ is the topological interior of S as a subset of its span.
7. An **affine subspace** of an affine space $\mathbb{A}^r$ is a nonempty intersection of a linear subspace with $\mathbb{A}^r$. A **signed affine hyperplane in** $\mathbb{A}^r$ is a triple $\mathcal{H} = (H^0 \cap \mathbb{A}^r, H^+ \cap \mathbb{A}^r, H^- \cap \mathbb{A}^r)$, where H^0 is a hyperplane, $H^0 \cap \mathbb{A}^r \neq \emptyset$, and H^+ and H^- are the two open half-spaces bounded by H^0.
8. S^d denotes the unit sphere in $\mathbb{R}^{d+1}$.
9. Let $S = \{\mathbf{v}_1, \ldots, \mathbf{v}_k\} \subset V$.

   1. The **open cone** on S is $\{\sum_{i=1}^k a_i \mathbf{v}_i : \forall i\ a_i > 0\}$.
   2. The **closed cone** on S is $\{\sum_{i=1}^k a_i \mathbf{v}_i : \forall i\ a_i \geq 0\}$.
   3. The **convex hull** of S is $\{\sum_{i=1}^k a_i \mathbf{v}_i : \forall i\ a_i \geq 0, \sum_{i=1}^k a_i = 1\}$.

   In particular, if S is contained in an affine space $\mathbb{A}^r$, then the convex hull of S is contained in $\mathbb{A}^r$.

   We also declare that $\mathbf{0} = \sum_{\mathbf{v} \in \emptyset} \mathbf{v}$, so $\{\mathbf{0}\}$ is the open cone on $\emptyset$, the closed cone on $\emptyset$, and the convex hull of $\emptyset$.

10. A **convex polytope** is the convex hull of a finite set of elements of $\mathbb{R}^r$. A subset Q of a convex polytope P is a **face** of P if there is a signed hyperplane $\mathcal{H}$ such that $Q = H^0 \cap P$ and $P \subset H^0 \cup H^+$. A **vertex** of a convex polytope P is a 0-dimensional face.

    A nontrivial theorem (cf. chapter 1 of Ziegler 1995) says that a set is a convex polytope if and only if it is a bounded intersection of closed halfspaces. A convex polytope P is the convex hull of its vertex set, and each face Q is the convex hull of the vertices of P in Q.

11. Let $\mathcal{O}$ be a set of geometric objects and E a finite set. An **arrangement of elements of $\mathcal{O}$ indexed by** E is an element of $\mathcal{O}^E$. An arrangement indexed by E is typically written as $\mathcal{A} = (A_e : e \in E)$. If $E = [n]$ then we may write an arrangement indexed by E as a list $(A_1, \ldots, A_n)$, as we have been doing in the case $\mathcal{O} = \mathbb{R}^r$. We will continue to denote a rank r arrangement of vectors in $\mathbb{R}^r$ indexed by $[n]$ by $M = (\mathbf{v}_1, \ldots, \mathbf{v}_n) \in$ Mat(r, n).

    This definition has one exception that will become prominent near the end of Chapter 2. We'll introduce the set $\mathcal{O}$ of *pseudospheres* in a fixed sphere and then define an *arrangement of pseudospheres* indexed by E to be an element of $\mathcal{O}^E$ satisfying certain additional properties.

1.3.2 Representing Arrangements

A signed hyperplane $\mathcal{H}$ in $\mathbb{R}^2$ will always be drawn with an arrow pointing from H^0 into H^+, as in Figure 1.1.

We'll represent a signed hyperplane arrangement $\mathcal{A}$ in $\mathbb{R}^3$ in two ways.

- Let $\mathbf{v}_\infty$ be a nonzero vector such that neither $\mathbf{v}_\infty^\perp$ nor $-\mathbf{v}_\infty^\perp$ is in $\mathcal{A}$. We will consider the affine plane $\mathbb{A} = \{\mathbf{x} \in \mathbb{R}^3 : \mathbf{x} \cdot \mathbf{v}_\infty = 1\}$. The set $\{\mathcal{H} \cap \mathbb{A} : \mathcal{H} \in \mathcal{A}\}$ is a signed affine hyperplane arrangement. Again, we draw an arrow from H^0 into H^+ to indicate the orientation. (See Figure 1.2 for an example.)

    Besides being easier to draw than an arrangement in $\mathbb{R}^3$, this kind of representation is useful for applying results in affine planar geometry to oriented matroids.

- We will draw the arrangement of equators $\{H^0 \cap S^2 : \mathcal{H} \in \mathcal{A}\}$ in S^2. We draw arrows from each equator $H^0 \cap S^2$ into $H^+ \cap S^2$. See Figure 1.3.

    This representation is preferred for two reasons: It saves us from worrying about the special plane $\mathbf{v}_\infty^\perp$, and, as we'll see in Chapter 4, it's the best route to representing arbitrary oriented matroids (not just realizable ones).

1.3 Arrangements

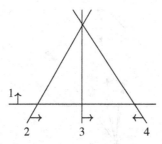

Figure 1.2 An affine representation of a signed hyperplane arrangement in $\mathbb{R}^3$.

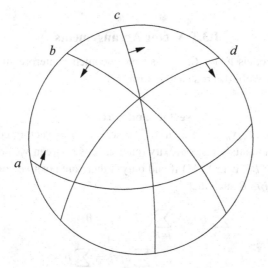

Figure 1.3 A spherical representation of a signed hyperplane arrangement in $\mathbb{R}^3$.

Consider a vector arrangement $M = (\mathbf{v}_1, \ldots, \mathbf{v}_n)$ in $\mathbb{R}^r$. If, for some $\mathbf{w} \in \mathbb{R}^r$, we have that $\mathbf{v}_i \cdot \mathbf{w} = 1$ for all i, then M can be viewed as a point arrangement in the affine space $\mathbb{A} := \{\mathbf{x} : \mathbf{x} \cdot \mathbf{w} = 1\}$. We will use this both to represent rank 3 vector arrangements in the plane and to imagine vector arrangements in higher dimensions.

Recall from Problem 1.6 that $\mathcal{V}(M)$, $\mathcal{V}^*(M)$, $\mathcal{C}(M)$, $\mathcal{C}^*(M)$, and $\chi(M)$ are all invariant under scaling the columns by positive scalars. Thus, if we have an arrangement $(\mathbf{v}_1, \ldots, \mathbf{v}_n)$ with all $\mathbf{v}_i$ in an open half-space $\{\mathbf{x} : \mathbf{x} \cdot \mathbf{w} > 0\}$, then we can scale each $\mathbf{v}_i$ to get an arrangement in the affine space $\{\mathbf{x} : \mathbf{x} \cdot \mathbf{w} = 1\}$, and this new arrangement can replace the original arrangement for our purposes. We call such an affine point arrangement an **affine representation** of M.

12   Realizable Oriented Matroids

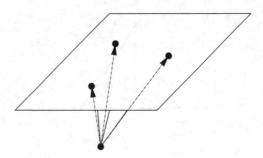

Figure 1.4 A point arrangement in an affine plane.

1.3.3 Vector Arrangements

Each of our models $\mathcal{V}, \mathcal{V}^*, \mathcal{C}, \mathcal{C}^*$ has a nice geometric interpretation for vector arrangements. We'll discuss each in turn.

Vectors and Circuits

Let $M = (\mathbf{v}_1, \ldots, \mathbf{v}_n)$ be a matrix. The set $\mathcal{V}(M)$ of vectors corresponding to M is useful in studying convexity relationships between $\mathbf{v}_i$. For instance, a sign vector $A^+ B^-$ is in $\mathcal{V}(M)$ if and only if there are positive constants a_i, b_j for all $i \in A, j \in B$ such that

$$\sum_{i \in A} a_i \mathbf{v}_i + \sum_{j \in B} (-b_j) \mathbf{v}_j = \mathbf{0},$$

$$\sum_{i \in A} a_i \mathbf{v}_i = \sum_{j \in B} b_j \mathbf{v}_j.$$

Thus, $A^+ B^- \in \mathcal{V}(M)$ if and only if the open cones spanned by $\{\mathbf{v}_i : i \in A\}$ and by $\{\mathbf{v}_i : i \in B\}$ intersect. (Recall that $\{\mathbf{0}\}$ is the open cone spanned by the empty set.)

If the columns of V lie in an affine space $\mathbb{A}$, then for nonempty sets A and B, $A^+ B^- \in \mathcal{V}(M)$ if and only if the convex hulls of the point sets corresponding to A and B intersect in their relative interiors. The following proposition tells us a bit more about $\mathcal{C}(M)$.

Proposition 1.15 *Let $\{\mathbf{v}_i : i \in A\}$ and $\{\mathbf{v}_i : i \in B\}$ be the vectors corresponding to independent nonempty subsets of an affine space $\mathbb{A}$. If $A^+ B^- \in \mathcal{C}(M)$ then the convex hulls of these two point sets intersect in a single point.*

(The converse is false: Consider the affine arrangement in Figure 1.5.)

1.3 Arrangements

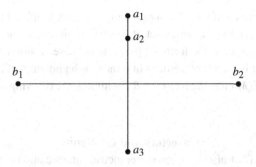

Figure 1.5 The convex hulls intersect in a single point, but this is not a minimal dependency.

Proof: By our preceding observations the intersection of the two convex hulls is nonempty. Assume by way of contradiction that it has two different elements $\sum_{i \in A} a_i \mathbf{v}_i = \sum_{i \in B} b_i \mathbf{v}_i$ and $\sum_{i \in A} \hat{a}_i \mathbf{v}_i = \sum_{i \in B} \hat{b}_i \mathbf{v}_i$. (Here all coefficients are nonnegative, and the sum of coefficients in each sum is 1.) Then for every ϵ

$$\sum_{i \in A}(a_i - \epsilon \hat{a}_i)\mathbf{v}_i = \sum_{i \in B}(b_i - \epsilon \hat{b}_i)\mathbf{v}_i.$$

Take $\epsilon > 0$ minimal such that some coefficient in this equation is 0. Then the other coefficients are all nonnegative, so this linear combination gives a smaller element of $\mathcal{V}(M)$. □

Here are further geometric insights about M that can be gleaned from $\mathcal{V}(M)$ and $\mathcal{C}(M)$:

- A subset S of $\{\mathbf{v}_1, \ldots, \mathbf{v}_n\}$ is linearly independent if and only if there is no $X \in \mathcal{V}(M)$ whose support is a subset of $\{i : \mathbf{v}_i \in S\}$. In particular, from $\mathcal{V}(M)$ (or just $\mathcal{C}(M)$) we see the rank of the arrangement. The **rank** of $\mathcal{V}(M)$ is defined to be the rank of M.
- From $\mathcal{V}(M)$, we can tell if M has an affine representation, as follows. The set $\{\mathbf{v}_1, \ldots, \mathbf{v}_n\}$ is contained in some open half-space $\{\mathbf{x} : \mathbf{x} \cdot \mathbf{w} > 0\}$ of $\mathbb{R}^n$ if and only if $\mathbf{0}$ is not in any of the relatively open cones spanned by nonempty subsets of $\{\mathbf{v}_1, \ldots, \mathbf{v}_m\}$.[2] $\mathbf{0}$ is in the open cone spanned by a subset S of $\{\mathbf{v}_1, \ldots, \mathbf{v}_n\}$ if and only if $\{i : i \in S\}^+ \in \mathcal{V}(M)$. Thus, M has an affine representation if and only if there is no nonempty S with $S^+ \in \mathcal{V}(M)$. In this case we say $\mathcal{V}(M)$ is *acyclic*. (A formal definition of acyclic is coming in Definition 3.7.)

---

[2] One direction of this assertion depends on the Farkas Lemma. See Section 1.5.2.

- If M does have an affine representation $(p_1, \ldots, p_n)$ with all p_i distinct, then $\{p_1, \ldots, p_n\}$ is the vertex set of a convex polytope if and only if no element of $\mathcal{V}(M)$ has the form $A^+\{b\}^-$. In this case, a subset $\{p_i : i \in S\}$ of the vertex set is the set of vertices of a face of the polytope if and only if no element of $\mathcal{V}(M)$ has the form A^+B^- with $B \subseteq S$. (Proving this requires some work.)

Covectors and Cocircuits

$\mathcal{V}^*(M)$ and $\mathcal{C}^*(M)$ also have good geometric interpretations in terms of the columns of M. This is another example of the oriented matroid trick brought up in Remark 1.12: $\mathcal{V}^*(M)$ is about the row space of M, but we can use it to talk about geometric properties of the set of columns.

Let $\mathbf{x} \in \text{row}(M)$, and so $\mathbf{x} = \mathbf{y}M$ for some $\mathbf{y} \in \mathbb{R}^r$. Thus $\text{sign}(\mathbf{x}) = (\text{sign}(\mathbf{y} \cdot \mathbf{v}_i) : i \in [n])$. Writing this in terms of the signed hyperplane $\mathcal{H} = \mathbf{y}^\perp$ and the sign vector $X = \text{sign}(\mathbf{x})$,

$$X(i) = \begin{cases} + & \text{if } \mathbf{v}_i \in H^+, \\ - & \text{if } \mathbf{v}_i \in H^-, \\ 0 & \text{if } \mathbf{v}_i \in H^0. \end{cases}$$

This proves the following.

Proposition 1.16 $A^+B^-C^0 \in \mathcal{V}^*(M)$ *if and only if there is a signed hyperplane $\mathcal{H}$ with* $\{\mathbf{v}_i : i \in A\} \subset H^+, \{\mathbf{v}_i : i \in B\} \subset H^-,$ *and* $\{\mathbf{v}_i : i \in C\} \subset H^0$.

Proposition 1.17 $A^+B^-C^0 \in \mathcal{C}^*(M)$ *if and only if there is an oriented hyperplane $\mathcal{H}$ with* $A = \{i : \mathbf{v}_i \in H^+\}$, $B = \{i : \mathbf{v}_i \in H^-\}$, *and H^0 the span of* $\{\mathbf{v}_i : i \in C\}$.

Proof: Let $A^+B^-C^0 \in \mathcal{V}^*(\mathcal{M}) - \{0\}$, and let $\mathcal{H}$ be a signed hyperplane with $\{\mathbf{v}_i : i \in A\} \subset H^+, \{\mathbf{v}_i : i \in B\} \subset H^-,$ and $\{\mathbf{v}_i : i \in C\} \subset H^0$. Then certainly the span of $\{\mathbf{v}_i : i \in C\}$ is contained in H^0. Further, this span is exactly H^0 if and only if $\text{rank}(\mathbf{v}_i : i \in C) = r - 1$. Proposition 1.14 says this holds if and only if $A^+B^-C^0 \in \mathcal{C}^*(\mathcal{M})$. □

Let's go back to some of the geometric ideas previously discussed for $\mathcal{V}$ and $\mathcal{C}$:

- A set $\{\mathbf{v}_i : i \in S\}$ contains a maximal independent set if and only if it is not contained in any hyperplane in $\mathbb{R}^r$, hence if and only if there is no element of $V^*(M)$ whose support is contained in the complement of S.

1.3 Arrangements

- The columns of M all lie in one open half-space (and hence M has an affine representation) if and only if $[n]^+ \in V^*(M)$.
- If M does have an affine representation, then the convex hull of the elements of an affine representation is a convex polytope P. A set $\{v_i : i \in A\}$ is the set of elements in a face of P if and only if there is a signed hyperplane $\mathcal{H}$ with $\{v_i : i \in A\} \subset H^0$ and $P \cap H^- = \emptyset$, hence if and only if $([n] \setminus A)^+ \in V^*(M)$.

Chirotopes

Definition 1.18 An arrangement $(v_1, \ldots, v_r)$ in $\mathbb{R}^r$ is a **positively oriented basis** of $\mathbb{R}^r$ if $\text{sign}(\det(v_1, \ldots, v_r)) = +$ and is a **negatively oriented basis** of $\mathbb{R}^r$ if $\text{sign}(\det(v_1, \ldots, v_r)) = -$.

For $M \in \text{Mat}(r, n)$, the chirotope $\chi(M)$ encodes for each ordered r-tuple of columns of M, whether that r-tuple is a positively oriented basis, a negatively oriented basis, or not an ordered basis at all.

To review the geometric meaning of this: The determinant function from GL_r to $\mathbb{R} - \{0\}$ is continuous and surjective, so GL_r has at least two connected components. With a little work one can see that GL_r has exactly two path components, namely $GL_r^+ = \{A \in GL_r : \det(A) > 0\}$ and $GL_r^- = \{A \in GL_r : \det(A) < 0\}$, and these components are homeomorphic. Thus an ordered basis $(v_{i_1}, \ldots, v_{i_r})$ is positively oriented if and only if we can continuously deform the matrix $(v_{i_1}, \ldots, v_{i_r})$ to the identity matrix while maintaining linear independence of columns, and similarly we can describe negatively oriented bases.

Another geometric way to think of this: Let $\{w_1, \ldots, w_{r-1}\}$ be a linearly independent set in $\mathbb{R}^r$, and let H be the span $\langle w_1, \ldots, w_{r-1}\rangle$. Then H is the zero locus of the continuous function $v \to \det(w_1, \ldots, w_{r-1}, v)$ from $\mathbb{R}^r$ to $\mathbb{R}$, and so the two connected components of $\mathbb{R}^r - H$ are the sets

$$\{v : \text{sign}(\det(w_1, \ldots, w_{r-1}, v)) = +\}$$

and

$$\{v : \text{sign}(\det(w_1, \ldots, w_{r-1}, v)) = -\}.$$

Now consider a vector arrangement $M = (v_1, \ldots, v_n)$, the associated chirotope χ, and $i_1, \ldots, i_{r-1}$ such that $\{v_{i_1}, \ldots, v_{i_{r-1}}\}$ is linearly independent. Let $H = \langle v_{i_1}, \ldots, v_{i_{r-1}}\rangle$. For each $j \in [n]$, we have that $\chi(i_1, \ldots, i_{nr1}, j) = 0$ if and only if $v_j \in H$, and for each $j, k \in [n]$, we have that $\chi(i_1, \ldots, i_{r-1}, j) = -\chi(i_1, \ldots, i_{r-1}, k) \neq 0$ if and only if v_j and v_k are on opposite sides of H.

Notice a difference between our geometric interpretation of chirotopes for vector arrangements and our geometric interpretations of $\mathcal{V}$, $\mathcal{V}^*$, $\mathcal{C}$, and $\mathcal{C}^*$.

Each of our earlier interpretations was coordinate-independent – given a vector arrangement in an arbitrary finite-dimensional real vector space (not necessarily $\mathbb{R}^r$), we could associate sets $\mathcal{V}$, $\mathcal{V}^*$, $\mathcal{C}$, and $\mathcal{C}^*$ by way of the interpretations we've described. By contrast, our interpretation of the chirotope references a preferred ordered basis for $\mathbb{R}^r$ (the list of columns of the identity matrix).

Given a general rank r vector space W over $\mathbb{R}$, we get an equivalence relation on the set of ordered bases of W, by saying $(\mathbf{v}_1, \ldots, \mathbf{v}_r) \sim (\mathbf{w}_1, \ldots, \mathbf{w}_r)$ if, for some (and therefore all) isomorphisms $f: W \to \mathbb{R}^r$, the matrices $(f(\mathbf{v}_1), \ldots, f(\mathbf{v}_r))$ and $(f(\mathbf{w}_1), \ldots, f(\mathbf{w}_r))$ are in the same connected component of GL_r. An *orientation* of W is a choice of one of these equivalence classes to be considered the set of positively oriented bases. So, given a rank r vector arrangement $(\mathbf{v}_1, \ldots, \mathbf{v}_n)$ in W and a choice of orientation of W, we get an associated chirotope $\chi: [n]^r \to \{0, +, -\}$. The opposite choice of orientation for W gives the chirotope $-\chi$.

We'll see in Section 1.5 that the same information about a vector arrangement M is encoded by each of $\mathcal{V}(M)$, $\mathcal{V}^*(M)$, $\mathcal{C}(M)$, $\mathcal{C}^*(M)$, or $\{\chi(M), -\chi(M)\}$.

1.3.4 Signed Hyperplane Arrangements

For a matrix $M = (\mathbf{v}_1 \cdots \mathbf{v}_n)$, consider the associated signed hyperplane arrangement $\mathcal{A} = (\mathbf{v}_1^\perp, \ldots, \mathbf{v}_n^\perp)$. Each triple $((\mathbf{v}_i^\perp)^0, (\mathbf{v}_i^\perp)^+, (\mathbf{v}_i^\perp)^-)$ with $\mathbf{v}_i \neq \mathbf{0}$ is a partition of $\mathbb{R}^r$ into three parts. The common refinement of these partitions is a decomposition of $\mathbb{R}^r$ into convex cones. Each cone can be specified by its relationship to each hyperplane: $A^+ B^- C^0$ represents the cone $\{\mathbf{y} \in \mathbb{R}^r : \mathbf{y} \cdot \mathbf{v}_i > 0 \text{ for all } i \in A; \mathbf{y} \cdot \mathbf{v}_i < 0 \text{ for all } i \in B; \mathbf{y} \cdot \mathbf{v}_i = 0 \text{ for all } i \in C\}$. (See Figure 1.6 for an example.) Thus we see a bijection from $\mathcal{V}^*(M)$ to cones of this partition, taking each X to $\{\mathbf{y} \in \mathbb{R}^r : \text{sign}(\mathbf{y}^\top M) = X\}$. The cones that are rays correspond to $\mathcal{C}^*(M)$ under this bijection.

Often it's more convenient to work with the arrangement of oriented equators $\mathbf{v}_i^\perp \cap S^{r-1}$ in S^{r-1}. This arrangement defines a decomposition of S^{r-1} into spherically convex cells that are in bijection with $\mathcal{V}^*(M)\setminus\{\mathbf{0}\}$.

The interpretations of $\mathcal{V}$, $\mathcal{C}$, and χ for signed hyperplane arrangements are not as illuminating and won't be discussed here. See Section 4.8 for some development of these interpretations.

1.4 Subspaces

Every rank r subspace W of $\mathbb{R}^n$ is the row space of some $M \in \text{Mat}(r,n)$, and by definition $\mathcal{V}(M) = \{\text{sign}(x) : x \in W^\perp\}$ and $\mathcal{V}^*(M) = \{\text{sign}(x) : x \in W\}$.

1.4 Subspaces

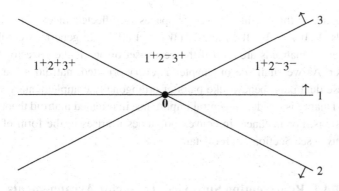

Figure 1.6 An arrangement of signed hyperplanes in $\mathbb{R}^2$ and some of the resulting sign vectors.

So our "discrete models for real matrices" could as well be thought of as "discrete models for real subspaces of $\mathbb{R}^n$," and these models nicely reflect orthogonality of vector spaces.

To recast things in these terms: The **real Grassmannian** $G(r, \mathbb{R}^n)$ is the set of all rank r subspaces of $\mathbb{R}^n$. Thus each element of $G(r, \mathbb{R}^n)$ is the row space of a matrix $M \in \mathrm{Mat}(r, n)$. Notice that two matrices $M, M' \in \mathrm{Mat}(r, n)$ have the same row space if and only if each row of M is a linear combination of rows of M', that is, $M = AM'$ for some invertible matrix A. Thus $G(r, \mathbb{R}^n)$ can be identified with the quotient of $\mathrm{Mat}(r, n)$ by the left action of GL_r. Our maps $\mathcal{V}: \mathrm{Mat}(r, n) \to \mathcal{P}(\{0, +, -\}^n)$ and $\mathcal{V}^*: \mathrm{Mat}(r, n) \to \mathcal{P}(\{0, +, -\}^n)$ quotient to maps

$$\begin{aligned} \mathcal{V}: \quad & G(r, \mathbb{R}^n) & \to & \quad \mathcal{P}(\{0, +, -\}^n) \\ & W & \to & \quad \{\mathrm{sign}(\mathbf{x}) : \mathbf{x} \in W^\perp\} \\ \mathcal{V}^*: \quad & G(r, \mathbb{R}^n) & \to & \quad \mathcal{P}(\{0, +, -\}^n) \\ & W & \to & \quad \{\mathrm{sign}(\mathbf{x}) : \mathbf{x} \in W\}, \end{aligned}$$

and we can define functions $\mathcal{C}$ and $\mathcal{C}^*$ on $G(r, \mathbb{R}^n)$ either as quotients of our maps $\mathcal{C}, \mathcal{C}^*$ on $\mathrm{Mat}(r, n)$, or in terms of $\mathcal{V}$ and $\mathcal{V}^*$ as before.

The view of vector and covector sets as simply the sets of sign vectors arising from subspaces of $\mathbb{R}^n$ is fundamental. We'll frequently bring up the idea of oriented matroids as combinatorial analogs to subspaces of $\mathbb{R}^n$. One useful point to notice is that the set of nonzero elements of $\{\mathrm{sign}(\mathbf{x}) : \mathbf{x} \in W\}$ is just the set $\{\mathrm{sign}(\mathbf{x}) : \mathbf{x} \in W, \|x\| = 1\}$ of sign vectors arising from the unit sphere in W. Loosely put, *an oriented matroid $\mathcal{M}$ is a combinatorial analog to a vector space, and $\mathcal{V}^*(\mathcal{M})\setminus\{0\}$ is a combinatorial analog to the unit sphere in that vector space.* This idea is central to Chapter 4.

Note that orthogonality of vector spaces is reflected nicely in oriented matroids: $\mathcal{V}(W) = \mathcal{V}^*(W^\perp)$ and $\mathcal{V}^*(W) = \mathcal{V}(W^\perp)$. In general, we will say two oriented matroids are *dual* if the vector set of one is the covector set of the other. As we shall see in Chapter 2, *every* oriented matroid has a dual, and these dual pairs behave like pairs of orthogonally complementary vector spaces. Duality is an idea of central importance in oriented matroid theory and applications. For instance, in convex polytopes it arises in the form of *Gale diagrams* – see Section 1.6 for details.

1.4.1 Representing Subspaces by Vector Arrangements

Let's tie the current discussion of subspaces together with our earlier discussion on vector arrangements. Given a subspace W of $\mathbb{R}^n$, choose a basis of W and make it the set of *rows* of a matrix M. Then the set of *columns* of M is a vector arrangement, and the vectors, covectors, and so on associated to W coincide with the vectors, covectors, and so on of this vector arrangement. Thus studying oriented matroid properties associated to subspaces of $\mathbb{R}^n$ is the same as studying oriented matroid properties of vector arrangements. This is often a useful way to address subspace issues – see, for instance, Exercise 1.6.

This transition from rows to columns may seem mysterious, but the following proposition shows that the arrangement of column vectors is essentially just the vector arrangement in W obtained by projecting the unit coordinate vectors in $\mathbb{R}^n$ onto W.

Proposition 1.19 *Let $M = (\mathbf{v}_1, \ldots, \mathbf{v}_n) \in \mathrm{Mat}(r,n)$ and let $W = \mathrm{row}(M)$. Let $\pi_W \colon \mathbb{R}^n \to \mathbb{R}^n$ denote the orthogonal projection onto W. Then there is an isomorphism $\mathbb{R}^r \to W$ sending each $\mathbf{v}_i$ to $\pi_W(\mathbf{e}_i)$.*

Proof: There is an automorphism of W sending the rows of M to an orthonormal basis for W, that is, there is an $A \in GL_r$ such that AM has orthonormal rows. The vector arrangement given by the columns of AM is the image under an isomorphism of the vector arrangement given by the columns of M. So without loss of generality we assume the rows of M to be orthonormal.

Let $\mathcal{B}$ be the vector arrangement in W consisting of the orthogonal projections $\pi_W(\mathbf{e}_i)$ of the unit coordinate vectors $\{\mathbf{e}_1, \ldots, \mathbf{e}_n\}$ onto W. We can think of $\mathcal{B}$ as the columns of the $n \times n$ matrix C_W representing orthogonal projection of $\mathbb{R}^n$ onto W, but in contrast to our usual identification of vector arrangements with matrices, the rank of C_W is typically less than the number of rows. Since the set of rows of M is an orthonormal basis for W, the columns of MC_W represent $\mathcal{B}$ in terms of this basis.

But, letting $C_{W^\perp}$ be the $n \times n$ matrix representing orthogonal projection of $\mathbb{R}^n$ onto $W^\perp$, we have

$$MC_W = M(C_W + C_{W^\perp}) \quad \text{since } MC_{W^\perp} = 0$$
$$= MI$$
$$= M,$$

and so the map $\mathbb{R}^r \to W$ sending the standard basis to the rows of M sends each $\mathbf{v}_i$ to $\pi_W(\mathbf{e}_i)$. □

1.4.2 Representing Subspaces by Signed Hyperplane Arrangements

Given a subspace W of $\mathbb{R}^n$, let $\mathcal{A}$ be the signed hyperplane arrangement $\{\mathbf{e}_i^\perp \cap W : i \in [n]\}$ in W. Then from the definition of $\mathcal{V}^*(W)$ and our earlier discussion of hyperplane arrangements, we see that the oriented matroid of $\mathcal{A}$ is the oriented matroid of W.

In fact, since $\mathbf{w} \cdot \mathbf{e}_i = \mathbf{w} \cdot \pi_W(\mathbf{e}_i)$ for each $\mathbf{w} \in W$, we see that $\mathcal{A}$ is the signed hyperplane arrangement corresponding to the vector arrangement $\mathcal{B} = \{\pi_W(\mathbf{e}_i) : i \in [n]\}$ of Section 1.4.1.

Problem 1.20 Consider the map $G(1, \mathbb{R}^n) \to \mathcal{P}(\{0, +, -\}^n)$ sending each W to $\mathcal{V}^*(W)$. Describe the partition of $\mathbb{RP}^{n-1} = G(1, \mathbb{R}^n)$ by preimages under this map.

1.4.3 Representing Subspaces by Points in Projective Space: The Plücker Embedding

Our discussion so far has put chirotopes in a different realm from the other oriented matroid characterizations. Each of those was given as the sign of a collection of vectors in $\mathbb{R}^n$, with each of these vectors having a nice geometric interpretation. In this section we'll see how to view the pair $\pm\chi(M)$ as giving the sign pair of a point in projective space, again with useful geometric meaning.

The key result is the following:

Proposition 1.21 *Let $M, M' \in \text{Mat}(r, n)$. Then $\text{row}(M) = \text{row}(M')$ if and only if there is a nonzero scalar c such that $|M_{i_1,\ldots,i_r}| = c|M'_{i_1,\ldots,i_r}|$ for each $i_1, \ldots, i_r \in [n]$.*

We can prove Proposition 1.21 in a naive way, or we can use the machinery of *exterior algebras*. The exterior algebra perspective has found powerful application (cf. Bokowski and Sturmfels 1989), but little of that has made it into this book. Here we'll give both the naive and fancier approaches: The reader may choose which to follow.

Naive proof: ($\Rightarrow$) If $\text{row}(M) = \text{row}(M')$ then $M = AM'$ for some $A \in GL_r$. For each $i_1 < \cdots < i_r$ in $[n]$, the submatrix $(AM')_{i_1,\ldots,i_r}$ is just $A(M'_{i_1,\ldots,i_r})$. Thus our scalar c is $|A|$.

($\Leftarrow$) Choose a $\{a_1, \ldots, a_r\} \subseteq [n]$ with $a_1 < \cdots < a_r$ such that $|M_{a_1,\ldots,a_r}| \neq 0$. Thus $\hat{M} := (M_{a_1,\ldots,a_r})^{-1} M$ is in reduced row-echelon form with respect to $\{a_1, \ldots, a_r\}$, and we can read off the entries of $\hat{M}$ as follows.

Let $\mathbf{v}_j$ denote the jth column of M, $\hat{\mathbf{v}}_j$ the jth column of $\hat{M}$, and $\hat{m}_{ij}$ denote individual entries of $\hat{M}$. Then

$$\hat{m}_{ij} = |\mathbf{e}_1, \ldots, \mathbf{e}_{i-1}, \hat{\mathbf{v}}_j, \mathbf{e}_{i+1}, \ldots, \mathbf{e}_r|$$
$$= |\hat{\mathbf{v}}_{a_1}, \ldots, \hat{\mathbf{v}}_{a_{i-1}}, \hat{\mathbf{v}}_j, \hat{\mathbf{v}}_{a_{i+1}}, \ldots, \hat{\mathbf{v}}_{a_r}|$$
$$= |\hat{M}_{a_1,\ldots,a_{i-1},j,a_{i+1},\ldots,a_r}|$$
$$= |(M_{a_1,\ldots,a_r})^{-1}||M_{a_1,\ldots,a_{i-1},j,a_{i+1},\ldots,a_r}|.$$

Our hypothesis tells us that

$$|(M_{a_1,\ldots,a_r})^{-1}||M_{a_1,\ldots,a_{i-1},j,a_{i+1},\ldots,a_r}| = |(M'_{a_1,\ldots,a_r})^{-1}||M'_{a_1,\ldots,a_{i-1},j,a_{i+1},\ldots,a_r}|.$$

Thus $\hat{M} = (M'_{a_1,\ldots,a_r})^{-1} M'$ as well, and so $M = M_{a_1,\ldots,a_r}(M'_{a_1,\ldots,a_r})^{-1} M'$. Since $M_{a_1,\ldots,a_r}(M'_{a_1,\ldots,a_r})^{-1} \in GL_r$, we have $\text{row}(M) = \text{row}(M')$. □

To give the exterior algebra perspective, we start with a quick review of tensor products.

Definition 1.22 Let $A_1, \ldots, A_r$, and V be vector spaces over a field $\mathbb{K}$. A function $f: A_1 \times \cdots \times A_r \to V$ is **multilinear** if it is linear in each coordinate.

One example of a multilinear map is the determinant, viewed as a function taking a sequence of row vectors to a field element. A generalization of this example: Let $r \leq n$, and for every $\{\mathbf{v}_1, \ldots, \mathbf{v}_r\} \subseteq \mathbb{K}^n$, let $R(\mathbf{v}_1, \ldots, \mathbf{v}_r)$ be the matrix with rows $\mathbf{v}_1, \ldots, \mathbf{v}_r$. We get a multilinear map $(\mathbb{K}^n)^r \to (\mathbb{K})^{\binom{n}{r}}$ by sending each $(\mathbf{v}_1, \ldots, \mathbf{v}_r)$ to the vector $(|R(\mathbf{v}_1, \ldots, \mathbf{v}_r)_{i_1,\ldots,i_r}| : 1 \leq i_1 < \cdots < i_r \leq n)$.

The **tensor product** $A \otimes B$ of vector spaces A and B over $\mathbb{K}$ is the essentially unique vector space with the following universal property: There is a map $A \times B \to A \otimes B$ such that for each vector space V, each multilinear map $m: A \times B \to V$ factors uniquely as $A \times B \to A \otimes B \xrightarrow{l} V$, where l is linear. Thus the tensor product is a gadget we use to move from multilinear maps to linear maps. The tensor product is constructed as a quotient of

1.4 Subspaces

the free vector space on $A \times B$: We simply quotient out by what we need to in order to get our universal property. That is, we identify formal sums $\sum_{\substack{\mathbf{a} \in A \\ \mathbf{b} \in B}} c_{\mathbf{a},\mathbf{b}}(\mathbf{a},\mathbf{b})$ and $\sum_{\substack{\mathbf{a} \in A \\ \mathbf{b} \in B}} d_{\mathbf{a},\mathbf{b}}(\mathbf{a},\mathbf{b})$ if, for every vector space V and multilinear map $f : A \times B \to V$, we have $\sum_{\substack{\mathbf{a} \in A \\ \mathbf{b} \in B}} c_{\mathbf{a},\mathbf{b}} f(\mathbf{a},\mathbf{b}) = \sum_{\substack{\mathbf{a} \in A \\ \mathbf{b} \in B}} d_{\mathbf{a},\mathbf{b}} f(\mathbf{a},\mathbf{b})$.

For each $\mathbf{a} \in A$ and $\mathbf{b} \in B$, let $\mathbf{a} \otimes \mathbf{b}$ denote the equivalence class of $(\mathbf{a}, \mathbf{b})$ under this equivalence. Hence every element of $A \otimes B$ can be expressed (usually not uniquely) as the equivalence class of a sum of terms of the form $k(\mathbf{a} \otimes \mathbf{b})$ with $k \in \mathbb{K}$, $\mathbf{a} \in A$, and $\mathbf{b} \in B$. The map $A \times B \to A \otimes B$ sends $(\mathbf{a}, \mathbf{b})$ to $\mathbf{a} \otimes \mathbf{b}$. It is easy to check that if $\{\mathbf{a}_1, \ldots, \mathbf{a}_\alpha\}$ is a basis for A and $\{\mathbf{b}_1, \ldots, \mathbf{b}_\beta\}$ is a basis for B then $A \otimes B$ has basis $\{\mathbf{a}_i \otimes \mathbf{b}_j : i \in [\alpha], j \in [\beta]\}$.

For a vector space A, the **exterior product** $\bigwedge^r A$ is the quotient of the tensor product $\otimes^r A$ by $\langle \mathbf{v}_1 \otimes \cdots \otimes \mathbf{v}_r - \mathrm{sign}(\sigma)\mathbf{v}_{\sigma(1)} \otimes \cdots \otimes \mathbf{v}_{\sigma(r)} : \mathbf{v}_i \in A, \sigma \in S_r \rangle$. It follows from the universal property for tensor products that the composition $A^r \to \otimes^r A \to \bigwedge^r A$ is universal for alternating multilinear maps. That is, each alternating multilinear map $A^r \to V$ factors through this map.

The image of $\mathbf{v}_1 \otimes \cdots \otimes \mathbf{v}_r$ under the quotient map $\otimes^r A \to \bigwedge^r A$ is denoted $\mathbf{v}_1 \wedge \cdots \wedge \mathbf{v}_r$. Since $\bigwedge^r A$ is the quotient of a vector space by a subspace, it inherits a vector space structure. If $\{\mathbf{a}_1, \ldots, \mathbf{a}_n\}$ is a basis for A then $\{\mathbf{a}_{i_1} \wedge \cdots \wedge \mathbf{a}_{i_r} : i_i < \cdots < i_r\}$ is a basis for $\bigwedge^r A$. Additionally, $\bigoplus_r \bigwedge^r A$ has a product operation, sending a pair $(\mathbf{x}_1 \wedge \cdots \wedge \mathbf{x}_r, \mathbf{x}_{r+1} \wedge \cdots \wedge \mathbf{x}_{r+s})$ to $\mathbf{x}_1 \wedge \cdots \wedge \mathbf{x}_{r+s}$.

Now let's focus on the vector space $\mathbb{R}^n$. The vector space $\bigwedge^r \mathbb{R}^n$ has basis $\{\mathbf{e}_{i_1} \wedge \mathbf{e}_{i_2} \wedge \cdots \wedge \mathbf{e}_{i_r} : 1 \leq i_1 < i_2 < \cdots < i_r \leq n\}$, hence is isomorphic as a vector space to $\mathbb{R}^{\binom{n}{r}}$.

Proposition 1.23 *Let* $\mathbf{v}_1, \mathbf{v}_2, \ldots, \mathbf{v}_r$ *be the rows of a* $r \times n$ *matrix* M. *Then*

$$\mathbf{v}_1 \wedge \cdots \wedge \mathbf{v}_r = \sum_{i_1 < \cdots < i_r} |M_{i_1, \ldots, i_r}| \mathbf{e}_{i_1} \wedge \cdots \wedge \mathbf{e}_{i_r}.$$

To see this, write each $\mathbf{v}_i$ as $\sum_j m_{i,j} \mathbf{e}_j$ and expand out $\mathbf{v}_1 \wedge \cdots \wedge \mathbf{v}_r$, remembering the formula for the determinant $|A| = \sum_{\sigma \in S_r} \mathrm{sign}(\sigma) a_{1,\sigma_1} \cdots a_{r,\sigma_r}$.

Corollary 1.24 *Let* $\mathbf{v}_1, \mathbf{v}_2, \ldots, \mathbf{v}_r \in \mathbb{R}^n$. *Then* $\mathbf{v}_1 \wedge \cdots \wedge \mathbf{v}_r = 0$ *if and only if* $\{\mathbf{v}_1, \mathbf{v}_2, \ldots, \mathbf{v}_r\}$ *has rank less than* r.

This happens when either the sequence $\mathbf{v}_1, \mathbf{v}_2, \ldots, \mathbf{v}_r$ has repeated elements or the set $\{\mathbf{v}_1, \mathbf{v}_2, \ldots, \mathbf{v}_r\}$ is dependent.

Proof: Let M be a matrix with rows $\mathbf{v}_1, \ldots, \mathbf{v}_r$. Then

$$\mathbf{v}_1 \wedge \cdots \wedge \mathbf{v}_r = 0 \Leftrightarrow |M_{i_1, \ldots, i_r}| = 0 \text{ for all } i_1, \ldots, i_r$$
$$\Leftrightarrow \mathrm{rank}(M_{i_1, \ldots, i_r}) < r \text{ for all } i_1, \ldots, i_r$$
$$\Leftrightarrow \mathrm{rank}(M) < r.$$

□

Exterior Algebra Proof of Proposition 1.21: The proof of ($\Rightarrow$) is the same as that for the naive proof. To see ($\Leftarrow$), let $\mathbf{v}_1, \ldots, \mathbf{v}_r$ be the rows of M and $\mathbf{w}_1, \ldots, \mathbf{w}_r$ be the rows of M'. Thus

$$\mathbf{v}_1 \wedge \cdots \wedge \mathbf{v}_r = \sum_{i_1 < \cdots < i_r} |M_{i_1,\ldots,i_r}| \mathbf{e}_{i_1} \wedge \cdots \wedge \mathbf{e}_{i_r}$$
$$= \sum_{i_1 < \cdots < i_r} |cM'_{i_1,\ldots,i_r}| \mathbf{e}_{i_1} \wedge \cdots \wedge \mathbf{e}_{i_r}$$
$$= c\mathbf{w}_1 \wedge \cdots \wedge \mathbf{w}_r.$$

Then for each i we have $(\mathbf{v}_1 \wedge \cdots \wedge \mathbf{v}_r) \wedge \mathbf{w}_i = c(\mathbf{w}_1 \wedge \cdots \wedge \mathbf{w}_r) \wedge \mathbf{w}_i$, which is 0 by Corollary 1.24. Thus each $\mathbf{w}_i$ is in $\langle \mathbf{v}_1, \ldots, \mathbf{v}_r \rangle$. Similarly, for each i we have $\mathbf{v}_i \in \langle \mathbf{w}_1, \ldots, \mathbf{w}_r \rangle$. $\square$

Notation 1.25 For a set S and an $r < n$, let $S^{\binom{n}{r}}$ denote the set of all vectors $(s_{i_1,\ldots,i_r} : 1 \leq i_1 < \cdots < i_r \leq n)$ with each component in S.

$\mathbb{R}P^{\binom{n}{r}-1}$ denotes the real projective space consisting of all one-dimensional subspaces of $\mathbb{R}^{\binom{n}{r}}$.

Proposition 1.21 gives us the **Plücker embedding**

$$P: G(r, \mathbb{R}^n) \to \mathbb{R}P^{\binom{n}{r}-1}$$

$$P(\text{row}(M)) = \mathbb{R}\left(\sum_{i_1 < \cdots < i_r} |M_{i_1,\ldots,i_r}| : 1 \leq i_1 < \cdots < i_r \leq n \right),$$

or, in terms of the exterior algebra,

$$P: G(r, \mathbb{R}^n) \to \mathbb{P}(\bigwedge^r \mathbb{R}^n) \cong \mathbb{R}P^{\binom{n}{r}-1}$$

$$\langle \mathbf{v}_1, \ldots, \mathbf{v}_r \rangle \mapsto \mathbb{R}\mathbf{v}_1 \wedge \cdots \wedge \mathbf{v}_r = \mathbb{R}\left(\sum_{i_1 < \cdots < i_r} |M_{i_1,\ldots,i_r}| \mathbf{e}_{i_1} \wedge \cdots \wedge \mathbf{e}_{i_r} \right).$$

This gives us a geometric interpretation of the chirotope as the sign of a vector: If χ is one of the two chirotopes arising from a space W and $i_1 < \cdots < i_r$, then $\chi(i_1, \ldots, i_r)$ is just the sign of the $(i_1, \ldots, i_r)$ coordinate of the Plücker embedding of W.

Remark 1.26 The naive proof of Proposition 1.21 pointed out how to find the entries of a matrix in reduced row-echelon form given just the maximal minors. Using this, we see how to recover the signs of entries of the matrix given only the chirotope. Let χ be the chirotope of a matrix $M = (m_{ij}) \in \text{Mat}(r, n)$

that is in reduced row-echelon form with respect to $\{a_1,\ldots,a_r\}$. Then for each i,j,

$$\text{sign}(m_{ij}) = \chi(a_1,\ldots,a_{i-1},j,a_{i+1},\ldots,a_r).$$

With this observation we have a small window into a thorny problem: *Given an oriented matroid, find the space of all matrices with this oriented matroid.* This is essentially the question of determining the *realization space* of an oriented matroid, the subject of Chapter 7.

1.4.4 Aside: Topology of the Grassmannian

We defined the Grassmannian $G(r,\mathbb{R}^n)$ as a set to be the quotient of $\text{Mat}(r,n)$ by the left action of GL_r. $\text{Mat}(r,n)$ has a topology as a subspace of $\mathbb{R}^{r\times n}$, and so $G(r,\mathbb{R}^n)$ has a topology as a quotient of $\text{Mat}(r,n)$. The Plücker embedding $P: G(r,\mathbb{R}^n) \to \mathbb{P}(\bigwedge^r \mathbb{R}^n)$ is a homeomorphism, so this topology on the Grassmannian coincides with the subspace topology in $\mathbb{P}(\bigwedge^r \mathbb{R}^n)$. It also coincides with the intuitive topology on the Grassmannian: If $V \in G(r,\mathbb{R}^n)$ then the set of r-dimensional subspaces of $\mathbb{R}^n$ that we get by wiggling V around just a bit is an open neighborhood of V.

The topology of the real Grassmannian is important for many reasons, for example in the theory of characteristic classes, and it will play a prominent role in later parts of this book. So we'll explore this topology briefly here. See, for instance, Milnor and Stasheff (1974) for more.

We'll show that $G(r,\mathbb{R}^n)$ is a manifold of dimension $r(n-r)$. First consider

$$U_{1,\ldots,r} = \{\text{row}(I|A) : (I|A) \in \text{Mat}(r,n)\}$$
$$= \{\text{row}(M) : M \in \mathbb{R}^{r\times n} \text{ and } M_{1,\ldots,r} \in GL_r\}.$$

This is an open subset of $G(r,\mathbb{R}^n)$, and the correspondence

$$U_{1,\ldots,r} \leftrightarrow \mathbb{R}^{r\times(n-r)}$$
$$\text{row}(I|A) \leftrightarrow A$$

is a homeomorphism. This homeomorphism is a coordinate chart for all $V \in G(r,\mathbb{R}^n)$ such that the $(1,\ldots,r)$-coordinate of $P(V)$ is nonzero.

Generalizing this, for each $\{i_1,\ldots,i_r\} \in [n]$ we define $U_{i_1,\ldots,i_r} = \{\text{row}(M) : M_{i_1,\ldots,i_r} \in GL_r\}$, and we see that $U_{i_1,\ldots,i_r}$ is homeomorphic to $\mathbb{R}^{r\times(n-r)}$. The collection of all such $U_{i_1,\ldots,i_r}$ is an open cover of the Grassmannian.

1.5 Cryptomorphisms for Realizable Oriented Matroids

The previous sections showed that the various sets $\mathcal{V}(M)$, $\mathcal{C}(M)$, $\mathcal{V}^*(M)$, and $\mathcal{C}^*(M)$ are geometrically interesting things to look at. As we are about to see, these sets encode exactly the same information about M, and χ encodes slightly more. Specifically,

1. There are bijections

$$\begin{array}{ccc} \mathcal{V}(\mathrm{Mat}(r,n)) & \longleftrightarrow & \mathcal{V}^*(\mathrm{Mat}(r,n)) \\ \updownarrow & & \updownarrow \\ \mathcal{C}(\mathrm{Mat}(r,n)) & & \mathcal{C}^*(\mathrm{Mat}(r,n)) \end{array}$$

commuting with the maps $\mathcal{V}$, $\mathcal{C}$, $\mathcal{V}^*$, and $\mathcal{C}^*$. That is, the diagram

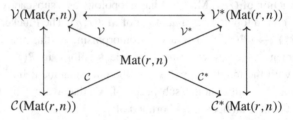

commutes.

2. For every $M \in \mathrm{Mat}(r,n)$, let $\tilde{\chi}(M) = \{\chi(M), -\chi(M)\}$. Then there is a bijection $\tilde{\chi}(\mathrm{Mat}(r,n)) \leftrightarrow \mathcal{C}(\mathrm{Mat}(r,n))$ so that

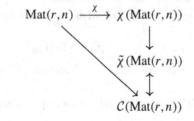

commutes.

The following sections will establish these two points. The phrase we'll use for these results is to say the models $\mathcal{V}$, $\mathcal{V}^*$, $\mathcal{C}$, $\mathcal{C}^*$, and $\tilde{\chi}$ are **cryptomorphic** – they encode the same data in different ways.

To phrase this in terms of the Grassmannian: Recall $G(r,\mathbb{R}^n)$ is essentially the quotient of $\mathrm{Mat}(r,n)$ by the left action of GL_r, and that $\mathcal{V}$, $\mathcal{V}^*$, $\mathcal{C}$, and $\mathcal{C}^*$ are all invariant under the action of GL_r on $\mathrm{Mat}(r,n)$. Additionally, define the **oriented Grassmannian** $OG(r,\mathbb{R}^n)$ to be the quotient of $\mathrm{Mat}(r,n)$ by the left action of the group GL_r^+ of matrices with positive determinant. An element

1.5 Cryptomorphisms for Realizable Oriented Matroids

of $OG(r, \mathbb{R}^n)$ can be thought of as an r-dimensional subspace of $\mathbb{R}^n$ equipped with a distinguished orientation. Thus the previous diagrams induce diagrams

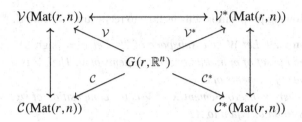

and

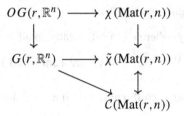

where the top two vertical maps $OG(r, \mathbb{R}^n) \to G(r, \mathbb{R}^n)$ and $\tilde{\chi}(\text{Mat}(r,n)) \to \chi(\text{Mat}(r,n))$ are double covers.

All of these results will be subsumed by results of Chapter 2, which will establish cryptomorphisms for general oriented matroids. We're doing the more limited results here to geometrically motivate our Chapter 2 results.

1.5.1 The Bijections $\mathcal{V}(\mathbf{Mat}(r,n)) \leftrightarrow \mathcal{C}(\mathbf{Mat}(r,n))$ and $\mathcal{V}^*(\mathbf{Mat}(r,n)) \leftrightarrow \mathcal{C}^*(\mathbf{Mat}(r,n))$

We'll get both of these bijections by a single argument. The forward maps $\mathcal{V}(\text{Mat}(r,n)) \to \mathcal{C}(\text{Mat}(r,n))$ and $\mathcal{V}^*(\text{Mat}(r,n)) \to \mathcal{C}^*(\text{Mat}(r,n))$ are obvious: Just take each set to its set of minimal nonzero elements. It remains to be seen how we can reconstruct elements of $\mathcal{V}(M)$ resp. $\mathcal{V}^*(M)$ from the minimal nonzero elements of these sets.

Definition 1.27 For two sign vectors $X, Y \in \{0, +, -\}^n$, define their composition $X \circ Y \in \{0, +, -\}^n$ by

$$X \circ Y(e) = \begin{cases} X(e) & \text{if } X(e) \neq 0, \\ Y(e) & \text{otherwise.} \end{cases}$$

We take the zero vector to be the composition of the empty sequence of elements of $\mathcal{C}$.

Problem 1.28 In Figure 1.6, in how many ways can the covector $(+, +, +)$ be expressed as a composition of two covectors?

The following proposition tells us how to reconstruct $\mathcal{V}$ from $\mathcal{C}$.

Proposition 1.29 *Let W be a subspace of $\mathbb{R}^n$. Let $\mathcal{V} = \{\text{sign}(\mathbf{x}) : \mathbf{x} \in W^{\perp}\}$, and let $\mathcal{C}$ be the set of minimal nonzero elements of $\mathcal{V}$. Then $\mathcal{V}$ is the set of all compositions of elements of $\mathcal{C}$.*

In fact, every nonzero element X of $\mathcal{V}$ is the composition of signed circuits that are less than or equal to X.

Lemma 1.30 *For every subspace V of $\mathbb{R}^n$ and $\mathbf{x}, \mathbf{y} \in V$, there exists $\mathbf{z} \in V$ such that $\text{sign}(\mathbf{x}) \circ \text{sign}(\mathbf{y}) = \text{sign}(\mathbf{z})$.*

Proof: Let $\mathbf{z} = \mathbf{x} + \epsilon \mathbf{y}$, where ϵ is a sufficiently small positive real number. □

Definition 1.31 $X, Y \in \{0, +, -\}^E$ are **conformal** if their separation set $S(X, Y)$ is empty.

A set of sign vectors is **conformal** if each pair of elements is conformal.

In particular, if $Y \leq X$ and $Y' \leq X$ then Y and Y' are conformal. A set $\{Y_1, \ldots, Y_k\} \subseteq \{0, +, -\}^E$ is conformal if and only if, for each e, $\max(Y_i(e) : i \in [k])$ exists. In this case $Y_1 \circ \cdots \circ Y_k(e) = \max(Y_i(e) : i \in [k])$.

Proof of Proposition 1.29: Lemma 1.30 shows that every composition of elements of $\mathcal{C}$ is in $\mathcal{V}$. Conversely, let $X \neq \mathbf{0}$ be an element of $\mathcal{V}$. Let $Y_1, \ldots, Y_k$ be the elements of $\mathcal{C}$ that are less than or equal to X, and let $Y = Y_1 \circ \cdots \circ Y_k$. Notice that Y does not depend on the order of Y_i and that $Y \leq X$. We'll induct on the height of X in the poset $\mathcal{V}$ to see that $Y = X$. The minimal case is when $X \in \mathcal{C}$, so that $X = Y_1 = Y$.

Above this, assume by way of contradiction that $\text{supp}(X) \setminus \text{supp}(Y) \neq \emptyset$. Let $\mathbf{x}, \mathbf{y} \in W^{\perp}$ with $X = \text{sign}(\mathbf{x})$ and $Y = \text{sign}(\mathbf{y})$. Consider the ray $\{\mathbf{x} - \lambda \mathbf{y} : \lambda \geq 0\}$ in $W^{\perp}$. All points on this ray have the same eth coordinate for each $e \in \text{supp}(X) \setminus \text{supp}(Y)$. If $f \in \text{supp}(Y)$ then $\text{sign}(x_f - \lambda y_f) = \text{sign}(x_f)$ for small values of λ, and $\text{sign}(x_f - \lambda y_f) = -\text{sign}(x_f)$ for large values of λ. Thus the ray leaves the orthant $\{\hat{\mathbf{x}} : \text{sign}(\hat{\mathbf{x}}) = X\}$ containing $\mathbf{x}$ at a point $\mathbf{z}$ with $z_e = x_e$ for each $e \in \text{supp}(X) \setminus \text{supp}(Y)$, and $z_f = 0$ for some $f \in \text{supp}(Y)$. Let $Z = \text{sign}(\mathbf{z})$. Then $\mathbf{0} \neq Z < X$, so by our induction hypothesis Z is a composition of circuits that are less than or equal to Z, and hence are less than or equal to X. But $Z \not\leq Y$, a contradiction. □

For another perspective on (co)vectors as compositions of signed (co)circuits, let $\mathcal{V}^*$ and $\mathcal{C}^*$ be the covector and signed cocircuit sets of a

1.5 Cryptomorphisms for Realizable Oriented Matroids

signed hyperplane arrangement. Consider a picture of the arrangement in an affine space $\mathbb{A}$. Then $X \in \mathcal{V}^*$ indexes a convex set c_X in the partition of $\mathbb{A}$ given by the arrangement, and a signed cocircuit $Y \leq X$ indexes a point p_Y in the closure of c_X. For covectors Y and Y' and points $p_Y \in c_Y$ and $p_{Y'} \in c_{Y'}$, the set $c_{Y \circ Y'}$ is the part of our partition containing a point obtained by starting at p_Y and moving a tiny step toward $p_{Y'}$. Here "tiny," means the step moves us off of every hyperplane of the arrangement that contains p_Y but not $p_{Y'}$, but is not so big that we cross any hyperplane of the arrangement. Proposition 1.29 says that if we start at some vertex of the closure of c_X and successively take tiny steps toward each remaining vertex, we will end in c_X.

1.5.2 $\mathcal{V}(\mathrm{Mat}(r,n)) \leftrightarrow \mathcal{V}^*(\mathrm{Mat}(r,n))$: Duality

Recall (Section 1.4) that for any linear subspace W of $\mathbb{R}^n$ we have $\mathcal{V}^*(W) = \mathcal{V}(W^\perp)$. Here we introduce a notion of orthogonality for sign vectors under which $\mathcal{V}(W)$ and $\mathcal{V}^*(W)$ are "orthogonal complements" to each other.

Algebra and Dot Product for Signs

Define an operation $\cdot$ on $\{0, +, -\}$ by

$$+ \cdot + = - \cdot - = +$$

$$+ \cdot - = - \cdot + = -$$

$$i \cdot 0 = 0 \cdot i = 0$$

for each i. (Often we'll suppress the "$\cdot$.") Thus for $s_1, s_2, s_3 \in \{0, +, -\}$ we have $s_1 \cdot s_2 = s_3$ if and only if there are elements $r_1, r_2, r_3 \in \mathbb{R}$ such that $\mathrm{sign}(r_i) = s_i$ for each i and $r_1 r_2 = r_3$.

A **hyperoperation** on a set S is a function $S \times S \to \mathcal{P}(S) - \{\emptyset\}$. We define a hyperoperation on $\{0, +, -\}$, called the **hypersum** and denoted $\boxplus$, by

$$+ \boxplus + = \{+\}$$

$$- \boxplus - = \{-\}$$

$$x \boxplus 0 = 0 \boxplus x = \{x\} \text{ for all } x$$

$$+ \boxplus - = - \boxplus + = \{0, +, -\}.$$

Thus for $s_1, s_2, s_3 \in \{0, +, -\}$ we have $s_3 \in s_1 \boxplus s_2$ if and only if there are elements $r_1, r_2, r_3 \in \mathbb{R}$ such that $\mathrm{sign}(r_i) = s_i$ for each i and $r_1 + r_2 = r_3$. When a set has a single element we will frequently omit the braces, for example, by denoting $x \boxplus 0$ by x. For every $S, T \subseteq \{0, +, -\}$ we define $S \boxplus T$ to be $\bigcup_{t \in T}^{s \in S} s \boxplus t$, and for $S \subseteq \{0, +, -\}$ and $x \in \{0, +, -\}$ we let $S \boxplus x = x \boxplus S := S \boxplus \{x\}$. With these definitions, $\boxplus$ is commutative and associative: In fact,

$$\boxplus_{s \in S} s = \begin{cases} \max(S) & \text{if } \{+,-\} \not\subseteq S \\ \{0, +, -\} & \text{otherwise.} \end{cases}$$

Also, $a \cdot (b \boxplus c) = (a \cdot b) \boxplus (a \cdot c)$.

Define the **inner product** of $X, Y \in \{0, +, -\}^m$ to be

$$X \cdot Y = \boxplus_{i=1}^{m} X(i) \cdot Y(i).$$

Define two sign vectors X and Y in $\{0, +, -\}^n$ to be **orthogonal**, written $X \perp Y$, if $0 \in X \cdot Y$. Thus $X \perp Y$ if $\{X(i) \cdot Y(i) : i \in [n]\}$ either is $\{0\}$ or contains both $+$ and $-$. For any set $\mathcal{F}$ of sign vectors, let $\mathcal{F}^\perp$ denote $\{Y : Y \perp X \,\forall X \in \mathcal{F}\}$.

Duality

If $\mathbf{x} \perp \mathbf{y}$ are elements of $\mathbb{R}^n$, then $\sum_i x_i y_i = 0$, and so either the terms in this sum are all zero or some terms are positive and some negative. Thus, sign($\mathbf{x}$) $\perp$ sign($\mathbf{y}$). Of course, from the hypothesis sign($\mathbf{x}$) $\perp$ sign($\mathbf{y}$) we can't conclude $\mathbf{x} \perp \mathbf{y}$. But kind of amazingly, orthogonality of sign vectors corresponds to orthogonality of vector spaces exactly as it should:

Proposition 1.32 *If W is a subspace of $\mathbb{R}^n$, then $\{\text{sign}(\mathbf{y}) : \mathbf{y} \in W^\perp\}^\perp = \{\text{sign}(\mathbf{x}) : \mathbf{x} \in W\}$.*

In other words, $\mathcal{V}(W)^\perp = \mathcal{V}(W^\perp) = \mathcal{V}^*(W)$.

We'll prove this in a moment.

This result generalizes to arbitrary oriented matroids (Section 2.3), giving a natural notion of "orthogonal pairs of oriented matroids." (The actual term used is *dual pairs of oriented matroids*.)

Proposition 1.32 follows from a result, important in linear programming, that does not initially look inspiring. We first state it in one form it is commonly seen:

Farkas Lemma 1[3] Let $A \in \mathbb{R}^{r \times n}$ and $\mathbf{b} \in \mathbb{R}^r$. Then exactly one of the following holds:

---

[3] Hungarian tutorial: The a's in "Farkas" are pronounced like the English long o, and the s is pronounced like the English *sh*. Farkas's name will be coming up a lot.

1.5 Cryptomorphisms for Realizable Oriented Matroids

1. There exists $x \in \mathbb{R}^n$ such that $Ax = b$ and $x \geq 0$.
2. There exists $z \in \mathbb{R}^r$ such that $zA \geq 0$ and $zb < 0$.

Geometrically, this is easy to believe: The first alternative says that b is in the closed cone, cone(A), consisting of nonnegative linear combinations of the columns of A, while the second alternative says that there is a hyperplane $z^\perp$ separating cone(A) and b.

Part of the proof is clear as well: Both alternatives can't be true simultaneously, because if $b \in $ cone(A) and $z \cdot a \geq 0$ for every column a of A then $z \cdot b \geq 0$. The remainder of the proof – showing that at least one of the two alternatives hold – is substantially trickier. See, for instance, Ziegler (1995) for a proof.

To interpret the Farkas Lemma in terms of subspaces, let W be the nullspace of the matrix $(A| - b)$. Then the two alternatives of the Farkas Lemma can be stated in terms of W: Either

1. there exists $x \in W$ such that $x \geq 0$ and $x_{n+1} > 0$, or
2. there exists $y \in W^\perp$ such that $y \geq 0$ and $y_{n+1} > 0$.

(Here $y = z(A| - b) \in$ row($A| - b$).) So, the Farkas Lemma says that for any subspace W of $\mathbb{R}^{n+1}$, exactly one of W, $W^\perp$ intersects the positive closed orthant $\{z : \forall i \text{ sign}(z_i) \in \{0, +\}\}$ in a point whose last coordinate is positive.

Let's adapt Farkas to consider more general orthants. Notice that the form of the second alternative changes when we consider nonmaximal orthants.

Farkas Lemma 2 For every linear subspace W of $\mathbb{R}^{n+1}$ and every $Z = A^+ B^- C^0 \in \{0, +, -\}^{n+1}$ with $Z(n+1) \neq 0$, exactly one of the following holds:

1. There exists $x \in W$ such that sign(x) $\leq Z$ and sign(x_{n+1}) $= Z(n+1)$.
2. There exists $y \in W^\perp$ such that sign(y_i) $\leq Z(i)$ for every $i \in A \cup B$ and sign(y_{n+1}) $= Z(n+1)$.

Proof: Both alternatives can't hold simultaneously because if sign(x) $\leq Z$ and sign(y_i) $\leq Z(i)$ for every $i \in A \cup B$ and sign(x_{n+1}) $=$ sign(y_{n+1}) $\neq 0$ then $x_i y_i \geq 0$ for all i and $x_{n+1} y_{n+1} > 0$, so $x \cdot y > 0$. Thus if $x \in W$ then $y \notin W^\perp$.

Now, say $M = (v_1, \ldots, v_{n+1})$ is a matrix such that $W =$ null(M). Let D denote the diagonal matrix with

$$D_{ii} = \begin{cases} 1 & \text{if } Z(i) = +, \\ 0 & \text{if } Z(i) = 0, \\ -1 & \text{if } Z(i) = -. \end{cases}$$

and let $W' = \text{null}(MD)$. Applying our previous version of Farkas to W', we have that exactly one of the following is true:

1. There exists $\mathbf{x}' \in W'$ such that $\mathbf{x}' \geq \mathbf{0}$ and $x'_{n+1} > 0$.
2. There exists $\mathbf{y}' \in (W')^\perp$ such that $\mathbf{y}' \geq \mathbf{0}$ and $y'_{n+1} > 0$.

If Alternative (1) holds then $\mathbf{x} := D\mathbf{x}'$ satisfies Alternative (1) in the statement of the proposition.

If Alternative (2) holds then $\mathbf{y}' = \mathbf{z}MD$ for some $\mathbf{z} \in \mathbb{R}^r$. Now consider $\mathbf{y} := \mathbf{z}M \in W^\perp$. For each i we have $y_i = \mathbf{z} \cdot \mathbf{v}_i$ and $y'_i = D_{ii}\mathbf{z} \cdot \mathbf{v}_i$. Thus

$$y'_i = \begin{cases} y_i & \text{if } Z(i) = +, \\ -y_i & \text{if } Z(i) = -, \\ 0 & \text{if } Z(i) = 0. \end{cases}$$

Since $y'_i \geq 0$ for each i, we see $\mathbf{y}$ is an element of $W^\perp$ satisfying Alternative (2) of the proposition. □

Of course, there is nothing special about the final coordinate: We have the following mild generalization.

Farkas Lemma 3 For every linear subspace W of $\mathbb{R}^{n+1}$, every $Z = A^+B^-C^0 \in \{0, +, -\}^{n+1}$, and every $j \in [n+1]$ with $Z(j) \neq 0$, exactly one of the following holds:

1. There exists $\mathbf{x} \in W$ such that $\text{sign}(x_i) \leq Z(i)$ for every i and $\text{sign}(x_j) = Z(j)$.
2. There exists $\mathbf{y} \in W^\perp$ such that $\text{sign}(y_i) \leq Z(i)$ for every $i \in A \cup B$ and $\text{sign}(y_j) = Z(j)$.

From this version of the Farkas Lemma we can prove Proposition 1.32.

Proof of Proposition 1.32: The discussion on orthogonality earlier in this section shows that $\{\text{sign}(\mathbf{x}) : \mathbf{x} \in W\} \subseteq \{X : X \perp Y \text{ for every } Y \in \text{sign}(W^\perp)\}$. Conversely, say $X \perp Y$ for every $Y \in \text{sign}(W^\perp)$. Consider a j such that $X(j) \neq 0$. Applying our final version of Farkas to X and j, we see that Alternative $\{2\}$ can't hold, and so there exists $\mathbf{x}^{(j)} \in W$ such that $\text{sign}(\mathbf{x}^{(j)}) \leq X$ and $\text{sign}(x_j^{(j)}) = X(j)$. Choose such an $\mathbf{x}^{(j)}$ for each j, and let $\mathbf{x} = \sum_j \mathbf{x}^{(j)}$. Then this $\mathbf{x}$ is the element of W we want. □

See Exercise 1.7 for a fourth version of the Farkas Lemma that removes the asymmetry between the two alternatives.

In Section 2.3.3 we will give a combinatorial version of the Farkas Lemma for families of sign vectors, rather than subspaces. As we shall see, the families

1.5 Cryptomorphisms for Realizable Oriented Matroids

that are vector sets of oriented matroids can be characterized as those which satisfy this combinatorial Farkas Lemma and a few other conditions. This will lead to a simple proof of cryptomorphism between vector sets and covector sets for general oriented matroids.

1.5.3 $\mathcal{C}(\text{Mat}(r,n)) \leftrightarrow \tilde{\chi}(\text{Mat}(r,n))$

This discussion will introduce some ideas that are useful in a lot of contexts:

- the Pivoting Property,
- the Dual Pivoting Property, and
- the Basis Exchange Principle.

The Dual Pivoting Property, which will be introduced in Problem 1.36, actually leads to a cryptomorphism $\mathcal{C}^*(\text{Mat}(r,n)) \leftrightarrow \tilde{\chi}(\text{Mat}(r,n))$ by a shorter argument than the one we'll give. We'll do the longer argument to highlight some geometric ideas.

Definition 1.33 An arrangement $(\mathbf{v}_1, \ldots, \mathbf{v}_n)$ is **linearly independent** if $\mathbf{v}_i \neq \mathbf{v}_j$ whenever $i \neq j$ and $\{\mathbf{v}_1, \ldots, \mathbf{v}_n\}$ is linearly independent. An arrangement is **linearly dependent** if it is not linearly independent.

Let $M = (\mathbf{v}_1, \ldots, \mathbf{v}_n) \in \text{Mat}(r,n)$. Our first step in showing that $\mathcal{C}(M)$ determines $\pm \chi(M)$ and vice versa is to show that $\{\text{supp}(X) : X \in \mathcal{C}(M)\}$ determines $\text{supp}(\chi(M))$ and vice versa.

Since $\mathcal{C}(M)$ is the set of elements of $\mathcal{V}(M)$ of minimal support, we see that a set S is $\text{supp}(X)$ for some $X \in \mathcal{C}(M)$ if and only if $(\mathbf{v}_s : s \in S)$ is a minimal dependent subarrangement of M. But the minimal dependent subarrangements of M determine the maximal independent subarrangements of M and vice versa, and $(\mathbf{v}_i : i \in S)$ is a maximal independent subarrangement of M if and only if $|S| = r$ and $\chi(M)(s_1, \ldots, s_r) \neq 0$ for all orderings $(s_1, \ldots, s_r)$ of S.

Our next step begins by noting that to determine a pair $X, -X$ of sign vectors it's enough to determine $\text{supp}(X)$ and the values $X(e)X(f)$ for each $e, f \in \text{supp}(X)$. We will show that $C(M)$ determines the product

$$\chi(M)(e, x_2, \ldots, x_r)\chi(M)(f, x_2, \ldots, x_r)$$

for each $e, f, x_2, \ldots, x_r$, and vice versa.

If $X \in \mathcal{C}(M)$ and $\text{supp}(X) = \{e\}$ then $\mathbf{v}_e = \mathbf{0}$, $\{X, -X\} = \{e^+, e^-\}$, and $\chi(M)(e, x_2, \ldots, x_r) = 0$ for all $x_2, \ldots, x_r$.

Now consider $X \in \mathcal{C}(M)$ with larger support, say $\text{supp}(X) = \{e, f, i_2, \ldots, i_k\}$. Since $\{\mathbf{v}_e, \mathbf{v}_{i_2}, \ldots, \mathbf{v}_{i_k}\}$ is linearly independent, it is contained in a basis $\{\mathbf{v}_e, \mathbf{v}_{i_2}, \ldots, \mathbf{v}_{i_r}\}$. Also, since $\{\mathbf{v}_e, \mathbf{v}_{i_2}, \ldots, \mathbf{v}_{i_k}\}$ and $\{\mathbf{v}_f, \mathbf{v}_{i_2}, \ldots, \mathbf{v}_{i_k}\}$ span the same subspace of $\mathbb{R}^n$, $\{\mathbf{v}_f, \mathbf{v}_{i_2}, \ldots, \mathbf{v}_{i_r}\}$ is also a basis.

Proposition 1.34 *Let $\{\mathbf{v}_e, \mathbf{v}_{i_2}, \ldots, \mathbf{v}_{i_r}\}$ and $\{\mathbf{v}_f, \mathbf{v}_{i_2}, \ldots, \mathbf{v}_{i_r}\}$ be distinct independent sets of columns of M. Let $X \in \mathcal{C}(M)$ such that the support of X is contained in $\{e, f, i_2, \ldots, i_r\}$. Then*

$$\chi(M)(e, i_2, \ldots, i_r) \cdot \chi(M)(f, i_2, \ldots, i_r) = -X(e) \cdot X(f).$$

This relationship between chirotopes and signed circuits will reappear for general oriented matroids in Theorem 2.54 as the *Pivoting Property*.

Proof: This just expresses a simple geometric idea: $\{\mathbf{v}_{i_2}, \ldots, \mathbf{v}_{i_r}\}$ spans a hyperplane H not containing $\mathbf{v}_e$ or $\mathbf{v}_f$, and the independence of the two sets tells us that neither $\mathbf{v}_e$ nor $\mathbf{v}_f$ lie on H. From our discussion of the chirotope in Section 1.3.3 we know that $\mathbf{v}_e$ and $\mathbf{v}_f$ lie on the same side of H if and only if $\chi(M)(e, i_2, \ldots, i_r) = \chi(M)(f, i_2, \ldots, i_r)$, that is, if $\chi(M)(e, i_2, \ldots, i_r) \cdot \chi(M)(f, i_2, \ldots, i_r) = +$. Also, they lie on the same side of H if and only if there is no positive linear combination of $\mathbf{v}_e$ and $\mathbf{v}_f$ lying on H, that is, there are no scalars a_e, a_f, b_2, b_r with $a_e > 0$, $a_f > 0$, and

$$a_e \mathbf{v}_e + a_f \mathbf{v}_f = \sum_{j=2}^{r} b_j \mathbf{v}_{i_j},$$

$$a_e \mathbf{v}_e + a_f \mathbf{v}_f + \sum_{j=2}^{r} b_j \mathbf{v}_{i_j} = \mathbf{0}.$$

Thus the signed circuit X with $\mathrm{supp}(X) \subseteq \{e, f, i_2, \ldots, i_r\}$ must satisfy $X(e) = -X(f)$. To summarize, if $\mathbf{v}_e$ and $\mathbf{v}_f$ lie on the same side of H then

$$\chi(M)(e, i_2, \ldots, i_r) \cdot \chi(M)(f, i_2, \ldots, i_r) = +$$

and $X(e)X(f) = -$. Likewise if $\mathbf{v}_e$ and $\mathbf{v}_f$ lie on opposite sides of H then

$$\chi(M)(e, i_2, \ldots, i_r) \cdot \chi(M)(f, i_2, \ldots, i_r) = -$$

and $X(e)X(f) = +$. □

Thus for each $X \in \mathcal{C}(M)$ for which we know $\mathrm{supp}(X)$, the pair $\{X, -X\}$ is determined by values $\chi(M)(e, i_2, \ldots, i_r) \cdot \chi(M)(f, i_2, \ldots, i_r)$. Conversely, if we know $\mathcal{C}(M)$ and we have $(e, i_2, \ldots, i_r), (f, i_2, \ldots, i_r) \in \mathrm{supp}(\chi(M))$ with $e \neq f$, the arrangement $(\mathbf{v}_e, \mathbf{v}_f, \mathbf{v}_{i_2}, \ldots, \mathbf{v}_{i_r})$ is dependent, and a minimal dependent subarrangement contains $\mathbf{v}_e$ and $\mathbf{v}_f$. A corresponding signed circuit X satisfies the hypothesis of Proposition 1.34, so from $X(e)X(f)$ we can find $\chi(M)(e, i_2, \ldots, i_r) \cdot \chi(M)(f, i_2, \ldots, i_r)$.

1.5 Cryptomorphisms for Realizable Oriented Matroids

The last step of the cryptomorphism is to show that from the values of products of the form

$$\chi(M)(e, i_2, \ldots, i_r)\chi(M)(f, i_2, \ldots, i_r),$$

we can determine $\chi(M)(b_1, \ldots, b_r)\chi(M)(b'_1, \ldots, b'_r)$ for all pairs of bases $B = \{\mathbf{v}_{b_1}, \ldots, \mathbf{v}_{b_r}\}$, $B' = \{\mathbf{v}_{b'_1}, \ldots, \mathbf{v}_{b'_r}\}$.

This follows from a linear algebra observation that's also a fundamental principle in the theory of ordinary matroids:

Proposition 1.35 (Basis Exchange Principle – realizable case) *If B and B' are bases for a vector space W and $\mathbf{x} \in B \backslash B'$ then there is a $\mathbf{y} \in B'$ such that $(B \cup \{\mathbf{y}\}) \backslash \{\mathbf{x}\}$ and $(B' \cup \{\mathbf{x}\}) \backslash \{\mathbf{y}\}$ are bases for W.*

Proof: $B \backslash \{\mathbf{x}\}$ spans a hyperplane H. Let $S = B' \cup \{\mathbf{x}\}$ and $T = (B' \cap H) \cup \{\mathbf{x}\}$. S spans W, and T is an independent subset of S, so T extends to a basis $(B' \cup \{\mathbf{x}\}) \backslash \{\mathbf{y}\}$ for W. Since $\mathbf{y} \notin H$, we also have that $(B \backslash \{\mathbf{x}\}) \cup \{\mathbf{y}\}$ is a basis for W. □

Using this, we can induct on $|B \backslash B'|$. Given B and B' with $|B \backslash B'| \geq 2$, we do a basis exchange to get a basis $\{\mathbf{v}_c : c \in (B \cup \{b'\}) \backslash \{b\}\}$ with $b' \in B'$. Without loss of generality assume $b = b_1$ and $b' = b'_1$. Then by our induction hypothesis we know the values of

$$\chi(M)(b_1, \ldots, b_r)\chi(M)(b'_1, b_2, \ldots, b_r)$$

and

$$\chi(M)(b'_1, b_2, \ldots, b_r)\chi(M)(b'_1, \ldots, b'_r).$$

Since $+ \cdot + = - \cdot - $ is the multiplicative identity in $\{0, +, -\}$ and the factor $\chi(M)(b'_1, b_2, \ldots, b_r)$ is in $\{+, -\}$, the product of these two values is $\chi(M)(b_1, \ldots, b_r)\chi(M)(b'_1, \ldots, b'_r)$.

Finally, there's a similar connection between χ and $\mathcal{C}^*$, whose abstract combinatorial analog will come up in Chapter 2.

Problem 1.36 If $\{\mathbf{v}_e, \mathbf{v}_{i_2}, \ldots, \mathbf{v}_{i_r}\}$ and $\{\mathbf{v}_f, \mathbf{v}_{i_2}, \ldots, \mathbf{v}_{i_r}\}$ are bases with $e \neq f$ and Y is a signed cocircuit with support contained in the complement of $\{i_2, \ldots, i_r\}$, show that

$$\chi(M)(e, i_2, \ldots, i_r)\chi(M)(f, i_2, \ldots, i_r) = Y(e) \cdot Y(f).$$

This relationship between chirotopes and signed cocircuits will reappear for general oriented matroids in Theorem 2.54 as the *Dual Pivoting Property*.

1.5.4 Conclusions on Cryptomorphism

To summarize: The functions $\mathcal{C}$, $\mathcal{V}$, $\mathcal{C}^*$, and $\mathcal{V}^*$ all encode the same data about matrices (while χ encodes this data plus an orientation of $\mathbb{R}^r$). This data about a matrix M is called the **oriented matroid corresponding to** M. We'll also refer to oriented matroids of vector or hyperplane arrangements, or of subspaces of $\mathbb{R}^n$. $\mathcal{C}(M)$ is called the set of **signed circuits** of the oriented matroid corresponding to M. $\mathcal{C}^*(M)$ is called the set of **signed cocircuits** of the oriented matroid corresponding to M. $\mathcal{V}(M)$ is called the set of **vectors** of the oriented matroid corresponding to M, and $\mathcal{V}^*(M)$ is called the set of **covectors** of the oriented matroid corresponding to M. (Ordinary (unoriented) matroid theory has (unsigned) circuits and cocircuits. See Section 1.8 for details.) $\chi(M)$ and $-\chi(M)$ are called the two **chirotopes** of the oriented matroid corresponding to M. We will use $\mathcal{M}(M)$ to denote the oriented matroid corresponding to M. The **dual** of $\mathcal{M}(M)$, denoted $\mathcal{M}^*(M)$, is the oriented matroid with vector set $\mathcal{V}^*(M)$ and covector set $\mathcal{V}(M)$.

This is not yet the definition of oriented matroids – it's only the special case of oriented matroids arising from matrices (these are called **realizable oriented matroids**). In Chapter 2 we will define oriented matroids in general, by first defining sets of signed circuits, sets of vectors, and so on. All of these objects will have purely combinatorial definitions inspired by the realizable case. We will see that in general signed circuits, vectors, and so on are cryptomorphic.

Every property we have described in terms of one of $\mathcal{C}(M)$, $\mathcal{V}(M)$, $\mathcal{C}^*(M)$, $\mathcal{V}^*(M)$, or $\pm\chi(M)$ can be thought of as a property of $\mathcal{M}(M)$. For instance, in Section 1.3.3 we defined the rank of $\mathcal{V}(M)$; henceforward we will call this the rank of $\mathcal{M}(M)$.

For ease of linear algebra, so far we have dealt with arrangements whose objects are indexed by $[n]$, resulting in oriented matroids defined in terms of $\{0, +, -\}^n$ and $\{0, +, -\}^{\binom{n}{r}}$. Going forward, this convention is unnecessary and occasionally inconvenient, so we will typically index the elements of an arrangement by a finite set E. Thus instead of working with a vector arrangement $M \in \text{Mat}(r, n)$, we will work with an arrangement $\mathcal{A} = (\mathbf{v}_e : e \in E)$. The resulting $\mathcal{C}(\mathcal{A})$, $\mathcal{V}(\mathcal{A})$, $\mathcal{C}^*(\mathcal{A})$, $\mathcal{V}^*(\mathcal{A})$ are subsets of $\{0, +, -\}^E$. We say E is the set of **elements** of $\mathcal{M}(\mathcal{A})$.

1.6 Convex Polytopes

Let's take a brief digression to see the usefulness of oriented matroids as a tool for studying convex polytopes. We'll examine the interaction between convex polytopes and oriented matroids in more depth in Chapter 8.

1.6 Convex Polytopes

Recall the definition of convex polytope from Section 1.3.1. It's not hard to see that a convex polytope P is the convex hull of its vertex set. The set of faces of P is a partially ordered set, ordered by inclusion. The **combinatorial type** of a convex polytope is the isomorphism class of its face poset.

Consider a convex polytope P in affine space with vertex set $V = \{p_e : e \in E\}$. As noted in Section 1.3.3, from the oriented matroid $\mathcal{M}$ of V we can read off the faces of P. For a subset F of V, F is the set of vertices of a face of V if and only if there is a signed hyperplane $\mathcal{H}$ with $H^0 \cap V = F$ and $H^- \cap V = \emptyset$. Further, this face is the convex hull of F. Thus we have a bijection between the covectors of $\mathcal{M}$ of the form A^+ and the faces of P, sending a covector A^+ to $\text{conv}(p_e : e \in E - A)$. This suggests realizable oriented matroids as a natural tool for studying face posets of convex polytopes.

Oriented matroid duality allows us to study high-dimensional convex polytopes with relatively few vertices using low-dimensional arrangements of points. Consider a d-dimensional convex polytope P in $\mathbb{A}^d \subset \mathbb{R}^{d+1}$ with vertex set $V = \{p_e : e \in E\}$, where $|V| \leq d + 4$. The oriented matroid $\mathcal{M}$ associated to V is rank $d + 1$ with $|V|$ elements, and so $\mathcal{M}^*$ is rank $|V| - (d + 1) \leq 3$. We can realize $\mathcal{M}^*$ in $\mathbb{R}^3$, but strictly speaking we can't realize it in $\mathbb{A}^2$: Since $E^+ \in \mathcal{V}^*(\mathcal{M}) = \mathcal{V}(\mathcal{M}^*)$, the elements of a realization can't all lie in a common plane not through the origin. We can get around this by introducing the notion of a *signed affine point arrangement*: This is an arrangement $((p_e, s_e) : e \in E)$, where each p_e is a point in $\mathbb{A}$ and each s_e is either $+$ or $-$. Given a realization $(\mathbf{w}_e : e \in E)$ of $\mathcal{M}^*$ in $\mathbb{R}^3$ and an affine plane $\mathbb{A} \subset \mathbb{R}^3$ that is not parallel to any $\mathbf{w}_e$, for each $e \in E$ there is a unique multiple $\lambda_e \mathbf{w}_e$ of $\mathbf{w}_e$ in $\mathbb{A}$: We let $(p_e, s_e) = (\lambda_e \mathbf{w}_e, \text{sign}(\lambda_e))$. The resulting signed affine point arrangement encodes our realization of $\mathcal{M}^*$, up to positive scaling: We call this arrangement an **affine Gale Diagram** for P. (This is actually weaker than the usual definition, which predates oriented matroids. Normally "Gale diagram" refers to a *dual vector arrangement* to V, obtained by treating V as the columns of a matrix M and finding a matrix N such that $\text{row}(N) = \text{row}(M)^\perp$.)

Problem 1.37 Find necessary and sufficient conditions on a realizable oriented matroid $\mathcal{M}$ for $\mathcal{M}$ to be the oriented matroid of a polytope.

Note that this also gives necessary and sufficient conditions for an oriented matroid to be the oriented matroid of the Gale diagram of a polytope.

Problem 1.38 Use your solution to Problem 1.37 to count the number of combinatorial types of four-dimensional convex polytopes on seven vertices.

To give a charming application, due to Perles, we consider the following question (cf. Ziegler 1995).

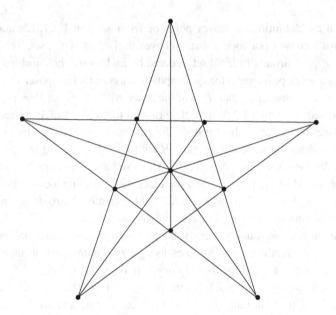

Figure 1.7 The "Betsy Ross" arrangement.

Given a convex polytope P in $\mathbb{R}^n$, is there some convex polytope P' in $\mathbb{Q}^n$ with the same face poset as P? That is, is every combinatorial type of real convex polytope **realizable over** $\mathbb{Q}$?

It's not hard to see that any combinatorial type of convex polytope of dimension less than or equal to three is realizable over $\mathbb{Q}$, as is any combinatorial type of convex polytope whose proper faces are all simplices. However, in general the answer to the question is No, and we'll give an example here.

Consider the affine point arrangement known as the *Betsy Ross arrangement*, shown in Figure 1.7. Make a signed affine point arrangement S with two elements $(p, +)$ and $(p, -)$ for each point p shown. Thus this arrangement describes 22 elements of $\mathbb{R}^3$ that come in pairs $\mathbf{v}, -\mathbf{v}$. Let $\mathcal{M}(S)$ be the oriented matroid of this arrangement.

Problem 1.39 Use the results of Problem 1.37 to verify that S is the Gale diagram of a convex polytope P in affine space. Determine the dimension of P. Show that any convex polytope in affine space of the same combinatorial type as P has oriented matroid isomorphic to $\mathcal{M}^*(S)$.

At the center of the argument are two observations.

- The face poset of P can be read off from the positive circuits of $\mathcal{M}(S)$.
- Any set T of vectors in $\mathbb{R}^3$ such that $\mathcal{M}(T)$ has the same positive circuits as $\mathcal{M}(S)$ (up to relabeling the elements of the oriented matroid) consists of

pairs $\{\mathbf{x}_e, -\mathbf{x}_e\}$, where, up to scaling, $T' := \{\mathbf{x}_e : e \in E\}$ is an affine point arrangement, and the set of positive circuits of $\mathcal{M}(T)$ determines the set of all circuits of $\mathcal{M}(T')$.

Note the many dependent triples among elements of S. Whenever a set $\{\mathbf{x}, \mathbf{y}, \mathbf{z}\}$ in $\mathbb{R}^3$ is dependent, the determinant $|\mathbf{x}, \mathbf{y}, \mathbf{z}| = 0$, and this equality is a polynomial equation in the coordinates for $\mathbf{x}$, $\mathbf{y}$, and $\mathbf{z}$. In the case of Betsy Ross, from the system of equations we can deduce that the only fields over which this collection of dependencies is realizable are fields containing $\sqrt{5}$. In particular, no vector arrangement in $\mathbb{Q}^3$ has oriented matroid $\mathcal{M}(S)$ – that is, $\mathcal{M}(S)$ is not realizable over $\mathbb{Q}$.

Problem 1.40 Show that an oriented matroid is realizable over $\mathbb{Q}$ if and only if its dual is as well.

Thus, P is not realizable over $\mathbb{Q}$!

Remark 1.41 The operation of replacing each element of our oriented matroid with two antiparallel copies is closely related to the *Lawrence construction*, which we will discuss more in Section 8.4. (The Lawrence construction performs this operation on $\mathcal{M}^*$ rather than on $\mathcal{M}$.)

Remark 1.42 For examples of nonrational combinatorial types of convex polytopes of dimension 4, see Richter-Gebert (1996a) and Dobbins (2011).

For much more on Gale diagrams, see chapter 6 of Ziegler (1995).

1.7 Deletion and Contraction

This section will discuss two operations on oriented matroids that are the basis for many inductive arguments.

From here on it will be convenient to index the columns of a matrix by an arbitrary finite set E, not necessarily $[n]$.

Given a matrix M with columns indexed by E and given $e \in E$, let $M \backslash e$ denote the matrix obtained from M by deleting the column indexed by e. For $X \in \{0, +, -\}^E$, let $X \backslash e \in \{0, +, -\}^{E \backslash e}$ be the restriction of X. Then from the definition of $\mathcal{V}^*$ it's clear that

$$\mathcal{V}^*(M \backslash e) = \{X \backslash e : X \in \mathcal{V}^*(M)\}.$$

It would be easy to jump to the conclusion that

$$\mathcal{C}^*(M \backslash e) = \{X \backslash e : X \in \mathcal{C}^*(M)\},$$

but this isn't quite true, for two reasons:

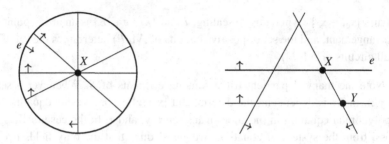

Figure 1.8 Deletions of circuits are not necessarily circuits of deletion.

1. If $X = \{e\}^+ \in \mathcal{V}^*(M)$ (i.e., row(M) has an element $\mathbf{x}$ whose only nonzero entry is x_e), then $X\backslash e = \mathbf{0} \notin \mathcal{C}^*$.
2. If there are $X, Y \in \mathcal{C}^*(M)$ so that $Y(e) \neq 0$, $X(e) = 0$, $Y(f) \leq X(f)$ for all $f \neq e$, and $Y(g) < X(g)$ for some g, then $\mathbf{0} \neq Y\backslash e < X\backslash e$, and thus $X\backslash e \notin \mathcal{C}^*$.

Illustrations of both issues are given in Figure 1.8. The figure on the left shows a representation of rank 3 oriented matroid by equators in S^2, and the figure on the right shows a representation of a rank 3 oriented matroid by hyperplanes in $\mathbb{A}^2$.

The correct description of the signed cocircuit set of the deletion is

$$\mathcal{C}^*(M\backslash e) = \min\{X\backslash e : X \in \mathcal{C}^*(M), X\backslash e \neq \mathbf{0}\}.$$

Problem 1.43 Describe $\mathcal{V}(M\backslash e)$ and $\mathcal{C}(M\backslash e)$ in terms of $\mathcal{V}(M)$ and $\mathcal{C}(M)$.

$\mathcal{M}(M\backslash e)$ is called the **deletion** of e from $\mathcal{M}(M)$, denoted $\mathcal{M}(M)\backslash e$. In Chapter 2 we'll define deletion for arbitrary oriented matroids in the way suggested by realizable oriented matroids.

In terms of vector and signed hyperplane arrangements, deletion of e corresponds to removing the vector resp. signed hyperplane corresponding to e from the arrangement.

The second operation to look at is *contraction*. The **contraction** of e from $\mathcal{M}(M)$, written $\mathcal{M}(M)/e$, is defined as $((\mathcal{M}^*(M))\backslash e)^*$. That is, $\mathcal{M}(M)/e$ is obtained from $\mathcal{M}(M)$ by deleting e in the dual. This operation has a direct interpretation in M as well, given in the following exercise.

Problem 1.44 (i) Let $\{\mathcal{H}_f : f \in E\}$ be an arrangement of signed hyperplanes with oriented matroid $\mathcal{M}$ and let $e \in E$. Show that the arrangement $\{(H_f^0 \cap H_e^0, H_f^+ \cap H_e^0, H_f^- \cap H_e^0) : f \in E\backslash e\}$ has oriented matroid $\mathcal{M}/e$.

(ii) Let $\{\mathbf{v}_f : f \in E\}$ be an arrangement of vectors in $\mathbb{R}^n$ with oriented matroid $\mathcal{M}$, and let $e \in E$. Let π_e be the orthogonal projection map from $\mathbb{R}^n$ to $(e^\perp)^0$. Show that the arrangement $\{\pi_e(\mathbf{v}_f) : f \in E\backslash e\}$ has oriented matroid $\mathcal{M}/e$.

This is an important idea to get used to: For oriented matroid purposes,

- deleting an element H_e in a signed hyperplane arrangement is equivalent to restricting to the subspace H_e in the dual, and
- deleting an element e in a vector arrangement is equivalent to projecting to the subspace $(e^\perp)^0$ in the dual.

1.8 A Few Words about Unoriented Matroids

For the reader who has never encountered (unoriented) matroids, here is a very, very brief introduction. Indeed, we won't get as far as a definition. For any real understanding, see, for instance, Oxley (1992). This section exists purely to satisfy a reader's mild curiosity – we will not use it elsewhere.

We have already seen that oriented matroids are modeling vector or hyperplane arrangements over *ordered* fields – fields with a natural partition into positive elements, negative elements, and 0. If we forget the data in an oriented matroid that arises from this order – by replacing each "+" and "−" by the word "nonzero" – we get a *matroid*. For instance:

- Instead of recording the signed circuit set of a finite arrangement $\mathcal{A}$ of vectors in $\mathbb{R}^r$, we look only at the supports of these signed circuits. This set of supports is called the set of *(unsigned) circuits* of the matroid of $\mathcal{A}$.
- Instead of recording the entire chirotope of $\mathcal{A}$, we record only which sets of elements are bases for $\mathbb{R}^r$.

This is a very oriented-matroid-chauvinist way to present the situation. In actuality, matroid theory (developed in Whitney 1935) much predates oriented theory (formalized independently in Folkman and Lawrence 1978 and Bland and Las Vergnas 1978, Bland and Las Vergnas 1979[4]), and was conceived of as a combinatorial abstraction of linear dependence over arbitrary fields. A matroid may be realizable as a vector arrangement over, for instance, a finite field without being realizable over $\mathbb{R}$.

Every oriented matroid has an underlying matroid, as described above, but not every matroid arises in this way – that is, not every matroid is *orientable*.

---

[4] See section 3.9 in Björner et al. (1999) and section 5.12 in Bachem and Kern (1992) for a more complete sketch of the origins of oriented matroid theory.

Examples can be found via vector arrangements over finite fields. See Ziegler (1991) for more discussion.

Exercises

1.1 Use the results of Section 1.2.1 to write (in a systematic way) a list of 2×4 rank 2 matrices giving each possible $\mathcal{C}(M)$ for this rank and size subject to the condition that the first two columns of the matrix are independent. Verify that these are also giving each possible $\chi(M)$, up to global change of sign.

1.2 (i) Let $\mathcal{C} = \{\{1,3\}^+\{2,4\}^-, \{1,3\}^-\{2,4\}^+, \{1,3\}^+\{5\}^-, \{1,3\}^-\{5\}^+, \{1,2,4\}^+\{5\}^-, \{1,2,4\}^-\{5\}^+, \{2,4\}^+\{3,5\}^-, \{2,4\}^-\{3,5\}^+\}$. Find a vector arrangement whose circuit set is $\mathcal{C}$. Then find another vector arrangement whose cocircuit set is $\mathcal{C}$.

(ii) Do the same for $\mathcal{C} = \{\{1,2,3\}^+\{4,5\}^-, \{1,2,3\}^-\{4,5\}^+\}$.

1.3 Let $p_1, \ldots, p_n$ be distinct points on an affine line, and let $\mathcal{M}$ be the rank 2 oriented matroid determined by these points. Prove that $\mathcal{M}$ depends only on the order of p_i along the line, and that $\mathcal{M}$ determines this order (up to global reversal).

1.4 Consider the vertex set of a regular pentagon in the plane. (There is nothing special about pentagons: We just want a concrete example.) By choosing a coordinate system for the plane, we can view these vertices as five vectors in $\mathbb{R}^2$ and get a rank 2 oriented matroid $\mathcal{M}^2$. On the other hand, by viewing our plane as an affine subspace of $\mathbb{R}^3$ we can view these vertices as five vectors in $\mathbb{R}^3$ and get a rank 3 oriented matroid $\mathcal{M}^3$. What can you say about the relationship between the oriented matroids $\mathcal{M}^2$ and $\mathcal{M}^3$ arising this way? For instance, what is the relationship between $\mathcal{V}(\mathcal{M}^2)$ and $\mathcal{V}(\mathcal{M}^3)$? What is the relationship between $\mathcal{V}^*(\mathcal{M}^2)$ and $\mathcal{V}^*(\mathcal{M}^3)$?

1.5 This is a generalization of Exercise 1.4. Let $M = (\mathbf{v}_1, \ldots, \mathbf{v}_n) \in \text{Mat}(r,n)$, and let $A \in \text{Mat}(k,r)$ for some $k \leq r \leq n$. What can you say about the relationship between the oriented matroid associated to M and the oriented matroid associated to AM?

1.6 (i) Consider a convex hexagon in the affine plane with vertices labeled $1, 2, \ldots, 6$ in cyclical order. The vertex set gives a rank 3 oriented matroid on elements [6]. Show that this oriented matroid $\mathcal{M}_{\text{hex}}$ is the same for all convex hexagons.

($\mathcal{M}_{hex}$ illustrates an important issue: The oriented matroid of an affine set of points tells you which *pairs* of subsets have intersecting convex hulls, but not which *triples* of subsets have intersecting convex hulls. For instance, a regular convex hexagon has three diagonals that intersect at a point, while a generic hexagon does not.)

(ii) If $W \subseteq V$ are subspaces of $\mathbb{R}^n$ then by definition $\mathcal{V}^*(W) \subseteq \mathcal{V}^*(V)$. Show that the converse fails by finding a rank 2 realizable oriented matroid $\mathcal{N}$ such that $\mathcal{V}^*(\mathcal{N}) \subseteq \mathcal{V}^*(\mathcal{M}_{hex})$ and a subspace V of $\mathbb{R}^6$ such that $\mathcal{V}^*(V) = \mathcal{V}^*(\mathcal{M}_{hex})$ but there is no subspace W of V such that $\mathcal{V}^*(W) = \mathcal{V}^*(\mathcal{N})$.

1.7 Prove the following version of the Farkas Lemma from the previous versions. (Here $\dot{\cup}$ denotes disjoint union.)

Farkas Lemma 4 For every subspace W of $\mathbb{R}^n$, every $Z = A^+ B^- (C \dot{\cup} D)^0 \in \{0, +, -\}^n$, and every $j \in A \cup B$, exactly one of the following holds.

1. There exists $\mathbf{x} \in W$ such that $\mathrm{sign}(x_i) \leq Z(i)$ for each $i \in A \cup B \cup C$ and $x_j \neq 0$.
2. There exists $\mathbf{y} \in W^\perp$ such that $\mathrm{sign}(y_i) \leq Z(i)$ for each $i \in A \cup B \cup D$ and $y_j \neq 0$.

1.8 Let $t_1 < t_2 < \cdots < t_n$ be real numbers. For each $i \in [n]$ let $\mathbf{v}_i = (1, t_i, t_i^2, \ldots, t_i^{r-1})$. The oriented matroid of $(\mathbf{v}_i : i \in [n])$ is called the **alternating oriented matroid** $\mathcal{M}_{alt}^{n,r}$. This oriented matroid is independent of the choice of t_j, as the first part of this exercise will show.

1. Show that $\mathcal{M}_{alt}^{n,r}$ has a chirotope χ such that $\chi(i_1, i_2, \ldots, i_r) = +$ whenever $i_1 < \cdots < i_r$. Once you've shown this, you know that $\mathcal{M}_{alt}^{n,r}$ is *uniform*, i.e., every r-tuple of elements is a basis, so you know the sizes of the supports of signed circuits and signed cocircuits. The following parts of the exercise fill out our knowledge.
2. Let $\{i_0, \ldots, i_r\} \subseteq [n]$ with $i_0 < \cdots < i_r$. Let X be the signed circuit of $\mathcal{M}_{alt}^{n,r}$ with support $\{i_0, \ldots, i_r\}$ and with $X(i_0) = +$. Prove that $X(i_j) = (-1)^j$ for each j.
3. Let $\{i_0, \ldots, i_{n-r}\} \subseteq [n]$ with $i_0 < \cdots < i_{n-r}$. Let Y be the signed cocircuit with support $\{i_0, \ldots, i_{n-r}\}$ and with $Y(i_0) = +$. For each $j \in \{0, \ldots, n-r\}$, let $\eta(j) = |\{k : i_0 < k < i_j \text{ and } k \notin \{i_0, \ldots, i_{n-r}\}\}|$. Prove that $Y(i_j) = (-1)^{\eta(j)}$.

Remark 1.45 Some sources use the term "alternating oriented matroid" for any oriented matroid on $[n]$ with a chirotope χ such that $\chi(i_1, \ldots i_r) \in \{0, +\}$ for all $i_1 < \cdots < i_r$. More recently, these have been called *positively oriented matroids*, or *positroids*. See Section 7.7 for more about these.

1.9 Use results from this chapter to prove *Carathéodory's Theorem*: If P is a set of points in a d-dimensional space and q is in the convex hull of P then q is in the convex hull of some subset of P of size at most $d + 1$.

2

Oriented Matroids

Now that we have some geometric motivation, we will generalize to a purely combinatorial theory of oriented matroids. As promised in Chapter 1, we'll introduce axiom systems for various representations of oriented matroids and show them to be cryptomorphic.

2.1 Scalar Multiplication and Hypersums of Sign Vectors

Throughout we'll make use of the hypersum operation and inner product introduced in Section 1.5.2. Hypersum notation was first used in matroid theory in 2016 (Baker and Bowler 2019), where it was used to describe a vast generalization of oriented matroids, called *matroids over hyperfields*. Those of us who were around before that wonder how we could have missed such an elegant and streamlined approach.

Definition 2.1 Define the scalar product of a sign and a sign vector in the obvious way: For $X \in \{0, +, -\}^E$ and $\alpha \in \{0, +, -\}$, αX is the sign vector with $(\alpha X)(e) = \alpha(X(e))$ for all $e \in E$.

For every $X, Y \in \{0, +, -\}^E$, the hypersum $X \boxplus Y$ is defined to be $\{Z \in \{0, +, -\}^E : \forall e \in E\ Z(e) \in X(e) \boxplus Y(e)\}$.

Example 2.2 If $\mathbf{x}, \mathbf{y} \in \mathbb{R}^E$, $a, b \in \mathbb{R}$, $X = \text{sign}(\mathbf{x})$, $Y = \text{sign}(\mathbf{y})$, $\alpha = \text{sign}(a)$, and $\beta = \text{sign}(b)$ then $\alpha X = \text{sign}(a\mathbf{x})$ and $\text{sign}(a\mathbf{x} + b\mathbf{y}) \in \alpha X \boxplus \beta Y$.

Not all elements of $\alpha X \boxplus \beta Y$ need arise as signs of linear combinations of $\mathbf{x}$ and $\mathbf{y}$. For instance, if $\mathbf{x} = (1,1)$ and $\mathbf{y} = (-1, -1)$ then $X \boxplus Y = (+,+) \boxplus (-,-) = \{0, +, -\}^2$ but $\{\text{sign}(a\mathbf{x} + b\mathbf{y}) : a > 0, b > 0\} = \{(+,+), (0,0), (-,-)\}$.

Example 2.3 Consider a signed hyperplane arrangement $\mathcal{A}$ in a vector space W over $\mathbb{R}$, $\mathcal{M}$ the oriented matroid associated to $\mathcal{A}$, and $X_1, \ldots, X_k \in \mathcal{V}^*(\mathcal{M})$.

43

Recall from Section 1.3.4 that the elements of $\mathcal{V}^*(\mathcal{M})$ index cones in the partition of W given by $\mathcal{A}$. The union of the cones indexed by $(\boxplus_{i=1}^k X_i) \cap \mathcal{V}^*(\mathcal{M})$ is a convex set: It's the intersection of

- all H_e^0 such that $X_i(e) = 0$ for each i,
- all H_e^+ such that $+ \in \{X_1(e), \ldots, X_k(e)\} \subseteq \{0, +\}$, and
- all H_e^- such that $- \in \{X_1(e), \ldots, X_k(e)\} \subseteq \{0, -\}$.

An observation we'll use often: If $Z \in X \boxplus Y$ then $\text{supp}(Z) \subseteq \text{supp}(X) \cup \text{supp}(Y)$ and $S(X, Z) \subseteq S(X, Y)$.

Notation 2.4 For every $\mathcal{F} \subseteq \{0, +, -\}^E$ we define $-\mathcal{F}$ to be $\{-X : X \in \mathcal{F}\}$.

2.2 Vectors and Signed Circuits

2.2.1 Vector and Signed Circuit Axioms

The following definition uses the separation set (Definition 1.7) and composition operation (Definition 1.27).

Definition 2.5 Let E be a finite set. A subset $\mathcal{V}$ of $\{0, +, -\}^E$ is the **set of vectors of an oriented matroid** on E if and only if all of the following hold.

1. $\mathbf{0} \in \mathcal{V}$.
2. (Symmetry) $\mathcal{V} = -\mathcal{V}$.
3. (Composition) For every $X, Y \in \mathcal{V}$ we have $X \circ Y \in \mathcal{V}$.
4. (Vector Elimination) For every $X, Y \in \mathcal{V}$ and $e \in S(X, Y)$ there exists $Z \in (X \boxplus Y) \cap \mathcal{V}$ such that $Z(e) = 0$.

For example, if W is a linear subspace of $\mathbb{R}^E$ and $\mathcal{V} = \{\text{sign}(\mathbf{x}) : \mathbf{x} \in W\}$ then linearity shows that $\mathcal{V}$ satisfies all of these axioms. In particular, if $\mathbf{x}, \mathbf{y} \in W$ then $\text{sign}(\mathbf{x}) \circ \text{sign}(\mathbf{y}) = \text{sign}(\mathbf{x} + \epsilon \mathbf{y})$ for all sufficiently small $\epsilon > 0$, and if $\text{sign}(x_e) = -\text{sign}(y_e) \neq 0$ then $\text{sign}(|y_e|\mathbf{x} + |x_e|\mathbf{y})$ is an element of $(\text{sign}(\mathbf{x}) \boxplus \text{sign}(\mathbf{y})) \cap \mathcal{V}$ whose value on e is 0.

Problem 2.6 Describe informally why the Vector Elimination and Composition Axioms hold for

1. the vector set of a vector arrangement,
2. the covector set of a vector arrangement,
3. the covector set of a signed hyperplane arrangement.

2.2 Vectors and Signed Circuits

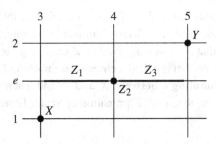

Figure 2.1 Three vector eliminations of e between X and Y.

Find your descriptions directly in each case, rather than going through subspaces of $\mathbb{R}^E$. The point here is to see the intuition for the axioms in each of these settings.

The Z of the Vector Elimination Axiom is called an **elimination of e between X and Y**. It is typically not unique. For example, consider the affine hyperplane arrangement shown in Figure 2.1, with the hyperplanes signed however you choose, and the indicated covectors of the resulting oriented matroid. Any of the three covectors Z_1, Z_2, Z_3 is an elimination of e between X and Y.

When we use Vector Elimination to deduce the existence of Z, we refer to the deduction as **eliminating e between X and Y**.

Definition 2.7 Let E be a finite set. A subset $\mathcal{C}$ of $\{0, +, -\}^E$ is the **set of signed circuits of an oriented matroid** on E if and only if all of the following hold:

1. $\mathbf{0} \notin \mathcal{C}$.
2. (Symmetry) $\mathcal{C} = -\mathcal{C}$.
3. (Incomparability) For every $X, Y \in \mathcal{C}$, if $\operatorname{supp}(X) \subseteq \operatorname{supp}(Y)$ then $X = Y$ or $X = -Y$.
4. (Circuit Elimination) For every $X, Y \in \mathcal{C}$ with $X \neq -Y$ and $e \in S(X,Y)$ there exists $Z \in \mathcal{C}$ such that $Z(f) \in (X(f) \boxplus Y(f)) \cup \{0\}$ for each f and $Z(e) = 0$.

Problem 2.8 Use results from Chapter 1 to explain why the Circuit Elimination Axiom holds for the signed circuit set of a realizable oriented matroid. Describe geometrically why the Circuit Elimination Axiom holds for the signed cocircuit set of an oriented hyperplane arrangement.

Again, the Z of the Circuit Elimination Axiom need not be unique, and it may or may not be an elimination in the sense of the vector axioms.

For instance, in Figure 2.1, either of the signed circuits that are less than Z_1 satisfies the conclusion of the Circuit Elimination Axiom, as does either of the signed circuits that are less than Z_3. Any Z satisfying the conclusion of the Circuit Elimination Axiom is called a **circuit elimination of e between X and Y**. We refer to **eliminating e between X and Y**, and when necessary we will clarify whether we mean Circuit Elimination or Vector Elimination.

2.2.2 Strong Circuit Elimination

The signed circuit axioms can be strengthened as follows.

Proposition 2.9 (Bland and Las Vergnas 1978, Folkman and Lawrence 1978) *Let $\mathcal{C}$ be a collection of sign vectors satisfying the first three signed circuit axioms. Then $\mathcal{C}$ satisfies the Circuit Elimination Axiom if and only if $\mathcal{C}$ satisfies the following* **Strong Elimination Condition**:

For all $X, Y \in \mathcal{C}$, $e \in S(X, Y)$ and $f \in \mathrm{supp}(X) \backslash S(X, Y)$, there is a $Z \in \mathcal{C}$ such that $Z(g) \in (X(g) \boxplus Y(g)) \cup \{0\}$ for each g, $Z(e) = 0$ and $Z(f) = X(f)$.

We'll prove Proposition 2.9 by induction at the end of this section. While we haven't defined contraction for arbitrary oriented matroids yet, our induction will work by considering the set we will eventually call the set of signed circuits of the contraction of an element p from $\mathcal{C}$.

Proposition 2.10 (Bland and Las Vergnas 1978) *Let $\mathcal{F} \subseteq \{0, +, -\}^E$ be such that $\mathbf{0} \notin \mathcal{F}$ and $\mathcal{F}$ satisfies the Symmetry and Circuit Elimination Axioms. Then:*

1. *An element Y of $\mathcal{F}$ is minimal in $\mathcal{F}$ if and only if $\mathrm{supp}(Y)$ is minimal in $\{\mathrm{supp}(X) : X \in \mathcal{F}\}$, and*
2. *the set $\min(\mathcal{F})$ of minimal elements of $\mathcal{F}$ is the set of signed circuits of an oriented matroid.*

Like several upcoming results, the proof of the first part of Proposition 2.10 uses Remark 1.8, that for $X, Y \in \{0, +, -\}^E$, $X \leq Y$ if and only if $\mathrm{supp}(X) \subseteq \mathrm{supp}(Y)$ and $S(X, Y) = \emptyset$.

Proof: 1. Certainly if $\mathrm{supp}(Y)$ is minimal then Y is minimal. If $\mathrm{supp}(Y)$ is not minimal, then choose an element of $\{X \in \mathcal{F} : \mathrm{supp}(X) \subset \mathrm{supp}(Y)\}$ with $S(X, Y)$ as small as possible. Assume by way of contradiction $S(X, Y) \neq \emptyset$. Then eliminating some $e \in S(X, Y)$ between X and Y gives $Z \in \mathcal{F}$ with $\mathrm{supp}(Z) \subseteq \mathrm{supp}(Y) \backslash \{e\}$ and $S(Y, Z) \subseteq S(X, Y) \backslash \{e\}$, a contradiction.

2.2 Vectors and Signed Circuits

2. We first verify that $\min(\mathcal{F})$ satisfies the Incomparability Axiom. Part (1) tells us that the elements of $\min(\mathcal{F})$ are exactly the elements of $\mathcal{F}$ with minimal support. Thus if $X, Y \in \min(\mathcal{F})$ and $\mathrm{supp}(X) \subseteq \mathrm{supp}(Y)$, then $\mathrm{supp}(X) = \mathrm{supp}(Y)$. If $X \neq \pm Y$, then there is some e we can eliminate between X and Y to get a new $Z \in \min(\mathcal{F})$ with $\mathrm{supp}(Z) \subset \mathrm{supp}(X)$, a contradiction.

To see that $\min(\mathcal{F})$ satisfies the Circuit Elimination Axiom, let X and Y be elements of $\min(\mathcal{F})$ with $X \neq \pm Y$, and let $e \in S(X, Y)$. Since $\mathcal{F}$ satisfies the Circuit Elimination Axiom, there is some circuit elimination $Z' \in \mathcal{F}$ of e between X and Y. Any minimal $Z \leq Z'$ in $\mathcal{F}$ is a circuit elimination in $\min(\mathcal{F})$. □

Given $\mathcal{C}$ the set of signed circuits of an oriented matroid on E and $p \in E$, let $\mathcal{C}_p = \{X \backslash p : X \in \mathcal{C}, X \backslash p \neq 0\}$. (Recall from Section 1.7 that in the realizable case the signed circuit set of the contraction of $\mathcal{C}$ by p is $\min(\mathcal{C}_p)$.)

Lemma 2.11 $\min(\mathcal{C}_p)$ *is the set of signed circuits of an oriented matroid.*

Proof: We apply Proposition 2.10. That $0 \notin \mathcal{C}_p$ and $\mathcal{C}_p$ satisfies Symmetry is clear.

To see $\mathcal{C}_p$ satisfies Circuit Elimination, let $X, Y \in \mathcal{C}$ with $X \backslash p, Y \backslash p \in \mathcal{C}_p$, $X \backslash p \neq -Y \backslash p$ and $e \in S(X \backslash p, Y \backslash p)$. Applying Circuit Elimination to X, Y, and e in $\mathcal{C}$, we get a circuit elimination Z.

We have that $Z(p) \in X(p) \boxplus Y(p)$, and so $\{e, p\}$ is a subset of either $\mathrm{supp}(X)$ or $\mathrm{supp}(Y)$. Since $\mathrm{supp}(Z)$ can't be a subset of either of these supports, we have that $Z \backslash p \neq 0$. Thus $Z \backslash p$ eliminates e between $X \backslash p$ and $Y \backslash p$. □

The next two lemmas tell us more about the elements of $\min(\mathcal{C}_p)$.

Lemma 2.12 *Let $\mathcal{C}$ be the set of signed circuits of an oriented matroid. If $X \in \mathcal{C}$ and $\{p\} \subset \mathrm{supp}(X)$ then $X \backslash p \in \min(\mathcal{C}_p)$.*

Proof: If not, there would be $Y \in \mathcal{C}$ with $Y \backslash p < X \backslash p$. But then $\mathrm{supp}(Y) \subset \mathrm{supp}(X)$, violating the Incomparability Axiom. □

Remark 2.13 If $X \in \mathcal{C}$ and $X(p) = 0$ then $X \backslash p$ may or may not be in $\min(\mathcal{C}_p)$, as the examples in Figure 2.2 show.

Lemma 2.14 *Let $\mathcal{C}$ be the set of signed circuits of an oriented matroid on E, and let $p \in E$. For every $X \in \mathcal{C}$ and $f \in \mathrm{supp}(X \backslash p)$ there is some $\hat{X} \in \min(\mathcal{C}_p)$ such that $f \in \mathrm{supp}(\hat{X})$ and $\hat{X} \leq X \backslash p$.*

Remark 2.15 Lemma 2.14 supports the idea that "an oriented matroid is like an oriented hyperplane arrangement." For example, consider the arrangement

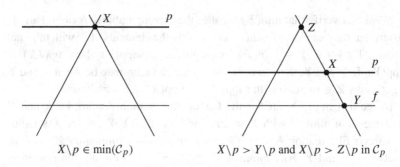

Figure 2.2 Illustration for Remarks 2.13 and 2.15.

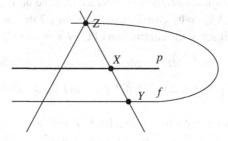

Figure 2.3 Lemma 2.14 rules this out.

on the right in Figure 2.2. The element of $\min(\mathcal{C}_p)$ indicated by Z is the $\hat{X}$ promised by Lemma 2.14. If the lemma were false, the "hyperplane" indicated by f would look more like the curve indicated in Figure 2.3.

The proof of Lemma 2.14 uses a combinatorial analog to a method we used in the proof of Proposition 1.29. There we followed a ray to the first point where it properly intersected a hyperplane of our arrangement. Looking at the arrangement on the left in Figure 2.2, one should imagine a ray from X to $-Y$. The point Z on that ray is our goal.

Proof: If $X\backslash p \in \min(\mathcal{C}_p)$ then $X\backslash p$ itself is the desired $\hat{X}$. So assume this is not the case – in particular, by Lemma 2.12, $p \notin \text{supp}(X)$. Then there is some $Y \in \mathcal{C}$ with $Y\backslash p \in \min(\mathcal{C}_p)$ and $Y\backslash p < X\backslash p$. Thus $\text{supp}(Y) \subset \text{supp}(X) \cup \{p\}$, and so by Incomparability in $\mathcal{C}$ we have $Y(p) \neq 0$.

If $Y(f) \neq 0$ then $Y\backslash p$ is the desired $\hat{X}$. Otherwise, consider

$$R = \{Z \in \mathcal{C} : \text{supp}(Z) \subseteq \text{supp}(X) \cup \{p\} \text{ and } Z(p) = -Y(p)\}.$$

This set is nonempty because $-Y \in R$. Consider $Y' \in R$ with $S(X, Y')$ minimal, and assume by way of contradiction that $S(X, Y') \neq \emptyset$. By eliminating

2.2 Vectors and Signed Circuits

some element of $S(X, Y')$ between X and Y' we get $Y'' \in \mathcal{C}$ with $S(X, Y'') \subset S(X, Y')$ and $\mathrm{supp}(Y'') \subset \mathrm{supp}(X) \cup \mathrm{supp}(Y') \subseteq \mathrm{supp}(X) \cup \{p\}$. Thus by Incomparability $Y''(p) \neq 0$, and so $Y''(p) \in X(p) \boxplus Y'(p) = -Y(p)$. Thus $Y'' \in R$, contradicting the minimality of $S(X, Y')$.

Thus $S(X, Y') = \emptyset$, and so $Y' \backslash p \leq X \backslash p$. We claim that $Y' \backslash p$ is our desired $\hat{X}$. It remains to show $Y'(f) \neq 0$.

Since $Y \backslash p \in \mathcal{C}_p$, we know that $\mathrm{supp}(Y)$ contains an element $g \neq p$, and since $Y \backslash p < X \backslash p$ and $S(X, Y') = \emptyset$, we have $Y(g) = X(g) \geq Y'(g)$. From this we conclude that $Y' \neq -Y$. Thus we can eliminate p between Y and Y' to get $X' \in \mathcal{C}$ with

$$\mathrm{supp}(X') \subseteq (\mathrm{supp}(Y) \cup \mathrm{supp}(Y')) \backslash p$$
$$\subseteq \mathrm{supp}(X)$$

and so by Incomparability $X' = \pm X$. Further

$$X'(g) \in (Y(g) \boxplus Y'(g)) \cup \{0\}) = \{X(g), 0\}$$

and so $X' = X$. In other words, X eliminates p between Y and Y'. Thus $X(f) \in Y(f) \boxplus Y'(f) = 0 \boxplus Y'(f)$. We conclude $Y'(f) = X(f)$. □

Now we're ready to prove Strong Circuit Elimination:

Proof: Consider $X, Y \in \mathcal{C}$, $e \in S(X, Y)$, and $f \in \mathrm{supp}(X) \backslash S(X, Y)$. We get the desired circuit elimination Z by induction on $|\mathrm{supp}(X) \cup \mathrm{supp}(Y)|$. Notice that there are no X and Y satisfying our hypotheses with $|\mathrm{supp}(X) \cup \mathrm{supp}(Y)| \in \{1, 2\}$, and if $|\mathrm{supp}(X) \cup \mathrm{supp}(Y)| = 3$ then Strong Circuit Elimination reduces to ordinary Circuit Elimination.

The inductive step has two cases:

Case 1: If $S(X, Y) \backslash e \neq \emptyset$, then let $p \in S(X, Y) \backslash e$. By Lemma 2.12 $X \backslash p$ and $Y \backslash p$ are elements of $\min(\mathcal{C}_P)$. By the induction hypothesis we can eliminate e between $X \backslash p$ and $Y \backslash p$ in $\min(\mathcal{C}_p)$ to get $Z \backslash p$, where $Z \in \mathcal{C}$ and Z satisfies all of the conditions to be a circuit elimination of e between X and Y except perhaps the condition that $Z(p) \in (X(p) \boxplus Y(p)) \cup \{0\}$. But $X(p) \boxplus Y(p) = \{0, +, -\}$, so this condition holds as well.

Case 2: If $S(X, Y) = \{e\}$, then let U be an elimination of e between X and Y. If $U(f) \neq 0$ then U is our desired Z. Otherwise, since

$$(\mathrm{supp}(X) \cup \mathrm{supp}(Y)) \backslash \{e\} \supseteq \mathrm{supp}(U) \nsubseteq \mathrm{supp}(X),$$

there is a p such that $X(p) = 0$ and $U(p) = Y(p) \neq 0$.

Now consider the element $\hat{X}$ of $\min(\mathcal{C}_p)$ given by Lemma 2.14. $\hat{X} = V \backslash \{p\}$ for some $V \in \mathcal{C}$. Thus $V(f) = X(f)$, for each $g \neq p$ we have

50   Oriented Matroids

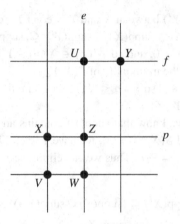

Figure 2.4 Final case of the proof of Proposition 2.9.

$V(g) \in \{0, X(g)\} \subseteq (X(g) \boxplus Y(g)) \cup \{0\}$, and $V(e) \in \{0, X(e)\}$. We have three subcases:

1. If $V(e) = 0$ and $V(p) \in \{0, Y(p)\}$ then V is our desired Z.
2. If $V(e) = 0$ and $V(p) = -Y(p)$ then since $\operatorname{supp}(U) \cup \operatorname{supp}(V) \subseteq (\operatorname{supp}(X) \cup \operatorname{supp}(Y)) \setminus \{e\}$, by the induction hypothesis we can do strong elimination of p between U and V to get our desired Z.
3. If $V(e) = X(e)$, then by the induction hypothesis we can do strong elimination of e in $\min(\mathcal{C}_p)$ between $\hat{X}$ and $Y \setminus \{p\}$ to get $W \setminus \{p\}$ with $W \in \mathcal{C}$, $W(e) = 0$, $W(f) = \hat{X}(f) = X(f)$, and for each $g \neq p$, $W(g) \in (X(g) \boxplus Y(g)) \cup \{0\}$. If $W(p) \leq Y(p)$ then W is our desired Z. Otherwise, we have a final case that is illustrated in the signed cocircuit set of an affine hyperplane arrangement in Figure 2.4. Since $\operatorname{supp}(U) \cup \operatorname{supp}(W) \subseteq (\operatorname{supp}(X) \cup \operatorname{supp}(Y)) \setminus \{e\}$, by the induction hypothesis we can do strong elimination of p between U and W to get our desired Z. □

2.2.3 Cryptomorphism $\mathcal{V} \leftrightarrow \mathcal{C}$

Let Vect denote the set of all collections of sign vectors satisfying the vector axioms, and let Circ denote the set of all collections satisfying the signed circuit axioms. As in the realizable case, there is a cryptomorphism between Vect and Circ.

Theorem 2.16 (cf. Björner et al. 1999, Bland and Las Vergnas 1978, Mandel 1982) *There is a bijection* b: Vect $\to$ Circ, *under which*

2.2 Vectors and Signed Circuits 51

- *for every* $\mathcal{V} \in$ Vect, $b(\mathcal{V}) = \min(\mathcal{V}\backslash\{0\})$, *and*
- *for every* $\mathcal{C} \in$ Circ, $b^{-1}(\mathcal{C})$ *is the set of all compositions of elements of* $\mathcal{C}$.

We have already done the work to see that b is well defined:

Proposition 2.17 *If* $\mathcal{V}$ *is the set of vectors of an oriented matroid, then* $\min(\mathcal{V}\backslash\{0\})$ *is the set of signed circuits of an oriented matroid.*

Proof: This follows immediately from Proposition 2.10. □

Proposition 2.18 *If* $\mathcal{C}$ *is the set of signed circuits of an oriented matroid then the set* $\mathcal{V}$ *of all compositions of elements of* $\mathcal{C}$ *satisfies the Signed Vector Axioms.*

Proof: All of the conditions except Vector Elimination are clear.

To prove Vector Elimination, we first induct on the height of $X \in \mathcal{V}$ in the poset $\mathcal{V}$ to see that X is a composition of elements of $\mathcal{C}$ conformal to X. Above the trivial cases $X = \mathbf{0}$ and $X \in \mathcal{C}$, let $X \in \mathcal{V}\backslash(\mathcal{C} \cup \{0\})$, and assume every element of $\mathcal{V}_{<X}$ is a composition of elements of $\mathcal{C}_{<X}$. Let X' be covered by X in $\mathcal{V}$. (This means that $X' < X$ and there is no $Z \in \mathcal{V}$ such that $X' < Z < X$.) Then $X = X' \circ Y$ for some $Y \in \mathcal{C}$ with $\mathrm{supp}(Y) \subset \mathrm{supp}(X)$. Note that $S(X, Y) = S(X', Y)$. Choose such a Y with $S(X, Y)$ minimal, and assume by way of contradiction there is an $e \in S(X, Y)$. Let $f \in \mathrm{supp}(X)\backslash\mathrm{supp}(X')$. By strong elimination of e between Y and some $X'' \in \mathcal{C}_{<X'}$, we get $Y' \in \mathcal{C}$ such that $Y'(f) = X(f), Y'(e) = 0$, and for every g, $Y'(g) \in (X''(g) \boxplus Y'(g)) \cup \{0\}$. Thus $\mathrm{supp}(Y') \subseteq \mathrm{supp}(X)\backslash e$, $S(X, Y') \subseteq S(X, Y)\backslash e$, and $X' < X' \circ Y' \leq X$. Because X covers X' we conclude $X' \circ Y' = X$, contradicting minimality of $S(X, Y)$. Thus Y is conformal to X, and $X = X' \circ Y$ is a composition of conformal elements of $\mathcal{C}$.

Now we prove Elimination. Let $X, Y \in \mathcal{V}$ and $e \in S(X, Y)$. Then X and Y can each be expressed as a composition of conformal elements of $\mathcal{C}$, say, $X = X_1 \circ \cdots \circ X_k$ and $Y = Y_1 \circ \cdots \circ Y_l$. We induct on $k + l$ to see that there is a vector elimination of e between X and Y.

In the base case X and Y are elements of $\mathcal{C}$. By Strong Elimination, for each $f \in (\mathrm{supp}(X) \cup \mathrm{supp}(Y))\backslash S(X, Y)$ there is a circuit elimination Z_f of e with $Z_f(f) \neq 0$. The composition of all such Z_f is our desired vector elimination.

Above this case, without loss of generality assume $k > 1$. We have cases:

- If $X_1 \circ \cdots \circ X_{k-1}(e) = 0$, then by the induction hypothesis there is a vector elimination Z_1 of e between X_k and Y, and $X_1 \circ \cdots \circ X_{k-1} \circ Z_1$ is a vector elimination of e between X and Y.

- If $X_1 \circ \cdots \circ X_{k-1}(e) \neq 0$ and $X_k(e) = 0$, then by the induction hypothesis there is a vector elimination Z_2 of e between $X_1 \circ \cdots \circ X_{k-1}$ and Y, and $Z_2 \circ X_k$ is a vector elimination of e between X and Y.
- If $X_1 \circ \cdots \circ X_{k-1}(e) \neq 0$ and $X_k(e) \neq 0$, then by the induction hypothesis there is a vector elimination Z_3 of e between $X_1 \circ \cdots \circ X_{k-1}$ and Y and a vector elimination Z_4 of e between X_k and Y, and $Z_3 \circ Z_4$ is a vector elimination of e between X and Y. □

The map $c \colon \text{Circ} \to \text{Vect}$ sending each $\mathcal{C}$ to the set of all compositions of elements of $\mathcal{C}$ is certainly a left inverse to the map b of Theorem 2.16. To complete the proof of Theorem 2.16, it remains to be seen that c is a right inverse to b, that is, every $\mathcal{V} \in \text{Vect}$ is the set of all compositions of elements of $\min(\mathcal{V}\setminus\{\mathbf{0}\})$. We will prove something stronger: a generalization of Proposition 1.29.

Proposition 2.19 (Bland and Las Vergnas 1978) *Let $\mathcal{V}$ be the set of vectors of an oriented matroid and $\mathcal{C} = \min(\mathcal{V}\setminus\{\mathbf{0}\})$. Each $X \in \mathcal{V}$ is a composition of elements of $\mathcal{C}$ that are less than or equal to X.*

Proof: Induct on the height of X in poset $\mathcal{V}$. The minimal cases, when $X = \mathbf{0}$ (the empty composition) and when $X \in \mathcal{C}$, come for free. For larger X, let X' be the composition of all elements of $\mathcal{C}$ that are less than or equal to X, so $X' \leq X$. Assume by way of contradiction $X' < X$. Let e be an element such that $X'(e) = 0$ and $X(e) \neq 0$.

We first check there exists $Y \in \mathcal{V}$ such that $\text{supp}(Y) \subset \text{supp}(X)$ and $Y(e) = X(e)$. This is easy: Let $f \in \text{supp}(X')$ and eliminate f between X and $-X'$ to get the desired Y.

So, take such a Y with $S(X, Y)$ minimal. Assume by way of contradiction there is a $g \in S(X, Y)$. Eliminate g between X and Y to get Y' with $\text{supp}(Y') \subset \text{supp}(X)$, $Y'(e) = X(e)$ and $S(X, Y') \subseteq S(X, Y)\setminus\{g\}$, a contradiction.

Thus, $Y < X$ and $Y(e) \neq 0$. By the induction hypothesis, Y is a composition of signed circuits less than or equal to Y, so one of these signed circuits is a signed circuit $Z < X$ with $Z(e) \neq 0$, contradicting $X'(e) = 0$. □

With the cryptomorphism from vector sets to circuit sets complete, going forward we will refer to "the oriented matroid $\mathcal{M}$ with vector set $\mathcal{V}$ and signed circuit set $\mathcal{C}$."

2.3 Duality

This section proves one of the most central results on oriented matroids – that oriented matroids come in dual pairs that behave like orthogonally complementary real vector spaces. The key to this is a "combinatorial Farkas Property" that holds for all oriented matroids and that plays a similar role here as the Farkas Lemma played for realizable oriented matroids.

Our discussion borrows heavily from Fukuda and Tamura (1990).

Recalling and expanding on notation from Chapter 1:

Notation 2.20 Let $S \subseteq E$.

1. For every $X \in \{0, +, -\}^E$, let $X \backslash S \in \{0, +, -\}^{E \backslash S}$ denote the restriction of X.
2. For every $\mathcal{F} \subseteq \{0, +, -\}^E$, let $\mathcal{F} \backslash S$ denote $\{X \backslash S : X \in \mathcal{F}\}$ and $\mathcal{F}/S$ denote $\{X \backslash S : X \in \mathcal{F}, X(S) = \{0\}\}$.
3. For $X \in \{0, +, -\}^{E \backslash S}$ and $\alpha \in \{0, +, -\}$, let XS^α denote the extension of X to E with value α on each $s \in S$.

For $e \in E$ we often write Xe^α rather than $X\{e\}^\alpha$.

Notation 2.21 Recall that a sign vector can be written as $A^+ B^- C^0$ or $A^+ B^-$. If one of A, B is empty then it can be omitted from the notation. If $X \in \{0, +, -\}^E$ is written in one of these ways, $e \notin E$, and $\alpha \in \{0, +, -\}$ then we denote the extension Xe^α by concatenating our expression for X with e^α. For instance if X is written $A^+ B^-$ then Xe^α is written $A^+ B^- e^\alpha$.

2.3.1 Orthogonality of Sets of Sign Vectors

Here we develop some lemmas to be used in the proof of duality. Recall the definition of orthogonality of sign vectors (Section 1.5.2).

Lemma 2.22 *For sign vectors X, Y, Z, if $X \perp Y$ and $X \perp Z$ then $X \perp Y \circ Z$.*

Proof: We have two cases:

- If $X \cdot Y = \{0\}$ then $X \cdot (Y \circ Z) = X \cdot Z$.
- If $X \cdot Y = \{0, +, -\}$ then there is some f such that $X(f)Y(f) = + = X(f)(Y \circ Z)(f)$ and there is some g such that $X(g)Y(g) = - = X(g)(Y \circ Z)(g)$. Hence $X \cdot (Y \circ Z) = \{0, +, -\}$. □

Proposition 2.23 *Let $\mathcal{V}$ and $\mathcal{C}$ be the sets of vectors resp. signed circuits of an oriented matroid. Then $\mathcal{V}^\perp = \mathcal{C}^\perp$.*

Proof: Because $\mathcal{V} \supseteq \mathcal{C}$, we have $\mathcal{V}^\perp \subseteq \mathcal{C}^\perp$. Conversely, since every vector is a composition of circuits, by Lemma 2.22 $\mathcal{C}^\perp \subseteq \mathcal{V}^\perp$. □

Lemma 2.24 *For every $\mathcal{F} \subseteq \{0, +, -\}^E$ and $S \subset E$,*

$$(\mathcal{F} \backslash S)^\perp = \mathcal{F}^\perp / S.$$

54   Oriented Matroids

Proof: For every $X \in \{0, +, -\}^{E\backslash S}$,

$$X \in (\mathcal{F}\backslash S)^\perp \Leftrightarrow X \perp Y\backslash S \text{ for all } Y \in \mathcal{F}$$
$$\Leftrightarrow XS^0 \perp Y \text{ for all } Y \in \mathcal{F}$$
$$\Leftrightarrow XS^0 \in \mathcal{F}^\perp$$
$$\Leftrightarrow X \in \mathcal{F}^\perp/S. \qquad \square$$

Problem 2.25   Give an example of an $\mathcal{F}$ and S such that $(\mathcal{F}/S)^\perp \neq \mathcal{F}^\perp\backslash S$.

2.3.2 Deletion and Contraction

Recall the notions of deletion and contraction in the realizable case (Section 1.7).

If $\mathcal{M}$ is an oriented matroid with vector set $\mathcal{V}$ and elements E and $S \subset E$, then one easily checks that both $\mathcal{V}\backslash S$ and $\mathcal{V}/S$ are the vector sets of oriented matroids.

Definition 2.26   For an oriented matroid $\mathcal{M}$ with vector set $\mathcal{V}$ and S a set of elements of $\mathcal{M}$, define the **contraction** $\mathcal{M}/S$ to be the oriented matroid with vector set $\mathcal{V}\backslash S$ and define the **deletion** $\mathcal{M}\backslash S$ to be the oriented matroid with vector set $\mathcal{V}/S$.

Define the **oriented submatroid** $\mathcal{M}(S)$ **induced by** S to be $\mathcal{M}\backslash(E\backslash S)$, where E is the set of elements of $\mathcal{M}$.

Theorem 2.34 will show that if $\mathcal{V}$ is the set of vectors of $\mathcal{M}$ then $\mathcal{V}^\perp$ satisfies the vector axioms, and that $\mathcal{V}^{\perp\perp} = \mathcal{V}$. As in Chapter 1 we denote $\mathcal{V}^\perp$ by $\mathcal{V}^*$ and call it the *covector set* of $\mathcal{M}$. Theorem 2.34 together with Lemma 2.24 tells us that $(\mathcal{V}\backslash S)^* = \mathcal{V}^*/S$ and $(\mathcal{V}/S)^* = (\mathcal{V}^{\perp\perp}/S)^* = \mathcal{V}^*\backslash S$.

The notation for deletion and contraction makes more sense in terms of $\mathcal{V}^*$: The contraction $\mathcal{M}/S$ is the oriented matroid with covector set $\mathcal{V}^*/S$, and the deletion $\mathcal{M}\backslash S$ is the oriented matroid with covector set $\mathcal{V}^*\backslash S$.

(For the record, the notation here for $\mathcal{V}\backslash S$ departs from that of Björner et al. (1999).)

Definition 2.27   A **minor** of $\mathcal{M}$ is an oriented matroid obtained from $\mathcal{M}$ by deletions and/or contractions.

Observe for every $\mathcal{F} \subseteq \{0, +, -\}^E$ and for every A and B disjoint subsets of E

$$(\mathcal{F}/A)/B = \mathcal{F}/(A \cup B),$$

$$(\mathcal{F}\backslash A)\backslash B = \mathcal{F}\backslash(A \cup B),$$

$$(\mathcal{F}/A)\backslash B = (\mathcal{F}\backslash B)/A.$$

Thus every minor can be described as obtained by a deletion followed by a contraction or vice versa.

2.3.3 The Farkas Property

Throughout the following $C \dot\cup D$ denotes the union of two disjoint sets C and D.

Exercise 1.7 suggests a "combinatorial Farkas Property" for collections of sign vectors.

Definition 2.28 Assume $\mathcal{F} \subseteq \{0, +, -\}^E$. Let $Z = A^+ B^- (C \dot\cup D)^0 \in \{0, +, -\}^E$, and let $j \in A \cup B$. We say $\mathcal{F}$ satisfies the (Z, C, D, j)-**Farkas Property** if exactly one of the following holds.

1. There exists $X \in \mathcal{F}$ such that $X(e) \leq Z(e)$ for every $e \in A \cup B \cup C$ and $X(j) \neq 0$.
2. There exists $Y \in \mathcal{F}^\perp$ such that $Y(e) \leq Z(e)$ for every $e \in A \cup B \cup D$ and $Y(j) \neq 0$.

We say $\mathcal{F}$ satisfies the **Farkas Property** if $\mathcal{F}$ satisfies the (Z, C, D, j)-Farkas Property for all (Z, C, D, j).

Remark 2.29 The Farkas Property is known elsewhere (cf. Fukuda and Tamura 1990, Bland and Las Vergnas 1978, Bland and Las Vergnas 1979) as the "4-Painting Property." It generalizes Minty's 4-Painting Property for directed graphs (Minty 1966). The special case when $A = E$ is called the Farkas Property in Fukuda and Tamura (1990).

We can simplify the conditions needed to check that a set $\mathcal{F}$ satisfies the Farkas Property. First of all, note that for every $\mathcal{F}$, at *most* one of the two alternatives of the Farkas Property holds. For if both held, the X of Alternative (1) and the Y of Alternative (2) would not be orthogonal. Thus, in the Farkas Property, "exactly one of the following" can be replaced with "at least one of the following."

Another simplification comes from the following lemma.

Lemma 2.30 *If $\mathcal{F}$ satisfies the (Z, C, D, j)-Farkas Property for all $Z = A^+ B^- (C \dot\cup D)^0$ with $|A \cup B| = 1$, then $\mathcal{F}$ satisfies the Farkas Property.*

Proof: Let $Z = A^+ B^- (C \dot\cup D)^0$, and let $j \in A \cup B$. We show $\mathcal{F}$ satisfies the (Z, C, D, j)-Farkas Property by induction on $|A \cup B|$, with the case $|A \cup B| = 1$ true by hypothesis.

For larger $|A \cup B|$, let f be an element of $(A \cup B) \setminus \{j\}$. Let

$$Z' = (A \setminus \{f\})^+ (B \setminus \{f\})^- (C \cup D \cup \{f\})^0.$$

The induction hypothesis tells us that $\mathcal{F}$ satisfies both the $(Z', C \cup \{f\}, D, j)$-Farkas Property and the $(Z', C, D \cup \{f\}, j)$-Farkas Property.

If Alternative (1) holds for $(Z', C \cup \{f\}, D, j)$ then the resulting X satisfies Alternative (1) for (Z, C, D, j). Likewise if Alternative (2) holds for $(Z', C, D \cup \{f\}, j)$ then the resulting Y satisfies Alternative (2) for (Z, C, D, j).

So assume Alternative (2) holds for $(Z', C \cup \{f\}, D, j)$ and Alternative (1) holds for $(Z', C, D \cup \{f\}, j)$, and consider the resulting $Y \in \mathcal{F}^\perp$ and $X \in \mathcal{F}$. Then the orthogonality of X and Y implies that $X(f) = -Y(f) \neq 0$. Thus either X satisfies Alternative (1) for (Z, C, D, j) or Y satisfies Alternative (2) for (Z, C, D, j). □

If $\mathcal{F} = -\mathcal{F}$, then we can simplify the Farkas Property even further, to the case $A = \{j\}$ and $B = \emptyset$. For if $X \in \mathcal{F}$ satisfies Alternative (1) for the $(\{j\}^-, C, D, j)$-Farkas property then $-X \in \mathcal{F}$ satisfies Alternative (1) for the $(\{j\}^+, C, D, j)$-Farkas property. Since $\mathcal{F}^\perp$ always satisfies Symmetry, we have a similar observation for Alternative (2).

Summing this up, we have:

Proposition 2.31 *Assume* $\mathcal{F} \subseteq \{0, +, -\}^E$ *and* $\mathcal{F} = -\mathcal{F}$. *Then* $\mathcal{F}$ *satisfies the Farkas Property if and only if $\mathcal{F}$ satisfies the following:*

For every $j \in E$ and sets C and D with $C \dot\cup D = E \setminus \{j\}$, at least one of the following holds:

1. *There exists $X \in \mathcal{F}$ such that $X(j) \neq 0$ and $X(C) \subseteq \{0\}$.*
2. *There exists $Y \in \mathcal{F}^\perp$ such that $Y(j) \neq 0$ and $Y(D) \subseteq \{0\}$.*

The following proposition is the critical element in our proof of duality.

Proposition 2.32 (Fukuda and Tamura 1990) *In the axioms for vector sets of oriented matroids, the Vector Elimination Property can be replaced with the Farkas Property.*

That is, $\mathcal{V} \subseteq \{0, +, -\}^E$ *is the vector set of an oriented matroid if and only if* $0 \in \mathcal{V}$ *and $\mathcal{V}$ satisfies the Symmetry Axiom, the Composition Axiom, and the Farkas Property.*

Proof: Let $\mathcal{V} \subseteq \{0, +, -\}^E$ satisfy $0 \in \mathcal{V}$ and the Symmetry and Composition Axioms. We first check that if $\mathcal{V}$ satisfies the Farkas Property then $\mathcal{V}$ satisfies Vector Elimination.

Given $X, Y \in \mathcal{V}$ and $e \in S(X, Y)$, let

$$A = \{e : X(e) \boxplus Y(e) = +\},$$

$$B = \{e : X(e) \boxplus Y(e) = -\},$$

$$C = \{e : X(e) \boxplus Y(e) = 0\} \cup \{e\},$$

$$D = \{e : X(e) \boxplus Y(e) = \{0, +, -\}\} \setminus \{e\} = S(X, Y) \setminus \{e\}.$$

For every $j \in A \cup B$, there is no $Z \in \mathcal{V}^\perp$ with $Z(j) \neq 0, Z(A) \subseteq \{0, +\}, Z(B) \subseteq \{0, -\}$, and $Z(D) \subseteq \{0\}$, because such a Z is not orthogonal to $X \circ Y$. Thus, by the Farkas Property, for every $j \in A \cup B$, there is $Z^{(j)} \in \mathcal{V}$ such that $Z^{(j)}(j) \neq 0, Z^{(j)}(A) \subseteq \{0, +\}, Z^{(j)}(B) \subseteq \{0, -\}$, and $Z^{(j)}(C) = \{0\}$. A composition of all $Z^{(j)}$ over all $j \in A \cup B$, in whatever order, is the desired elimination of e.

To see that the Farkas Property holds if Vector Elimination holds, it's enough to check the equivalent formulation of the Farkas Property given in Proposition 2.31. Thus we wish to show: For each $j \in E$ and sets C and D with $C \dot\cup D = E \setminus \{j\}$, at least one of the following holds:

1* There exists $X \in \mathcal{V}$ such that $X(j) \neq 0$ and $X(C) \subseteq \{0\}$.
2* There exists $Y \in \mathcal{V}^\perp$ such that $Y(j) \neq 0$ and $Y(D) \subseteq \{0\}$.

We prove this by induction on E. In the minimal case, $E = \{j\}$ and the result is trivial. Above this, we have two cases:

Case 1: $D \neq \emptyset$. Let $f \in D$, and consider the vector set $\mathcal{V} \setminus f$. By Lemma 2.24, $(\mathcal{V} \setminus f)^\perp = \mathcal{V}^\perp / f$. Thus by the induction hypothesis one of the two alternatives holds for $\mathcal{V} \setminus f, \mathcal{V}^\perp / f$ and the sets C and $D \setminus f$, and either of these alternatives gives the corresponding alternative for $\mathcal{V}, \mathcal{V}^\perp, C$, and D.

Case 2: $D = \emptyset$. Let $e \in C$. Then by the induction hypothesis one of the two alternatives of Proposition 2.31 holds for $\mathcal{V} \setminus e, (\mathcal{V} \setminus e)^\perp = \mathcal{V}^\perp / e$, and the sets $C \setminus e$ and $D = \emptyset$. If Alternative (2) holds, then this $Y \setminus e \in \mathcal{V}^\perp / e$ gives $Y \in \mathcal{V}^\perp$ satisfying Alternative ($2*$).

So, assume that for every $e \in C$ there is no such Y. Thus, for every $e \in C$ there is some $X \in \mathcal{V}$ with $\{j\} \subseteq \operatorname{supp}(X) \subseteq \{e, j\}$. If $\operatorname{supp}(X) = \{j\}$ for some such X, then X satisfies Alternative ($1*$).

It remains to consider the possibility that $j^+ \notin \mathcal{V}$, and for every $e \in C$, there exists $X \in \mathcal{V}$ with $\operatorname{supp}(X) = \{e, j\}$. Thus for each $e \in C$ there is an $\alpha_e \in \{+, -\}$ such that $j^+ e^{\alpha_e} \in \mathcal{V}$. Since $j^+ \notin \mathcal{V}$ and every vector is a composition of signed circuits, we have $j^+ e^{\alpha_e} \in \mathcal{C}$.

From this we can deduce all of $\mathcal{C}$. For every $e_1 \neq e_2$, we can eliminate j between $j^+ e^{\alpha_{e_1}}$ and $j^- e^{-\alpha_{e_2}}$ to get $e_1^{\alpha_{e_1}} e_2^{-\alpha_{e_2}} \in \mathcal{V}$. Since each $j^+ e_i^{\alpha_{e_i}} \in \mathcal{C}$, we see $e_i^{\alpha_{e_i}} \notin \mathcal{C}$, so in fact $e_1^{\alpha_{e_1}} e_2^{-\alpha_{e_2}} \in \mathcal{C}$. Thus we have shown that every pair of elements of E is the support of some signed circuit, so by Incomparability we have found all of the signed circuits: $\mathcal{C} = \{\pm j^+ e^{\alpha_e} : e \in C\} \cup \{e_1^{\alpha_{e_1}} e_2^{-\alpha_{e_2}} : e_1, e_2 \in C, e_1 \neq e_2\}$.

Define $Z \in \{0, +, -\}^E$ by $Z(j) = +$ and $Z(e) = -\alpha_e$ for each $e \in C$. Then $Z \in C^\perp$. But $C^\perp = \mathcal{V}^\perp$ by Proposition 2.23, and so Alternative (2*) holds. □

The operator on the power set of $\{0, +, -\}^E$ taking each $\mathcal{F}$ to $\mathcal{F}^\perp$ is a Galois connection (cf. section 1.5 in Kung et al. 2009). In particular, $(\mathcal{F}^\perp)^\perp \supseteq \mathcal{F}$ and $((\mathcal{F}^\perp)^\perp)^\perp = \mathcal{F}^\perp$. If $\mathcal{F}$ satisfies the Farkas Property then we can say more about these sets.

Lemma 2.33 *Let $\mathcal{F} \subseteq \{0, +, -\}^E$ satisfy the Farkas Property.*

1. *$\mathcal{F}^\perp$ also satisfies the Farkas Property.*
2. *$(\mathcal{F}^\perp)^\perp$ consists of all compositions of elements of $\mathcal{F}$.*

Proof: 1. To see that the (Z, C, D, j)-Farkas Property holds for $\mathcal{F}^\perp$, consider the (Z, D, C, j)-Farkas Property for $\mathcal{F}$. If Alternative (1) of the (Z, D, C, j)-Farkas Property for $\mathcal{F}$ holds, then since $\mathcal{F} \subseteq (\mathcal{F}^\perp)^\perp$, Alternative (2) of the (Z, C, D, j)-Farkas Property for $\mathcal{F}^\perp$ holds. If Alternative (2) of the (Z, D, C, j)-Farkas Property for $\mathcal{F}$ holds, then Alternative (1) of the (Z, C, D, j)-Farkas Property for $\mathcal{F}^\perp$ holds. Thus at least one of the two alternatives of the (Z, C, D, j)-Farkas Property for $\mathcal{F}^\perp$ holds, and by the same argument we have seen before both alternatives cannot simultaneously hold.

2. By Lemma 2.22, every composition of elements of $\mathcal{F}$ is in $(\mathcal{F}^\perp)^\perp$. Conversely, consider $Z \in (\mathcal{F}^\perp)^\perp$. For each $j \in \text{supp}(Z)$, we apply the $(Z, Z^0, \emptyset, j)$-Farkas Property of $\mathcal{F}$. The second alternative for this property would give a $Y \in \mathcal{F}^\perp$ with $Y \cdot Z = \{+\}$, a contradiction, and so the first alternative holds: There is an $X_j \in \mathcal{F}$ such that $X_j \leq Z$ and $X_j(j) \neq 0$. The composition of the X_j over all $j \in \text{supp}(Z)$ is Z. □

2.3.4 Cryptomorphism $\mathcal{V} \leftrightarrow \mathcal{V}^*$

Now we use the Farkas Property to prove duality.

Theorem 2.34 *If $\mathcal{V}$ is the set of vectors of an oriented matroid then:*

1. *$\mathcal{V}^\perp$ is also the set of vectors of an oriented matroid, and*
2. *$(\mathcal{V}^\perp)^\perp = \mathcal{V}$.*

Proof: The second statement follows from the second part of Lemma 2.33.

It's clear that $\mathcal{V}^\perp$ satisfies Symmetry, and by Lemma 2.22 $\mathcal{V}^\perp$ satisfies Composition. Applying Proposition 2.3.3, we see $\mathcal{V}$ satisfies the Farkas Property, so by Lemma 2.33 $\mathcal{V}^\perp$ also satisfies the Farkas Property, and so $\mathcal{V}^\perp$ satisfies Elimination. □

2.3 Duality

Definition 2.35 If an oriented matroid $\mathcal{M}$ has vector set $\mathcal{V} \subset \{0, +, -\}^E$, then $\mathcal{V}^\perp$ is denoted $\mathcal{V}^*$ and is called the set of **covectors** of $\mathcal{M}$. The oriented matroid with vector set $\mathcal{V}^*$ and covector set $\mathcal{V}$ is called the **dual** to $\mathcal{M}$, denoted $\mathcal{M}^*$. The set $\mathcal{C}^*$ of minimal nonzero elements of $\mathcal{V}^*$ is called the set of **signed cocircuits** of $\mathcal{M}$. The set E is called the set of **elements** of $\mathcal{M}$.

We may also denote the vector set, covector set, signed circuit set, and signed cocircuit set of $\mathcal{M}$ by $\mathcal{V}(\mathcal{M})$, $\mathcal{V}^*(\mathcal{M})$, $\mathcal{C}(\mathcal{M})$, and $\mathcal{C}^*(\mathcal{M})$.

2.3.5 Dual Pair Axioms for Signed Circuits and Signed Cocircuits

We introduce here another interesting characterization of signed circuits and signed cocircuits that will be used in our cryptomorphism between signed circuit sets and chirotopes.

Proposition 2.36 *Let $\mathcal{F} \subseteq \{0, +, -\}^E$ such that $\mathcal{F} = -\mathcal{F}$ and every element X of $\mathcal{F}$ resp. $\mathcal{F}^\perp$ can be written as a composition of elements of $\min(\mathcal{F}\backslash\{0\})$ resp. $\min(\mathcal{F}^\perp\backslash\{0\})$ that are conformal to X. Then $\mathcal{F}$ satisfies the Farkas Property if and only if $\mathcal{F}$ satisfies the following.*

For every $j \in E$ and sets C, D with $C \dot\cup D = E\backslash\{j\}$, at least one of the following holds:

1. *There exists $X \in \min(\mathcal{F}\backslash\{0\})$ such that $X(j) \neq 0$ and $X(C) \subseteq \{0\}$.*
2. *There exists $Y \in \min(\mathcal{F}^\perp\backslash\{0\})$ such that $Y(j) \neq 0$ and $Y(D) \subseteq \{0\}$.*

This follows immediately from Proposition 2.31.

Proposition 2.37 (Dual Pair Axioms) *Let $\mathcal{D}$, $\mathcal{D}' \subseteq \{0, +, -\}^E$ such that $\mathcal{D} \perp \mathcal{D}'$ and $\mathcal{D}$ and $\mathcal{D}'$ each satisfies the Incomparability and Symmetry Axioms. Then $\mathcal{D}$, $\mathcal{D}'$ are the signed circuit set and signed cocircuit set of an oriented matroid if and only if for every $j \in E$ and sets C, D with $C \dot\cup D = E\backslash\{j\}$, at least one of the following holds:*

1. *There exists $X \in \mathcal{D}$ such that $X(j) \neq 0$ and $X(C) \subseteq \{0\}$.*
2. *There exists $Y \in \mathcal{D}'$ such that $Y(j) \neq 0$ and $Y(D) \subseteq \{0\}$.*

Proof: The forward direction follows from the Farkas Property for vectors and covectors and Proposition 2.36.

Conversely, recall (Proposition 2.22) that for every $\mathcal{F} \subseteq \{0, +, -\}^E$, every composition of elements of $\mathcal{F}^\perp$ is in $\mathcal{F}^\perp$. So let $\mathcal{V}$ be the set of all compositions of elements of $\mathcal{D}$, and let $\mathcal{V}'$ be the set of all compositions of elements of $\mathcal{D}'$.

Then $\mathcal{V} \perp \mathcal{V}'$ and $\mathcal{V}$ and $\mathcal{V}'$ satisfy the Symmetry and Composition Axioms and the Farkas Property. Thus $\mathcal{V}$ and $\mathcal{V}'$ are the vectors and covectors of an oriented matroid, and by Proposition 2.10 $\mathcal{D}$ is the set of signed circuits of $\mathcal{V}$, and $\mathcal{D}'$ is the set of signed circuits of $\mathcal{V}'$. □

There are various formulations of dual pair axioms (cf. Bland and Las Vergnas 1978), and dual pair axioms have an interesting role beyond oriented matroids. While our proof used vector and covector sets, one can prove cryptomorphisms between circuit sets, dual pairs, and chirotopes directly (cf. chapter 3 of Björner et al. 1999). This is a better approach in contexts beyond oriented matroids, including ordinary matroids and matroids over hyperfields, in which vectors and covectors are less natural objects.

2.3.6 Notes on Duality

1. The Farkas Lemma plays a central role in linear programming. Indeed, interest in duality from linear programming was a driving force in the development of oriented matroids. For further discussion of this – indeed, an entire introduction to oriented matroids from the perspective of linear programming – see Bachem and Kern (1992).
2. The only difference between the Farkas Lemma and Minty's "4-painting" is one of approach: 4-painting is a means to describe oriented matroids in terms of ordinary matroids, while the Farkas Lemma describes oriented matroids in terms of orthogonal families of sign vectors. For more on 4-painting, see Minty's original discussion (Minty 1966) – which developed painting on a restricted collection of matroids – or Bland and Las Vergnas's treatment for general oriented matroids (Bland and Las Vergnas 1978, Bland and Las Vergnas 1979).
3. It can be proved (Kung 1983, Bland and Dietrich 1988) that duality is the only "good" involution on the set of all oriented matroids. By "good" we mean an involution that preserves the set of elements and interchanges deletion and contraction.

2.4 Chirotopes

2.4.1 The Grassmann–Plücker Relations

Recall the definition of the chirotope associated to a matrix $M \in \text{Mat}(r, n)$ as $\chi(i_1, \ldots, i_r) = \text{sign}(|M_{i_1, \ldots, i_r}|)$. Our chirotope axioms should thus be

2.4 Chirotopes

motivated by conditions satisfied by the signs of the maximal minors of M. Two such conditions are easy: χ should be nonzero and alternating. Not surprisingly, these two conditions are not enough to give chirotope axioms cryptomorphic to the other oriented matroid axioms. We need a more subtle property of determinants, given by the following theorem.

Theorem 2.38 *Let $\mathbb{K}$ be a field and $f : [n]^r \to \mathbb{K}$ be a nonzero alternating function. The following are equivalent.*

1. There is a $r \times n$ matrix M of rank r over $\mathbb{K}$ such that for each $i_1, \ldots, i_r \in [n]$,

$$f(i_1, \ldots, i_r) = |M_{i_1, \ldots, i_r}|.$$

2. For each $i_1, \ldots, i_r, j_1, \ldots, j_r \in [n]$,

$$f(i_1, \ldots, i_r) f(j_1, \ldots, j_r) = \qquad (2.1)$$

$$\sum_{k=1}^{r} f(j_k, i_2, \ldots, i_r) f(j_1, \ldots, j_{k-1}, i_1, j_{k+1}, \ldots, j_r).$$

Lemma 2.39 *Let $\mathbb{K}$ be a field and $\mathbf{x}_1, \ldots, \mathbf{x}_r, \mathbf{y}_1, \ldots, \mathbf{y}_r \in \mathbb{K}^r$. Then*

$$|\mathbf{x}_1 \cdots \mathbf{x}_r||\mathbf{y}_1 \cdots \mathbf{y}_r| = \sum_{k=1}^{r} |\mathbf{y}_k \mathbf{x}_2 \cdots \mathbf{x}_r||\mathbf{y}_1 \cdots \mathbf{y}_{k-1} \mathbf{x}_1 \mathbf{y}_{k+1} \cdots \mathbf{y}_r|. \qquad (2.2)$$

An elegant proof of Lemma 2.39 comes from the exterior algebra (Section 1.4.3).

Proof: For a fixed $\mathbf{x}_2, \ldots, \mathbf{x}_r \in \mathbb{K}^r$, consider the function from $(\mathbb{K}^r)^{r+1}$ to $\mathbb{K}$ sending each $(\mathbf{x}_1, \mathbf{y}_1, \ldots, \mathbf{y}_r)$ to

$$|\mathbf{x}_1 \cdots \mathbf{x}_r||\mathbf{y}_1 \cdots \mathbf{y}_r| - \sum_{k=1}^{r} |\mathbf{y}_k \mathbf{x}_2 \cdots \mathbf{x}_r||\mathbf{y}_1 \cdots \mathbf{y}_{k-1} \mathbf{x}_1 \mathbf{y}_{k+1} \cdots \mathbf{y}_r|.$$

This is an alternating multilinear function in $r + 1$ variables, so it factors into a composition $(\mathbb{K}^r)^{r+1} \to \bigwedge^{r+1} \mathbb{K}^r \to \mathbb{K}$. But $\bigwedge^{r+1} \mathbb{K}^r = \{0\}$, by Corollary 1.24. □

A more naive approach better fits the spirit of our proof of Theorem 2.38. For both proofs we use an observation we first used in proving Proposition 1.21: For a square matrix $(x_{i,j} : i, j \in [r])$, given by column vectors as $(\mathbf{x}_1, \ldots, \mathbf{x}_r)$,

$$|\mathbf{e}_1, \ldots, \mathbf{e}_{i-1}, \mathbf{x}_j, \mathbf{e}_{i+1}, \ldots, \mathbf{e}_r| = x_{i,j}.$$

Additionally, $(-1)^{i-1}|\mathbf{e}_i,\mathbf{x}_2,\ldots,\mathbf{x}_r|$ is the determinant of the submatrix of $(x_{i,j} : i,j \in [r])$ obtained by deleting the first column and the ith row. Thus

$$\sum_{i=1}^r |\mathbf{e}_i,\mathbf{x}_2,\ldots,\mathbf{x}_r||\mathbf{e}_1,\ldots,\mathbf{e}_{i-1},\mathbf{x}_1,\mathbf{e}_{i+1},\ldots,\mathbf{e}_r|$$

is the cofactor expansion of $|\mathbf{x}_1,\ldots,\mathbf{x}_r|$ along its first column.

Naive Proof of Lemma 2.39: If $\{\mathbf{y}_1,\ldots,\mathbf{y}_r\}$ has rank less than r, then the left-hand side of Equation (2.2) is 0, and by expressing one of the $\mathbf{y}_i$ as a linear combination of the others, one can easily check that the right-hand side is 0 as well.

If $(\mathbf{y}_1,\ldots,\mathbf{y}_r) = (\mathbf{e}_1,\ldots,\mathbf{e}_r)$, then the left-hand side of Equation 2.2 is $|\mathbf{x}_1\cdots\mathbf{x}_r|$, and by our preceding observations the right-hand side is the cofactor expansion of $|\mathbf{x}_1\cdots\mathbf{x}_r|$ along its first column.

For general invertible $(\mathbf{y}_1,\ldots,\mathbf{y}_r)$, let $A = (\mathbf{y}_1,\ldots,\mathbf{y}_r)^{-1}$. Since for every $r \times r$ matrix $(\mathbf{z}_1,\ldots,\mathbf{z}_r)$ we have $A(\mathbf{z}_1,\ldots,\mathbf{z}_r) = (A\mathbf{z}_1,\ldots,A\mathbf{z}_r)$, we apply the previous paragraph to $A\mathbf{x}_i,\ldots,A\mathbf{x}_r,A\mathbf{y}_1,A\mathbf{y}_r$ to see

$$|A(\mathbf{x}_1\cdots\mathbf{x}_r)||A(\mathbf{y}_1\cdots\mathbf{y}_r)| = \sum_{k=1}^r |A(\mathbf{y}_k\mathbf{x}_2\cdots\mathbf{x}_r)||A(\mathbf{y}_1\cdots\mathbf{y}_{k-1}\mathbf{x}_1\mathbf{y}_{k+1}\cdots\mathbf{y}_r)|.$$

Dividing both sides by $|A|^2$ completes the proof. □

Proof of Theorem 2.38: Lemma 2.39 proves that (1) implies (2). To see the converse, notice that f satisfies (1) if and only if every nonzero scalar multiple of f satisfies (1), and likewise for (2). So given an f satisfying (2) and a choice of a particular $(b_1,\ldots,b_r)$ in the support of f with $b_1 < \cdots < b_r$, let $\hat{f} = f(b_1,\ldots,b_r)^{-1}f$. We'll find a matrix $\hat{M}$ in reduced row-echelon form with respect to $\{b_1,\ldots,b_r\}$ such that $\hat{f}(i_1,\ldots,i_r) = |\hat{M}_{i_1,\ldots,i_r}|$ for all $i_1,\ldots,i_r$. Multiplying the first row of $\hat{M}$ by $f(b_1,\ldots,b_r)$ gives the matrix M we need. For convenience we'll carry this out when $(b_1,\ldots,b_r) = (1,\ldots,r)$.

Let

$$\hat{M} = \left(\begin{array}{c|ccc} & \hat{m}_{1,r+1} & \cdots & \hat{m}_{1,n} \\ I & \vdots & & \vdots \\ & \hat{m}_{r,r+1} & \cdots & \hat{m}_{r,n} \end{array}\right),$$

where $\hat{m}_{i,j} = \hat{f}(1,\ldots,i-1,j,i+1,\ldots,r)$. Thus $\hat{M}$ is defined so that $\hat{f}(i_1,\ldots,i_r) = |\hat{M}_{i_1,\ldots,i_r}|$ whenever all but at most one element of $\{i_1,\ldots,i_r\}$ is in $[r]$. In particular, $\hat{f}(i_1,\ldots,i_r) = |\hat{M}_{i_1,\ldots,i_r}|$ if $\max(i_1,\ldots,i_r) \leq r+1$. We induct on $\max(i_1,\ldots,i_r)$ to see that $\hat{f}(i_1,\ldots,i_r) = |\hat{M}_{i_1,\ldots,i_r}|$ for all $(i_1,\ldots,i_r)$.

2.4 Chirotopes

Since $\hat{f}$ and the determinant function are both alternating, we may assume that this maximum is achieved only at i_1. Applying the definition of $\hat{f}$ and our assumption that $\hat{f}(1,\ldots,r) = 1$, we have

$$\hat{f}(i_1,\ldots,i_r) = \hat{f}(i_1,\ldots,i_r)\hat{f}(1,\ldots,r)$$

$$= \sum_{k=1}^{r} \hat{f}(k,i_2,\ldots,i_r)\hat{f}(1,\ldots,k-1,i_1,k+1,\ldots,r).$$

By our induction hypothesis $\hat{f}(k,i_2,\ldots,i_r) = |\hat{M}_{k,i_2,\ldots,i_r}|$, and by the definition of $\hat{M}$ we have $\hat{f}(1,\ldots,k-1,i_1,k+1,\ldots,r) = |\hat{M}_{1,\ldots,k-1,i_1,k+1,\ldots,r}|$. Thus

$$\hat{f}(i_1,\ldots,i_r) = \sum_{k=1}^{r} |\hat{M}_{k,i_2,\ldots,i_r}||\hat{M}_{1,\ldots,k-1,i_1,k+1,\ldots,r}|,$$

which is the cofactor expansion of $\hat{M}_{i_1,\ldots,i_r}$ along its first column.   □

From Theorem 2.38 we get a condition on the chirotope associated to M: The sign of the left-hand side of Expression (2.1) must be in the hypersum of the signs of the summands on the right-hand side. Thus for the chirotope χ asociated to row(M) we have for each $i_1,\ldots,i_r, j_1,\ldots,j_r \in [n]$

$$\chi(i_1,\ldots,i_r)\chi(j_1,\ldots,j_r) \in \boxplus_{k=1}^{r} \chi(j_k,i_2,i_3,\ldots,i_r)\chi(j_1,\ldots,j_{k-1},i_1,j_{k+1},\ldots,j_r)$$

(2.3)

For general chirotopes, we take each expression of the form given in 2.3 as an axiom.

Remark 2.40 Had we only proved the implication (1)⇒(2) in Theorem 2.38, we would still have arrived at Expression (2.3) as a potential chirotope axiom, but we might have wondered whether there were still more relations between maximal minors that should be reflected in our chirotope axioms. The implication (2)⇒(1) should lead the reader to hope that this is not the case. This hope will be fulfilled when we justify the cryptomorphism between chirotopes, defined via Expression (2.3), and oriented matroids.

Remark 2.41 The equations (2.1) are called the *Grassmann–Plücker relations*. They can be taken as conditions on the projective coordinates of the Plücker embedding of the Grassmannian introduced in Section 1.4.3. In other words, $P(G(r,\mathbb{R}^n))$ is the algebraic variety defined by the Grassmann–Plücker relations.

2.4.2 General Chirotopes

Definition 2.42 A rank r **chirotope** on a set E is a function $\chi : E^r \to \{0, +, -\}$ satisfying:

1. χ is not identically 0.
2. χ is alternating. That is, for every $i_1, \ldots, i_r \in E$ and every permutation σ,

$$\chi(i_{\sigma_1}, \ldots, i_{\sigma_r}) = \mathrm{sign}(\sigma)\chi(i_1, \ldots, i_r).$$

3. (Combinatorial Grassmann–Plücker relations) For all $i_1, \ldots, i_r, j_1, \ldots, j_r \in E$

$$\chi(i_1, \ldots, i_r) \cdot \chi(j_1, \ldots, j_r) \in$$

$$\boxplus_{k=1}^{r} \chi(j_k, i_2, \ldots, i_r)\chi(j_1, \ldots, j_{k-1}, i_1, j_{k+1}, \ldots, j_r).$$

We can see several equivalent versions of the combinatorial Grassmann–Plücker relations by way of a few observations:

1. Because χ is alternating,

$$\chi(j_1, \ldots, j_{k-1}, i_1, j_{k+1}, \ldots, j_r) = (-1)^{k-1}\chi(i_1, j_1, \ldots, \hat{j}_k, \ldots, j_r)$$

(Here $j_1, \ldots, \hat{j}_k, \ldots, j_r$ denotes $j_1, \ldots, j_{k-1}, j_{k+1}, \ldots, j_r$.)

2. For $a, b_1, \ldots, b_r \in \{0, +, -\}$,

   1. $a \in \boxplus_{k=1}^{r} b_k$ if and only if $0 \in a \boxplus \boxplus_{k=1}^{r}(-b_k)$, and
   2. if $a \neq 0$ then $a \in \boxplus_{k=1}^{r} b_k$ if and only if $a = b_k$ for some k.

Proposition 2.43 *In the definition of chirotope, the combinatorial Grassmann–Plücker relations can be replaced by any of the following:*

1. For all $i_1, \ldots, i_r, j_1, \ldots, j_r \in E$

$$\chi(i_1, \ldots, i_r) \cdot \chi(j_1, \ldots, j_r) \in$$

$$\boxplus_{k=1}^{r} (-1)^{k-1}\chi(j_k, i_2, \ldots, i_r)\chi(i_1, j_1, \ldots, \hat{j}_k, \ldots, j_r).$$

2. Let $j \in [r]$. For all $i_1, \ldots, i_r, j_1, \ldots, j_r \in E$

$$\chi(i_1, \ldots, i_r) \cdot \chi(j_1, \ldots, j_r) \in$$

$$\boxplus_{k=1}^{r} (-1)^{k-1}\chi(i_1, \ldots, i_{j-1}, j_k, i_{j+1}, \ldots, i_r)\chi(i_j, j_1, \ldots, \hat{j}_k, \ldots, j_r).$$

3. For all $i_2, \ldots, i_r, j_0, \ldots, j_r \in E$,

$$0 \in \boxplus_{k=0}^{r} (-1)^k \chi(j_k, i_2, \ldots, i_r) \chi(j_0, \ldots, \hat{j_k}, \ldots, j_r).$$

4. *Signed Basis Exchange Axiom: For every* $i_1, \ldots, i_r, j_1, \ldots, j_r \in E$ *such that* $\chi(i_1, \ldots, i_r) \chi(j_1 \ldots, j_r) \neq 0$, *there is a k such that*

$$\chi(j_k, i_2, \ldots, i_r) \chi(j_1, \ldots, j_{k-1}, i_1, j_{k+1}, \ldots, j_r) = \chi(i_1, \ldots, i_r) \chi(j_1 \ldots, j_r).$$

By ignoring signs in Signed Basis Exchange, we get the Basis Exchange Principle, an axiom in unoriented matroid theory that we noted for realizable oriented matroids in Section 1.5.3. (See Definition 2.52 for the definition of a *basis* of a chirotope.)

Corollary 2.44 (Basis Exchange Principle) *Let B and B' be bases of a chirotope χ. Let $i \in B$. There is a $j \in B'$ such that $(B \backslash \{i\}) \cup \{j\}$ and $(B' \backslash \{j\}) \cup \{i\}$ are bases of χ.*

2.4.3 Chirotopes, Circuits, and Cocircuits

Our next goal is Theorem 2.54, the cryptomorphism between chirotopes and oriented matroids. We'll describe the cryptomorphism in two ways: as a map between signed circuit sets and chirotopes, and as a map between signed cocircuit sets and chirotopes.

Before we even begin to associate a chirotope to an oriented matroid, we need to determine the domain of the chirotope. That is, we need to show that an oriented matroid has a well-defined rank. So before we describe our cryptomorphism, we will take care of this and other "unoriented" matters:

- Given an oriented matroid $\mathcal{M}$ on elements E, we will define the bases of $\mathcal{M}$ and show that all bases have the same size r. We call r the *rank*. Then from either $\mathcal{C}$ or $\mathcal{C}^*$ we will obtain $\chi : E^r \to \{+, 0, -\}$.
- Given a chirotope χ we will define the unsigned circuits and the unsigned cocircuits associated to χ, that is, the sets that will eventually be the supports of circuits and the supports of cocircuits of an oriented matroid. Then we use χ to put signs on the unsigned circuits and unsigned cocircuits.

The cryptomorphism of Theorem 2.54 is more intricate than our previous cryptomorphisms. One aim of the first part of our discussion, before Theorem 2.54, is to point out how to translate ideas between the chirotope perspective and the circuit/cocircuit perspective without working out the entire cryptomorphism. To that end, we prove a few results that are not strictly needed for the cryptomorphism.

Bases of Oriented Matroids

Definition 2.45 Let $\mathcal{M}$ be an oriented matroid with elements E. A set $S \subseteq E$ is **independent** in $\mathcal{M}$ if there is no signed circuit of $\mathcal{M}$ with support contained in S. A **basis** of $\mathcal{M}$ is a maximal independent set.

Proposition 2.46 *Let $\mathcal{M}$ be an oriented matroid with elements E.*

1. *All bases of $\mathcal{M}$ have the same number of elements.*
2. *If B is a basis of $\mathcal{M}$ then $E\backslash B$ is a basis of $\mathcal{M}^*$.*

Lemma 2.47 *If B is a basis of $\mathcal{M}$ and $e \in E\backslash B$ then there is a unique pair $\{X, -X\} \in \mathcal{C}(\mathcal{M})$ such that $\mathrm{supp}(X) \subseteq B \cup \{e\}$.*

Proof: Since B is a maximal independent set, $B \cup \{e\}$ is dependent, and so there is some $X \in \mathcal{C}(\mathcal{M})$ such that $\mathrm{supp}(X) \subseteq B \cup \{e\}$. Assume by way of contradiction there is a $Y \in \mathcal{C}(\mathcal{M})\backslash\{X, -X\}$ such that $\mathrm{supp}(Y) \subseteq B \cup \{e\}$. Since $Y(e) \neq 0$ and $X(e) \neq 0$, we can eliminate e between Y and either X or $-X$ to get a signed circuit with support contained in B, contradicting independence of B. □

Proof of Proposition 2.46: We show (1) by induction on $|E|$, with the base case $|E| = 1$.

Above the minimal case, consider distinct bases B_1 and B_2. We have two cases.

Case 1: If there is an $e \in B_1 \cap B_2$, then since $\mathcal{V}(\mathcal{M}/e) = \mathcal{V}(\mathcal{M})\backslash e$, we see that $B_1\backslash\{e\}$ and $B_2\backslash\{e\}$ are bases of $\mathcal{M}/e$. Thus by the induction hypothesis $|B_1\backslash\{e\}| = |B_2\backslash\{e\}|$, and $|B_1| = |B_2|$.

Case 2: If $B_1 \cap B_2 = \emptyset$, then let $e \in B_1$. Lemma 2.47 tells us there is an $X \in \mathcal{C}(\mathcal{M})$ with support contained in $B_2 \cup \{e\}$. Since $\{e\}$ is independent, there is an $f \in \mathrm{supp}(X)\backslash\{e\}$. Uniqueness of $\pm X$ tells us that there is no signed circuit of $\mathcal{M}$ with support contained in $(B_2 \cup \{e\})\backslash\{f\}$, and so $(B_2 \cup \{e\})\backslash\{f\}$ is independent. Let B_3 be a basis containing $(B_2\cup\{e\})\backslash\{f\}$. Then $e \in B_1 \cap B_3$, and so by Case (2) $|B_1| = |B_3| \geq |B_2|$. But by the same argument, $|B_1| \geq |B_1|$.

To see (2), first assume by way of contradiction there is some $Y \in \mathcal{C}^*(\mathcal{M})$ with support contained in $E\backslash B$. Let e be an element of $\mathrm{supp}(Y)$. Then $B \cup \{e\}$ is dependent in $\mathcal{M}$, so there exists $X \in \mathcal{C}(\mathcal{M})$ such that $\mathrm{supp}(X) \subseteq B\cup\{e\}$ and $X(e) \neq 0$. But then $X \cdot Y = 0 \boxplus X(e)Y(e) \neq \{0\}$, contradicting orthogonality of $\mathcal{C}$ and $\mathcal{C}^*$. Thus, $E\backslash B$ is independent in $\mathcal{M}^*$.

To see $E\backslash B$ is a basis of $\mathcal{M}^*$, we check that for every $j \in B$ there is some $Y \in \mathcal{V}^*(\mathcal{M})$ with support contained in $(E\backslash B) \cup \{j\}$. This follows from the Dual Pair Axioms (Proposition 2.37) applied to j and the sets $C = E\backslash B$ and $D = B\backslash\{j\}$. Since B is independent, the $\mathcal{C}$ alternative fails. Thus, the $\mathcal{C}^*$ alternative holds. □

2.4 Chirotopes

Definition 2.48 Define the **rank** of an oriented matroid to be the size of a basis. Define the rank of a set S of elements to be the rank of $\mathcal{M}(S)$.
Define the **corank** of an oriented matroid $\mathcal{M}$ to be the rank of $\mathcal{M}^*$.

Thus if $\mathcal{M}$ is rank r on elements E then the corank of $\mathcal{M}$ is $|E| - r$. Recall that if $\mathcal{M}$ is realized by a subspace W of $\mathbb{R}^n$ then rank$(\mathcal{M})$ = dim(W) and rank$(\mathcal{M}^*)$ = dim$(W^\perp)$. Thus the second part of Proposition 2.46 generalizes the statement dim(W) + dim$(W^\perp)$ = n.

Definition 2.49 Define a **hyperplane** of $\mathcal{M}$ to be a maximal set not containing a basis.

In other words, a hyperplane is maximal among sets whose rank is rank$(\mathcal{M}) - 1$. If $\mathcal{M}$ has a realization $(\mathbf{v}_e : e \in E)$ in a vector space V then $H \subset E$ is a hyperplane of $\mathcal{M}$ if and only if there is a hyperplane (in the linear algebra sense) $W \subset V$ spanned by elements of $\{\mathbf{v}_e : e \in H\}$ such that $W \cap \{\mathbf{v}_e : e \in E\} = \{\mathbf{v}_e : e \in H\}$. Recall that signs on such hyperplanes W gave us cocircuits of $\mathcal{M}$. The following proposition generalizes this to nonrealizable $\mathcal{M}$. (Recall that for $Y \in \{0, +, -\}^E$, Y^0 denotes $Y^{-1}(0)$.)

Proposition 2.50 *Let $\mathcal{M}$ be an oriented matroid on E and $H \subseteq E$. Then H is a hyperplane of $\mathcal{M}$ if and only if $H = Y^0$ for some $Y \in \mathcal{C}^*(\mathcal{M})$.*

Proof: For every subset H of E,

H is a hyperplane of $\mathcal{M}$;

$\Leftrightarrow H$ does not contain a basis of $\mathcal{M}$ but for every $e \in E \backslash H$,

$H \cup \{e\}$ contains a basis of $\mathcal{M}$;

$\Leftrightarrow E \backslash H$ is not contained in a basis of $\mathcal{M}^*$ but for every $e \in E \backslash H$,

$E \backslash (H \cup \{e\})$ is contained in a basis of $\mathcal{M}^*$;

$\Leftrightarrow E \backslash H$ is dependent in $\mathcal{M}^*$ but for every $e \in E \backslash H$,

$E \backslash (H \cup \{e\})$ is independent in $\mathcal{M}^*$;

$\Leftrightarrow E \backslash H$ is the support of some $Y \in \mathcal{C}^*$. □

Thus we have two equivalent characterizations of bases of an oriented matroid:

1. A basis is a maximal element of

$\{A \subseteq E : \forall X \in \mathcal{C} \ \text{supp}(X) \not\subseteq A\}$.

2. A basis is a minimal element of

$\{A \subseteq E : \forall Y \in \mathcal{C}^* \ E \backslash \text{supp}(Y) \not\supseteq A\}$.

The following lemma paves the way for a Pivoting Property (Proposition 1.34) and a Dual Pivoting Property (Problem 1.36) for general oriented matroids.

Lemma 2.51 *Let $\mathcal{M}$ be an oriented matroid on elements E, and let $\{e, x_2, \ldots, x_r\}$ and $\{f, x_2, \ldots, x_r\}$ be bases of $\mathcal{M}$ with $e \neq f$.*

1. *There are exactly two signed circuits X, $-X$ of $\mathcal{M}$ with $\operatorname{supp}(X) \subseteq \{e, f, x_2, \ldots, x_r\}$, and for this pair we have $\{e, f\} \subseteq \operatorname{supp}(X)$.*
2. *There are exactly two signed cocircuits Y, $-Y$ of $\mathcal{M}$ with $\operatorname{supp}(Y) \subseteq E \setminus \{x_2, \ldots, x_r\}$, and for this pair we have $\{e, f\} \subseteq \operatorname{supp}(Y)$.*

Proof: 1. This follows from Lemma 2.47 applied to $\{e, x_2, \ldots, x_x\}$ and f.
2. Apply the first part of the lemma to $\mathcal{M}^*$ and its bases $\{e\} \cup (E \setminus \{x_2, \ldots, x_r\})$ and $\{f\} \cup (E \setminus \{x_2, \ldots, x_r\})$. □

Unsigned Circuits and Cocircuits of a Chirotope

Definition 2.52 A **basis of a chirotope** χ is a set $\{x_1, \ldots, x_r\}$ such that $\chi(x_1, \ldots, x_r) \neq 0$. A subset S of E is **independent** in χ if S is contained in a basis and is **dependent** otherwise. The **unsigned circuits** of χ are the minimal dependent sets of χ. The **hyperplanes** of χ are the maximal sets not containing a basis, and the **unsigned cocircuits** of χ are the complements of hyperplanes.

Lemma 2.53 *Let χ be a rank r chirotope on E.*

1. *For every unsigned circuit $\tilde{X}$ of χ there is an $S \subseteq E \setminus \tilde{X}$ such that $(\tilde{X} \cup S) \setminus \{e\}$ is a basis for each $e \in \tilde{X}$.*
   In fact, for each $S \subseteq E \setminus \tilde{X}$, if $(\tilde{X} \cup S) \setminus \{e\}$ is a basis for some $e \in \tilde{X}$ then $(\tilde{X} \cup S) \setminus \{e\}$ is a basis for all $e \in \tilde{X}$.
2. *For every unsigned cocircuit $\tilde{Y}$ of χ there is a $T \subseteq E \setminus \tilde{Y}$ such that $\{e\} \cup T$ is a basis for each $e \in \tilde{Y}$.*
   In fact, for each $T \subseteq E \setminus \tilde{Y}$, if $\{e\} \cup T$ is a basis for some $e \in \tilde{Y}$ then $\{e\} \cup T$ is a basis for all $e \in \tilde{Y}$.

Proof: 1. Fix $e \in \tilde{X}$. Since $\tilde{X} \setminus \{e\}$ is independent, there is an $S \subseteq E \setminus \tilde{X}$ such that $(\tilde{X} \cup S) \setminus \{e\}$ is a basis. For any other element f of $\tilde{X}$, there is at least one $S_f \subseteq E \setminus \tilde{X}$ such that $(\tilde{X} \cup S_f) \setminus \{f\}$ is a basis. Choose S_f with $S \cap S_f$ as large as possible, and assume by way of contradiction that $S \neq S_f$. Let $g \in S_f \setminus S$. By Basis Exchange there is an $h \in (\tilde{X} \cup S) \setminus \{e\}$ such that $(\tilde{X} \cup S_f \cup \{h\}) \setminus \{f, g\}$

2.4 Chirotopes

is a basis. Certainly $h \notin \tilde{X}$, so $S'_f := (S_f \cup \{h\}) \setminus \{g\} \subseteq E \setminus \tilde{X}$. But $S \cap S'_f = (S \cap S_f) \cup \{h\}$, contradicting our maximality assumption.

2. $\tilde{Y}$ is a cocircuit, so $E \setminus \tilde{Y}$ is a maximal set not containing a basis of χ. Fix an $e \in \tilde{Y}$. Then $(E \setminus \tilde{Y}) \cup \{e\}$ contains a basis $T \cup \{e\}$. For any other element f of $\tilde{Y}$, there is at least one $T_f \subseteq E \setminus \tilde{Y}$ such that $T_f \cup \{f\}$ is a basis. Applying Basis Exchange to $T \cup \{e\}$ and $T_f \cup \{f\}$, we see there is a $g \in T_f \cup \{f\}$ such that $T \cup \{g\}$ is a basis. Since $T \cup T_f \subseteq E \setminus \tilde{Y}$ and $E \setminus \tilde{Y}$ does not contain a basis, we see $g \notin T_f$. Thus $g = f$, and so $T \cup \{f\}$ is a basis. □

Cryptomorphism from Chirotopes to Oriented Matroids
Theorem 2.54 *1. Let χ be a rank r chirotope on elements E.*

i. *There is a unique $\mathcal{C} \subseteq \{0, +, -\}^E$ such that*

 a. $\mathcal{C} = -\mathcal{C}$;
 b. $\{\mathrm{supp}(X) : X \in \mathcal{C}\}$ *is the set of unsigned circuits of χ; and*
 c. **(Pivoting Property)** *For every $X \in \mathcal{C}$, $\{e, f\} \subseteq \mathrm{supp}(X)$, and bases $\{e, x_2, \ldots, x_r\}$ and $\{f, x_2, \ldots, x_r\}$ of χ with $\mathrm{supp}(X) \subseteq \{e, f, x_2, \ldots, x_r\}$, we have*

$$\chi(e, x_2, \ldots, x_r) \chi(f, x_2, \ldots, x_r) = -X(e)X(f).$$

ii. *There is a unique $\mathcal{C}^* \subseteq \{0, +, -\}^E$ such that*

 a. $\mathcal{C}^* = -\mathcal{C}^*$;
 b. $\{\mathrm{supp}(X) : X \in \mathcal{C}^*\}$ *is the set of unsigned cocircuits of χ; and*
 c. **(Dual Pivoting Property)** *For every $Y \in \mathcal{C}^*$, $\{e, f\} \subseteq \mathrm{supp}(Y)$, and bases $\{e, x_2, \ldots, x_r\}$ and $\{f, x_2, \ldots, x_r\}$ of χ with $\{x_2, \ldots, x_r\} \subseteq E \setminus \mathrm{supp}(Y)$ we have*

$$\chi(e, x_2, \ldots, x_r) \chi(f, x_2, \ldots, x_r) = Y(e)Y(f).$$

Further $\mathcal{C}$ and $\mathcal{C}^*$ are the signed circuit set and signed cocircuit set of a rank r oriented matroid with the same bases as χ.

2. *Let $\mathcal{M}$ be a rank r oriented matroid on elements E. There are exactly two alternating functions $\chi, -\chi$ from E^r to $\{0, +, -\}$ satisfying:*

 (i) *The support of χ is the set of orderings of bases of $\mathcal{M}$, and*
 (ii) *The Pivoting Property holds for χ and $\mathcal{C}(\mathcal{M})$.*

Further χ is a chirotope whose unsigned circuits are the supports of the signed circuits of $\mathcal{M}$ and whose unsigned cocircuits are the supports of the signed cocircuits of $\mathcal{M}$.

70   Oriented Matroids

3. In (2), the Pivoting Property for χ and $C(\mathcal{M})$ can be replaced with the Dual Pivoting Property for χ and $C^*(\mathcal{M})$.

(The term "Pivoting Property" comes from the oriented matroid interpretation of the Simplex Algorithm – see chapter 10 of Björner et al. (1999) for details. The geometric idea of the Pivoting Property can be seen in the proof of Proposition 1.34.)

The first part of Theorem 2.54 shows that there is a well-defined map from the set of rank r chirotopes on E to the set of rank r oriented matroids on E. The second part shows that the preimage of an oriented matroid under this map is a pair $\chi, -\chi$. The following corollary completes the cryptomorphism.

Corollary 2.55 *Let* $\mathrm{Chir}(r, E)$ *be the set of pairs* $\pm \chi$ *of rank r chirotopes on E, and let* $\mathrm{MacP}(r, E)$ *be the set of rank r oriented matroids on E. The maps* $m \colon \mathrm{Chir}(r, E) \to \mathrm{MacP}(r, E)$ *and* $c \colon \mathrm{MacP}(r, E) \to \mathrm{Chir}(r, E)$ *given by Theorem 2.54 are inverse to each other.*

Remark 2.56 $\mathrm{MacP}(r, E)$, with a partial order we'll introduce in Chapter 5, is called the *MacPhersonian*.

Proof: Notice that for an object in either $\mathrm{Chir}(r, E)$ or $\mathrm{MacP}(r, E)$, the set of bases (maximal independent sets) determines the set of unsigned circuits (minimal dependent sets), and vice versa. Thus for each $\mathcal{M} \in \mathrm{MacP}(r, E)$, $\mathcal{M}$ and $c(\mathcal{M})$ have the same unsigned circuits, which are also the unsigned circuits of $m \circ c(\mathcal{M})$. Let $\tilde{X}$ be an unsigned circuit of $\mathcal{M}$, let $e \in \tilde{X}$, let X be the signed circuit of $\mathcal{M}$ with $\mathrm{supp}(X) = \tilde{X}$ and $X(e) = +$, and let X' be the signed circuit of $m \circ c(\mathcal{M}')$ with $\mathrm{supp}(X') = \tilde{X}$ and $X'(e) = +$. By Lemma 2.53 applied to $c(\mathcal{M})$ there are bases $\{e, x_2, \ldots, x_r\}$ and $\{f, x_2, \ldots, x_r\}$ to which we can apply the Pivoting Property to see that

$$X(f) = -c(\mathcal{M})(e, x_2, \ldots, x_r) c(\mathcal{M})(f, x_2, \ldots, x_r) = X'(f),$$

and so $X = X'$. Thus $\mathcal{M} = m \circ c(\mathcal{M})$.

Likewise, for each chirotope χ, χ and $c \circ m(\{\pm \chi\})$ have the same bases, and by a similar argument, this time using Basis Exchange and Lemma 2.51, we see that $\{\pm \chi\} = c \circ m(\{\pm \chi\})$. □

Lemma 2.57 *Let χ be a chirotope and $\tilde{X}$ be an unsigned circuit of χ.*

1. Let $e \neq f$ be elements of $\tilde{X}$. Let S be a set such that $(\tilde{X} \cup S) \setminus \{x\}$ is a basis for each $x \in \tilde{X}$, and let $(\tilde{X} \cup S) \setminus \{e, f\} = \{x_2, \ldots, x_r\}$. Define $\sigma(e, f) = -\chi(e, x_2, \ldots, x_r) \chi(f, x_2, \ldots, x_r)$. Then $\sigma(e, f)$ is independent of the choice of S.

2.4 Chirotopes

2. σ induces an equivalence relation on $\tilde{X}$ with at most two equivalence classes, with $e \sim f$ if and only if $e = f$ or $\sigma(e, f) = +$.

Remark 2.58 Lemma 2.53 shows the existence of sets S as in this lemma.

Proof: (1) Assume S and T are sets such that $(\tilde{X} \cup S)\setminus\{e\}$, $(\tilde{X} \cup S)\setminus\{f\}$, $(\tilde{X} \cup T)\setminus\{e\}$, and $(\tilde{X} \cup T)\setminus\{f\}$ are bases. Let $(\tilde{X} \cup S)\setminus\{e, f\} = \{x_2, \ldots, x_r\}$ and $(\tilde{X} \cup T)\setminus\{e, f\} = \{y_2, \ldots, y_r\}$. Then by Signed Basis Exchange, one of the following is true.

1. There is an $i \in \{2, \ldots, r\}$ such that

$$\chi(e, x_2, \ldots, x_r)\chi(f, y_2, \ldots, y_r) = \chi(y_i, x_2, \ldots, x_r)$$
$$\chi(f, y_2, \ldots, y_{i-1}, e, y_{i+1}, \ldots, y_r). \quad (2.4)$$

2. $\chi(e, x_2, \ldots, x_r)\chi(f, y_2, \ldots, y_r) = \chi(f, x_2, \ldots, x_r)\chi(e, y_2, \ldots, y_r)$.

Assume by way of contradiction that Equation (2.4) holds. Since the left-hand side of (2.4) is nonzero, we have that neither factor on the right-hand side is 0. Thus we get two contradictory statements:

- $y_i \notin \{x_2, \ldots, x_r\}$, and since also $y_i \notin \{e, f\}$, we have $y_i \notin \tilde{X}$.
- $\{e, f, y_2, \ldots, \hat{y}_i, \ldots, y_r\}$ is a basis, and so does not contain $\tilde{X}$. But $\{e, f, y_2, \ldots, y_r\} \supseteq \tilde{X}$, so $y_i \in \tilde{X}$.

Thus we conclude the second statement,

$$\chi(e, x_2, \ldots, x_r)\chi(f, y_2, \ldots, y_r) = \chi(f, x_2, \ldots, x_r)\chi(e, y_2, \ldots, y_r).$$

Multiplying both sides by $\chi(f, x_2, \ldots, x_r)\chi(f, y_2, \ldots, y_r)$, we get

$$\chi(e, x_2, \ldots, x_r)\chi(f, x_2, \ldots, x_r) = \chi(e, y_2, \ldots, y_r)\chi(f, y_2, \ldots, y_r).$$

(2) Certainly the relation is reflexive and symmetric. To complete the proof we show for each $e, f, g \in \tilde{X}$, that $\sigma(e, f) \cdot \sigma(f, g) = \sigma(e, g)$. Let $\{x_3, \ldots, x_r\} = (\tilde{X} \cup S)\setminus\{e, f, g\}$.

$$\sigma(e, g) = -\chi(e, f, x_3, \ldots, x_r)\chi(g, f, x_3, \ldots, x_r)$$
$$= -\chi(f, e, x_3, \ldots, x_r)\chi(f, g, x_3, \ldots, x_r)(-\chi(e, g, x_3, \ldots, x_r)\chi(g, e, x_3, \ldots, x_r))$$
$$= (-\chi(e, g, x_3, \ldots, x_r)\chi(f, g, x_3, \ldots, x_r))(-\chi(f, e, x_3, \ldots, x_r)\chi(g, e, x_3, \ldots, x_r))$$
$$= \sigma(e, f)\sigma(f, g).$$

□

If the equivalence relation σ of Lemma 2.57 has two equivalence classes A and B, we call A^+B^- and B^+A^- the **signed circuits induced by** $\tilde{X}$. If σ has only one equivalence class then $\tilde{X}^+$ and $\tilde{X}^-$ are the signed circuits induced by $\tilde{X}$. Let $\mathcal{C}$ denote the set of signed circuits induced by unsigned circuits of χ.

A similar proof gives an analogous result to Lemma 2.57 about cocircuits.

Lemma 2.59 *Let χ be a chirotope and $\tilde{Y}$ be an unsigned cocircuit of χ. Let $\{x_2, \ldots, x_r\}$ be a maximal independent subset of the hyperplane complementary to $\tilde{Y}$.*

1. *Let $e \neq f$ be elements of $\tilde{Y}$. Define $\sigma(e, f) = \chi(e, x_2, \ldots, x_r)\chi(f, x_2, \ldots, x_r)$. Then $\sigma(e, f)$ is independent of the choice of $x_2, \ldots, x_r$.*
2. *σ induces an equivalence relation on $\tilde{Y}$ with at most two equivalence classes, with e and f in the same part if and only if $\sigma(e, f) = +$.*

As with circuits, we use the equivalence classes of σ to define, for each unsigned cocircuit $\tilde{Y}$, two sign vectors $Y, -Y$ with support $\tilde{Y}$ and call these the **signed cocircuits induced by** $\tilde{Y}$. Let $\mathcal{C}^*$ denote the set of signed cocircuits induced by unsigned cocircuits of χ.

We now prove Part (1) of Theorem 2.54.

Proof: Consider an unsigned circuit $\tilde{X}$ of χ. By Lemma 2.53, for each $\{e, f\} \in \tilde{X}$ there are bases $\{e, x_2, \ldots, x_r\}$ and $\{f, x_2, \ldots, x_r\}$ as in the hypothesis of the Pivoting Property. Thus a sign vector $X = A^+ B^-$ with support $\tilde{X}$ is consistent with the Pivoting Property if and only if the set of nonempty elements of $\{A, B\}$ is the set of equivalence classes of the relation given in Lemma 2.57. Thus the set $\mathcal{C}$ defined after Lemma 2.57 is the unique set of sign vectors satisfying the three conditions of Part (1a) of Theorem 2.54.

Likewise, by Lemma 2.53, for each unsigned cocircuit $\tilde{Y}$ of χ and $\{e, f\} \in \tilde{Y}$ there are bases $\{e, x_2, \ldots, x_r\}$ and $\{f, x_2, \ldots, x_r\}$ as in the hypothesis of the Dual Pivoting Property. The set $\mathcal{C}^*$ that we defined after Lemma 2.59 is the unique subset of $\{0, +, -\}$ satisfying the three conditions of Part (1b) of Theorem 2.54.

To see that $\mathcal{C}$ and $\mathcal{C}^*$ are the signed circuits and signed cocircuits of an oriented matroid, we use the Dual Pair Axioms (Proposition 2.37). Certainly $\mathcal{C}$ and $\mathcal{C}^*$ satisfy the Incomparability and Symmetry Axioms.

To see $\mathcal{C} \perp \mathcal{C}^*$: Let $X \in \mathcal{C}$ and $Y \in \mathcal{C}^*$ such that $X \cdot Y \neq \{0\}$. Let $e \in \text{supp}(X) \cap \text{supp}(Y)$. Let $B_X = \{x_1, \ldots, x_r\}$ be a basis with $\text{supp}(X) \backslash e \subseteq B_X \subseteq E \backslash e$. Let $B_Y = \{e, y_2, \ldots, y_r\}$ be a basis with $B_Y \cap \text{supp}(Y) = \{e\}$. By Signed Basis Exchange applied to B_X and B_Y, there is an i such that

$$\chi(e, y_2, \ldots, y_r)\chi(x_1, \ldots, x_r) = \chi(x_i, y_2, \ldots, y_r)$$
$$\chi(x_1, \ldots, x_{i-1}, e, x_{i+1}, \ldots, x_r), \quad (2.5)$$

2.4 Chirotopes

and these products are nonzero. For the first factor on the right-hand side to be nonzero, x_i must be in the support of Y. For the second factor on the right-hand side to be nonzero, x_i must be in the support of X.

Multiplying both sides of (2.5) by $\chi(x_i, y_2, \ldots, y_r)\chi(x_1, \ldots, x_r)$, we get

$$\chi(e, y_2, \ldots, y_r)\chi(x_i, y_2, \ldots, y_r) = \chi(x_1, \ldots, x_{i-1}, e, x_{i+1}, \ldots, x_r)\chi(x_1, \ldots, x_r)$$
$$= \chi(e, x_1, \ldots, \hat{x}_i, \ldots, x_r)\chi(x_i, x_1, \ldots, \hat{x}_i, \ldots, x_r)$$
$$Y(e)Y(x_i) = -X(e)X(x_i)$$
$$Y(e)X(e) = -Y(x_i)X(x_i).$$

By the definition of e, the left-hand side of the final equation is nonzero. Thus, $X \cdot Y = + \boxplus - = \{0, +, -\}$.

To see the final condition of the Dual Pair Axioms: Let $j \in E$, and let $C \dot\cup D = E\backslash\{j\}$. Suppose there is no $X \in \mathcal{C}$ such that $X(j) \neq 0$ and $X(C) \subseteq \{0\}$. It then suffices to check that D is contained in a hyperplane of χ not containing j.

Let S be a maximal independent subset of D. Then since the $\mathcal{C}$ alternative of the Dual Pair Axiom fails, $S \cup \{j\}$ is also independent, so it is contained in a basis $S \cup \{j\} \cup T$ of χ. Thus, $S \cup T$ has rank $r - 1$, and so $D \cup T$ has rank greater than or equal to $r - 1$. But also, an independent subset of $D \cup T$ must have the form $S' \cup T'$ with $S' \subseteq D$ and $T' \subseteq T$, and by the maximality of $|S|$, $|S' \cup T'| = |S'| + |T'| \leq |S| + |T| = r - 1$. Thus, the rank of $D \cup T$ is $r - 1$, so $D \cup T$ is contained in a hyperplane H. Since $S \cup T \subseteq H$, we conclude that $j \notin H$.

Finally, since χ and $\mathcal{M}$ have the same minimal dependent sets, they have the same independent sets. In particular, they have the same bases, and rank$(\mathcal{M}) = n$. □

We prove the last two parts of Theorem 2.54 together.

Proposition 2.60 *Let $\mathcal{M}$ be a rank r oriented matroid on E. Let $\chi: E^r \to \{0, +, -\}$ be a function whose support is the set of orderings of bases of $\mathcal{M}$. Then χ and $\mathcal{C}(\mathcal{M})$ satisfy the Pivoting Property if and only if χ and $\mathcal{C}^*(\mathcal{M})$ satisfy the Dual Pivoting Property.*

Proof: By Lemma 2.51, for each pair of bases $\{e, x_2, \ldots, x_r\}$ and $\{f, x_2, \ldots, x_r\}$ of $\mathcal{M}$ with $e \neq f$ there is a pair $X, -X \in \mathcal{C}(\mathcal{M})$ with support contained in $\{e, f, x_2, \ldots, x_r\}$ and there is a pair $Y, -Y \in \mathcal{C}^*(\mathcal{M})$ with support contained in $E\backslash\{x_2, \ldots, x_r\}$. By orthogonality of circuits and cocircuits,

$$0 \in X \cdot Y = X(e)Y(e) \boxplus X(f)Y(f) \neq \{0\}.$$

Thus $X(e)Y(e) = -X(f)Y(f) \neq 0$. Multiplying both sides by $X(f)Y(f)$, we get $X(e)X(f) = -Y(e)Y(f)$. □

74   Oriented Matroids

Lemma 2.61 *Let $\mathcal{M}$ be an oriented matroid on elements E. Let $S \subset E$ and $\{a,b,c\} \subseteq E\backslash S$ such that $\{a\} \cup S$, $\{b\} \cup S$, and $\{c\} \cup S$ are all bases. Let $X, Y, Z \in \mathcal{C}(\mathcal{M})$ such that*

$$\text{supp}(X) \subseteq \{a,b\} \cup S,$$
$$\text{supp}(Y) \subseteq \{a,c\} \cup S,$$
$$\text{supp}(Z) \subseteq \{b,c\} \cup S,$$

and $Y(c) = -Z(c)$. Then either X or $-X$ is the unique circuit elimination of c between Y and Z. If X is this circuit elimination then $X(a) = Y(a)$ and $X(b) = Z(b)$.

Proof: A circuit elimination of c between Y and Z is a signed circuit with support contained in $\{a,b\} \cup S$, so by Lemma 2.47 it must be $\pm X$. Assuming it's X, since $X(a) \neq 0$ we have $X(a) \in Y(a) \boxplus Z(a) = \{Y(a)\}$, and similarly $X(b) = Z(b)$. □

Now we prove the last two parts of Theorem 2.54.

Proof: Let $\{b_1, \ldots, b_r\}$ be a basis of $\mathcal{M}$. We first prove by induction on $|E|$ that there is a unique alternating $\chi: E^r \to \{0, +, -\}$ such that $\text{supp}(\chi)$ is the set of orderings of bases of $\mathcal{M}$, $\chi(b_1, \ldots, b_r) = +$, and χ satisfies the Pivoting Property. In the base case, $E = \{b_1, \ldots, b_r\}$, $\mathcal{C}(\mathcal{M}) = \emptyset$, and $\mathcal{C}^*(\mathcal{M}) = \{\pm\{b_i\}^+ : i \in [r]\}$, so both the Pivoting Property and Dual Pivoting Property are vacuous.

For larger E, let $a \in E\backslash\{b_1, \ldots, b_r\}$. The induction hypothesis applied to $\mathcal{M}\backslash a$ gives us a unique $\chi_a: (E\backslash\{a\})^r \to \{0, +, -\}$. Note that $\mathcal{M}\backslash a$ has rank r, and that a set is independent in a deletion of $\mathcal{M}$ if and only if it is independent in $\mathcal{M}$. Thus the bases of $\mathcal{M}\backslash a$ are the bases of $\mathcal{M}$ not containing a, and for two such bases $\{e, x_2, \ldots, x_r\}$ and $\{f, x_2, \ldots, x_r\}$, the signed circuits $\pm X$ of $\mathcal{M}\backslash a$ with support contained in $\{e, f, x_2, \ldots, x_r\}$ extend to signed circuits of $\mathcal{M}$ by setting $X(a) = 0$. Thus every extension of χ_a to a function $E^r \to \{0, +, -\}$ will satisfy the Pivoting Property with respect to pairs of bases not containing a.

To extend χ_a to a function $\chi: E^r \to \{0, +, -\}$, we first define χ to have value 0 on all nonbases containing a. Now consider a basis $\{a, x_2, \ldots, x_r\}$. Because $\{x_2, \ldots, x_r\}$ is independent in $\mathcal{M}\backslash a$ and bases of $\mathcal{M}\backslash a$ are also bases of $\mathcal{M}$, we see there is at least one $e \neq a$ such that $\{e, x_2, \ldots, x_r\}$ is a basis. Let $\pm X$ be the two signed circuits of $\mathcal{M}$ with $\text{supp}(X) \subseteq \{a, e, x_2, \ldots, x_r\}$. In order for the Pivoting Property to hold, we would like to define

$$\chi(a, x_2, \ldots, x_r) = -X(a)X(e)\chi(e, x_2, \ldots, x_r),$$

2.4 Chirotopes

but we need to show that this is independent of our choice of e. Let $\{e', x_2, \ldots, x_r\}$ be a basis with $e' \notin \{a, e\}$. Let Y be the signed circuit with $\operatorname{supp}(Y) \subseteq \{a, e', x_2, \ldots, x_r\}$ and $Y(a) = X(a)$. We wish to show that

$$X(a)X(e)\chi(e, x_2, \ldots, x_r) = Y(a)Y(e')\chi(e', x_2, \ldots, x_r).$$

Let Z be the signed circuit with $\operatorname{supp}(Z) \subseteq \{e, e', x_2, \ldots, x_r\}$ and $Z(e') = -Y(e')$. The induction hypothesis tells us that

$$Z(e)Z(e') = -\chi(e, x_2, \ldots, x_r)\chi(e', x_2, \ldots, x_r).$$

Multiplying both sides by $Z(e')\chi(e, x_2, \ldots, x_r)$,

$$Z(e)\chi(e, x_2, \ldots, x_r) = -Z(e')\chi(e', x_2, \ldots, x_r).$$

Lemma 2.61 tells us that $X(e) = Z(e)$. Thus we have

$$\begin{aligned}
X(a)X(e)\chi(e, x_2, \ldots, x_r) &= Y(a)Z(e)\chi(e, x_2, \ldots, x_r) \\
&= -Y(a)Z(e')\chi(e', x_2, \ldots, x_r) \\
&= Y(a)Y(e')\chi(e', x_2, \ldots, x_r),
\end{aligned}$$

so indeed we have a definition for $\chi(a, x_2, \ldots, x_r)$. We define χ on permutations of $(a, x_2, \ldots, x_r)$ so that χ is alternating.

The only remaining step in showing that this χ satisfies the Pivoting Property is to consider bases $\{e, y_2, \ldots, y_r\}$ and $\{f, y_2, \ldots, y_r\}$ such that $a \in \{y_2, \ldots, y_r\}$. Since χ is alternating, we may assume $a = y_2$. Let W be a signed circuit with $\operatorname{supp}(W) \subseteq \{e, f, a, y_3, \ldots, y_r\}$. We have two cases.

Case 1: $a \in \operatorname{supp}(W)$. Then

$$\begin{aligned}
\chi(e, a, y_3, \ldots, y_r)\chi(f, a, y_3, \ldots, y_r) &= \chi(a, e, y_3, \ldots, y_r)\chi(a, f, y_3, \ldots, y_r) \\
&= \chi(a, e, y_3, \ldots, y_r)\chi(e, f, y_3, \ldots, y_r) \\
&\quad \chi(e, f, y_3, \ldots, y_r)\chi(a, f, y_3, \ldots, y_r) \\
&= -\chi(a, e, y_3, \ldots, y_r)\chi(f, e, y_3, \ldots, y_r) \\
&\quad (-W(e)W(a)) \\
&= -W(a)W(f)(W(e)W(a)) \\
&= -W(e)W(f).
\end{aligned}$$

Case 2: $a \notin \operatorname{supp}(W)$. Because $\{e, y_3, \ldots, y_r\}$ is independent in $\mathcal{M}\backslash a$, there is a $b \neq a$ such that $\{e, b, y_3, \ldots, y_r\}$ is a basis. Additionally, since $\pm W$ is the unique pair of signed circuits with support contained in $\{e, f, b, y_3, \ldots y_r\}$, we have that $\{f, b, y_3, \ldots, y_r\}$ is independent, and hence is a basis. By the induction hypothesis we have that

$$W(e)W(f) = -\chi(e, b, y_3, \ldots, y_r)\chi(f, b, y_3, \ldots, y_r). \tag{2.6}$$

Now we have two subcases.

If $\{a,b,y_3,\ldots,y_r\}$ is dependent then there is a signed circuit U with $\{a,b\} \subseteq \operatorname{supp}(U) \subseteq \{a,b,y_3,\ldots,y_r\}$, and by our definition of χ we have both

$$U(a)U(b) = -\chi(a,e,y_3,\ldots,y_r)\chi(b,e,y_3,\ldots,y_r)$$

and

$$U(a)U(b) = -\chi(a,f,y_3,\ldots,y_r)\chi(b,f,y_3,\ldots,y_r).$$

Thus

$$\chi(e,a,y_3,\ldots,y_r)\chi(e,b,y_3,\ldots,y_r) = \chi(f,a,y_3,\ldots,y_r)\chi(f,b,y_3,\ldots,y_r).$$

Multiplying both sides by $\chi(e,b,y_3,\ldots,y_r)\chi(f,a,y_3,\ldots,y_r)$,

$$\chi(e,a,y_3,\ldots,y_r)\chi(f,a,y_3,\ldots,y_r) = \chi(e,b,y_3,\ldots,y_r)\chi(f,b,y_3,\ldots,y_r)$$
$$= -W(e)W(f) \quad \text{by Equation (2.6)}.$$

If $\{a,b,y_3,\ldots,y_r\}$ is independent, let Y be the unique signed circuit with support contained in $\{a,b,e,y_3,\ldots,y_r\}$ and with $Y(b) = +$, and let Z be the unique signed circuit with support contained in $\{a,b,f,y_3,\ldots,y_r\}$ and with $Z(b) = -$. By Lemma 2.61, either $(X(e) = Y(e)$ and $X(f) = Z(f))$ or $(-X(e) = Y(e)$ and $-X(f) = Z(f))$. Thus

$$X(e)X(f) = Y(e)Z(f)$$
$$= Y(e)Y(b)Y(b)Z(f)$$
$$= -Y(e)Y(b)Z(b)Z(f)$$
$$= -(-\chi(e,a,y_3,\ldots,y_r)\chi(b,a,y_3,\ldots,y_r))$$
$$\quad (-\chi(b,a,y_3,\ldots,y_r)\chi(f,a,y_3,\ldots,y_r))$$
$$= -\chi(e,a,y_3,\ldots,y_r)\chi(f,a,y_3,\ldots,y_r)).$$

To see that the function χ we've defined is a chirotope, we verify that it satisfies Signed Basis Exchange. Consider bases $\{x_1,\ldots,x_r\}$ and $\{y_1,\ldots,y_r\}$. If $x_1 = y_i$ for some i, then the result is clear. Otherwise, there is a signed circuit X with support contained in $\{x_1,y_1,\ldots,y_r\}$ and a signed cocircuit Y with support contained in $E\backslash\{x_2,\ldots,x_r\}$ with $x_1 \in \operatorname{supp}(Y)$. Assume without loss of generality $X(x_1)Y(x_1) = +$. Then for some i we have $X(y_i)Y(y_i) = -$, so $-X(x_1)X(y_i) = Y(x_1)Y(y_i) \neq 0$.

Since $\pm X$ is the unique pair of signed circuits with support contained in $\{x_1,y_1,\ldots,y_r\}$, and $X(y_i) \neq 0$, we have that $\{x_1,y_1,\ldots,\hat{y}_i,\ldots,y_r\}$ is independent, and so it's a basis. Applying the Pivoting Property, we see

$$-X(x_1)X(y_i) = \chi(x_1,y_1,\ldots,\hat{y}_i,\ldots,y_r)\chi(y_i,y_1,\ldots,\hat{y}_i,\ldots,y_r)$$
$$= \chi(y_1,\ldots,y_{i-1},x_1,y_{i+1},\ldots,y_r)\chi(y_1,\ldots,y_r).$$

Since rank($\{x_2, \ldots, x_r\}$) = $r - 1$ and $Y(\{x_2, \ldots, x_r\}) = \{0\}$, Proposition 2.50 tells us that for every c, $\{c, x_2, \ldots, x_r\}$ is a basis if and only if $Y(c) \neq 0$. In particular, $\{y_i, x_2, \ldots, x_r\}$ is a basis, and we can apply the Dual Pivoting Property to see

$$Y(x_1)Y(y_i) = \chi(x_1, \ldots, x_r)\chi(y_i, x_2, \ldots, x_r).$$

Setting these equal gives Signed Basis Exchange.

Finally, since $\mathcal{M}$ and χ have the same bases, they also have the same dependent sets and hyperplanes. Thus the unsigned circuits of χ are the supports of signed circuits of $\mathcal{M}$, and by Lemma 2.50, the unsigned cocircuits of χ are the supports of signed cocircuits of $\mathcal{M}$. □

Remark 2.62 The base case of our induction introduced the simplest oriented matroids – for each finite E, the rank $|E|$ oriented matroid on elements E. We call this the **coordinate oriented matroid** on E. We denote a coordinate oriented matroid on $[r]$ by $\mathcal{M}_{\text{coord}}$ or $\mathcal{M}_{\text{coord}}^r$. Thus $\mathcal{C}(\mathcal{M}_{\text{coord}}) = \emptyset$, $\mathcal{C}^*(\mathcal{M}_{\text{coord}}^r) = \{\pm\{i\}^+ : i \in [r]\}$, and $\mathcal{M}_{\text{coord}}^r$ is realized as a hyperplane arrangement by the coordinate hyperplanes in $\mathbb{R}^r$.

2.5 Summary on Cryptomorphism

We've now generalized the conclusions we summarized for realizable oriented matroids in Section 1.5.4. We can define an oriented matroid $\mathcal{M}$ to be the information encoded in any of the following equivalent ways:

- by a set $\mathcal{V}(\mathcal{M})$ satisfying Definition 2.5,
- by the dual $\mathcal{V}(\mathcal{M})^\perp$, which also satisfies Definition 2.5 and is denoted $\mathcal{V}^*(\mathcal{M})$,
- by a set $\mathcal{C}(\mathcal{M})$ satisfying Definition 2.7, which is the set of minimal elements of $\mathcal{V}(\mathcal{M})\backslash\{\mathbf{0}\}$,
- by a set $\mathcal{C}^*(\mathcal{M})$ satisfying Definition 2.7, which is the set of minimal elements of $\mathcal{V}^*(\mathcal{M})\backslash\{\mathbf{0}\}$, or
- by a pair $\pm\chi$ each satisfying Definition 2.42, which are related to $\mathcal{C}(\mathcal{M})$ and $\mathcal{C}^*(\mathcal{M})$ by the Pivoting Property and Dual Pivoting Property.

The dual of $\mathcal{M}$, denoted $\mathcal{M}^*$, is the oriented matroid with $\mathcal{V}(\mathcal{M}^*) = \mathcal{V}^*(\mathcal{M})$. An oriented matroid and its dual behave like a linear subspace of $\mathbb{R}^n$ and its complement in various ways. For instance, if $\mathcal{M}$ is rank r and has n elements then $\mathcal{M}^*$ is rank $n - r$ (Proposition 2.46). We saw in Section 2.3.2

that deletion and contraction are dual operations: $(\mathcal{M}/A)^* = \mathcal{M}^*\backslash A$ and $(\mathcal{M}\backslash A)^* = \mathcal{M}^*/A$.

Finally, we note two dual properties that make an element of $\mathcal{M}$ uninteresting. In the realizable case, one property amounts to an element of a vector arrangement being **0**, while the other amounts to one vector in an arrangement lying outside the span of the remaining vectors.

Proposition 2.63 *Let e be an element of an oriented matroid $\mathcal{M}$. Let $\chi_\mathcal{M}: E^r \to \{0, +, -\}$ be a chirotope for $\mathcal{M}$ and $\chi_{\mathcal{M}^*}: E^{n-r} \to \{0, +, -\}$ be a chirotope for $\mathcal{M}^*$. The following are equivalent.*

1. $e^+ \in \mathcal{V}(\mathcal{M})$.
2. $X(e) = 0$ *for each* $X \in \mathcal{V}^*(\mathcal{M})$.
3. $\chi_\mathcal{M}(e, x_2, \ldots, x_r) = 0$ *for all* $x_2, \ldots x_r$.
4. $\chi_{\mathcal{M}^*}(y_1, \ldots, y_{n-r}) = 0$ *if* $e \notin \{y_1, \ldots, y_r\}$.

Problem 2.64 Prove Proposition 2.63.

Definition 2.65 We say that e is a **loop** of $\mathcal{M}$ if the equivalent conditions of Proposition 2.63 hold.

A **coloop** of $\mathcal{M}$ is a loop of $\mathcal{M}^*$.

If e is a loop of $\mathcal{M}$ then $\mathcal{V}^*(\mathcal{M}) = \{Xe^0 : X \in \mathcal{V}^*(\mathcal{M}\backslash e)\} \cong \mathcal{V}^*(\mathcal{M})\backslash e$, and if e is a coloop of $\mathcal{M}$ then $\mathcal{V}^*(\mathcal{M}) = \{Xe^\epsilon : X \in \mathcal{V}^*(\mathcal{M}\backslash e), \epsilon \in \{0, +, -\}\}$.

Problem 2.66 Find similar characterizations of $\mathcal{C}(\mathcal{M})$, $\mathcal{V}(\mathcal{M})$, $\mathcal{C}^*(\mathcal{M})$, and $\chi_\mathcal{M}$ in terms of $\mathcal{C}(\mathcal{M}\backslash e)$, $\mathcal{V}(\mathcal{M}\backslash e)$, $\mathcal{C}^*(\mathcal{M}\backslash e)$, and $\chi_{\mathcal{M}\backslash e}$ when e is a loop and when e is a coloop. Conclude that for many purposes we may limit our attention to oriented matroids with no loops or coloops.

2.6 Other Axiomatizations

This section offers some interesting alternative perspectives but will not be needed in later chapters.

The signed circuit axioms can be weakened considerably, if one is willing to build on matroid theory. The papers Las Vergnas (1978), Las Vergnas (1984) proved a collection of results on the theme "assuming our signed structure is built on a matroid, we can reduce to considering rank 2 contractions." One of these results is the subject of Sections 6.2 and 6.3, and in the exercises for Chapter 6 we'll use this result to prove the following oriented matroid axiomatizations.

2.6 Other Axiomatizations

Definition 2.67 Let E be a finite set and Γ^* a collection of subsets of E. We say Γ^* is the **cocircuit set of an (unoriented) matroid** if $\Gamma^* \neq \{\emptyset\}$, no element of Γ^* properly contains another element of Γ^*, and, for each S, T distinct elements of Γ^*, if $e \in S \cap T$ then some element of Γ^* is contained in $(S \cup T) \setminus \{e\}$. We say $A \subseteq E$ is **independent** if, for each $a \in A$, there is an element S of Γ^* such that $S \cap A = \{a\}$. For each $A \subseteq E$, we define rank(A) to be the largest size of an independent subset of A. The **rank** of the matroid with cocircuit set Γ^* is defined to be the rank of E.

If C^* is the set of signed cocircuits of an oriented matroid then $\Gamma^* := \{\text{supp}(X) : X \in C^*\}$ is the set of cocircuits of a matroid, and the definitions of independence and rank for C^* and Γ^* coincide.

Theorem 2.68 (Las Vergnas 1984) $C^* \subseteq \{0, +, -\}^E$ *is the set of signed cocircuits of a rank r oriented matroid if and only if all of the following hold:*

1. $\Gamma^* := \{\text{supp}(X) : X \in C^*\}$ *is the cocircuit set of a matroid,*
2. $C^* = -C^*$, *and*
3. *for each $A \subseteq E$ of rank $r - 2$ with respect to Γ^*, C^*/A is the signed cocircuit set of an oriented matroid.*

Las Vergnas phrased the result differently, in terms of *comodular pairs*.

Definition 2.69 Let Γ^* be the cocircuit set of a rank r matroid and A, B be distinct elements of Γ^*. We say that the pair A, B is **comodular** if rank$(E \setminus (A \cup B)) = r - 2$.

Let C^* be the set of signed cocircuits of an oriented matroid, $C := \{\text{supp}(X) : X \in C^*\}$, and $X, Y \in C^*$. We say the pair X, Y is **comodular** if supp(X), supp(Y) is comodular in C. In other words, X, Y is comodular if rank$(X^0 \cap Y^0) = 2$.

Theorem 2.68 can be restated as follows:

Theorem 2.70 (Modular Cocircuit Elimination; Las Vergnas 1984) *Let $C^* \subseteq \{0, +, -\}^E$ such that $C^* = -C^*$ and $\{\text{supp}(X) : X \in C^*\}$ is the set of cocircuits of a matroid. Then C^* is the set of signed cocircuits of an oriented matroid if, for each comodular pair $X, Y \in C^*$ and each $e \in S(X, Y)$, there is a $Z \in C^*$ such that $Z(f) \in (X(f) \boxplus Y(f)) \cup \{0\}$ for each f and $Z(e) = 0$.*

Definition 2.71 $\mathcal{B} \subseteq \mathcal{P}(E)$ is the set of **bases of a matroid** if $\mathcal{B} \neq \emptyset$, and for each $B, B' \in \mathcal{B}$ and $b \in B \setminus B'$ there is a $b' \in B' \setminus B$ such that $(B \setminus \{b\}) \cup \{b'\} \in \mathcal{B}$.

As the reader surely suspects, there is a cryptomorphism between matroid cocircuit sets and matroid basis sets, as well as a duality theory, none of which we will go into here. Neither will we discuss Las Vergnas's characterization of signed circuit sets of oriented matroids, a close analog to Theorem 2.68.

Theorem 2.72 (Las Vergnas 1978, 1984) *Let E be a finite set, let $r > 2$, and let $\chi : E^r \to \{0, +, -\}$ be a nonzero alternating function. Then χ is a chirotope if and only if χ satisfies both of the following:*

1. *The set $\mathcal{B}$ of bases of χ is the set of bases of a matroid.*
2. *(**3-Term Grassmann–Plücker Relations**) For each $a, b, c, d, x_3, \ldots, x_r \in E$,*

$$0 \in \chi(a, b, x_3, \ldots, x_r)\chi(c, d, x_3, \ldots, x_r)$$
$$\boxplus -\chi(a, c, x_3, \ldots, x_r)\chi(b, d, x_3, \ldots, x_r)$$
$$\boxplus \chi(a, d, x_3, \ldots, x_r)\chi(b, c, x_3, \ldots, x_r).$$

Together with the chirotope characterization of contraction (Exercise 2.4), this says that to determine whether an alternating function χ is a chirotope, it suffices to check that χ has an underlying matroid and that χ induces a chirotope on each rank 2 contraction. The condition that χ has an underlying matroid is essential: For instance, the alternating function $\chi : [6]^3 \to \{0, +, -\}$ with $\chi(1, 2, 3) = \chi(4, 5, 6) = +$ and $\chi(i, j, k) = 0$ whenever $\{i, j, k\} \notin \{\{1, 2, 3\}, \{4, 5, 6\}\}$ satisfies the 3-term Grassmann–Plücker relations but is not a chirotope.

Another flavor of axiomatization for oriented matroids centers on maximal covectors. A maximal covector of $\mathcal{M}$ is called a *tope* of $\mathcal{M}$; these are discussed more in Chapters 3 and 4. As a "top-down" analog to Proposition 2.19, we have the following.

Theorem 2.73 (da Silva 1987, chapter 6, theorem 1) *If $\mathcal{T}$ is the set of topes of an oriented matroid on elements E, then the set of covectors is*

$$\{X \in \{0, +, -\}^E : X \circ T \in \mathcal{T} \text{ for all } T \in \mathcal{T}\}.$$

In particular, the set of topes determines the set of covectors. As a "top-down" analog to the signed circuit axioms, we have the following.

Theorem 2.74 (da Silva 1987, chapter 6, theorem 1) *Let $\mathcal{T} \subset \{0, +, -\}^E$, and let $\mathcal{T}_\downarrow = \{X \in \{0, +, -\}^E : X \leq T \text{ for some } T \in \mathcal{T}\}$. $\mathcal{T}$ is the set of topes of an oriented matroid if and only if $\mathcal{T}$ satisfies all of the following:*

1. $\mathcal{T} \neq \emptyset$.
2. $\mathcal{T} = -\mathcal{T}$.

2.6 Other Axiomatizations
81

3. *For each* $X \in \mathcal{T}_\downarrow$, *either there is a* $T \in \mathcal{T}$ *such that* $X \circ T \in \mathcal{T}$ *and* $X \circ (-T) \notin \mathcal{T}$ *or for all* $T \in \mathcal{T}$ $X \circ T \in \mathcal{T}$.

(For other characterizations of the tope set, see da Silva (1988), Lawrence (1983), Björner et al. (1999)).

Still another flavor of axiomatization centers on convexity. The notion of convexity we use for these purposes is motivated by affine point arrangements. If $\mathcal{M}$ is an oriented matroid on elements E and $A \subseteq E$, define the **convex hull** of A with respect to $\mathcal{M}$ to be

$$\widehat{\text{conv}}(A) = A \cup \{e \in E : \hat{A}^+ e^- \in \mathcal{C}(\mathcal{M}) \text{ for some } \hat{A} \subseteq A\}.$$

(The notation $\widehat{\text{conv}}$ is to distinguish this notion of convexity from convexity of covector sets, which will play a larger role in later chapters.) The function $\widehat{\text{conv}} \colon \mathcal{P}(E) \to \mathcal{P}(E)$ does not determine $\mathcal{M}$ by itself. For instance, if $\mathcal{M}$ is a rank 3 oriented matroid on elements $[n]$ realized by the set of vertices of a convex n-gon in an affine plane, then $\widehat{\text{conv}}$ is the identity function, and from $\widehat{\text{conv}}$ we can't deduce the cyclical order of the vertex labels, whereas we can deduce it from $\mathcal{M}$. We get convexity axioms motivated by a vector arrangement consisting of vectors in an affine subspace of $\mathbb{R}^n$, together with all of their negatives.

Theorem 2.75 (Folkman and Lawrence 1978, Büchi and Fenton 1988) *Let E be a finite set and $*$ an involution on E. Let $\widehat{\text{conv}} \colon \mathcal{P}(E) \to \mathcal{P}(E)$ satisfy all of the following:*

1. $\widehat{\text{conv}}(\emptyset) = \emptyset$.
2. $\widehat{\text{conv}}$ *is a* closure operator, *that is, for each $A \subseteq E$ we have $A \subseteq \widehat{\text{conv}}(A) = \widehat{\text{conv}}(\widehat{\text{conv}}(A))$, and for each $A \subseteq B \subseteq E$ we have $\widehat{\text{conv}}(A) \subseteq \widehat{\text{conv}}(B)$.*
3. $\widehat{\text{conv}}(A)^* = \widehat{\text{conv}}(A^*)$ *for each $A \subseteq E$.*
4. *If $e \in \widehat{\text{conv}}(A \cup \{e^*\})$ then $e \in \widehat{\text{conv}}(A)$.*
5. *If $e \in \widehat{\text{conv}}(A \cup \{f^*\})$ and $e \notin \widehat{\text{conv}}(A)$ then $f \in \widehat{\text{conv}}((A \cup \{e^*\}) \setminus \{f\})$.*

Then there is a unique oriented matroid $\mathcal{M}$ on E such that

- $\{e, e^*\}^+ \in \mathcal{V}(\mathcal{M})$ *for each $e \in E$, and*
- *for each $A \subset E$ and $b \in E \setminus A$*

$$\hat{A}^+ b^- \in \mathcal{V}(\mathcal{M}) \text{ for some } \hat{A} \subseteq A \Leftrightarrow b \in \widehat{\text{conv}}(A).$$

Of course, the oriented matroids arising from Theorem 2.75 are quite special, but every oriented matroid can be obtained from one of these by deletion.

2.7 Realizable vs. Nonrealizable Oriented Matroids

Every oriented matroid of rank 1 or 2 is realizable (see Exercise 2.2). An oriented matroid is realizable (by a subspace W of $\mathbb{R}^n$) if and only if its dual is realizable (by $W^\perp$). So the above exercise proves that every rank r oriented matroid with at most $r + 2$ elements is realizable. With some work (Goodman and Pollack 1980b, Richter-Gebert 1989) one can see that all rank 3 oriented matroids on at most eight elements are realizable.

Thus a nonrealizable oriented matroid is given by either a rather large set of sign vectors or by a function χ with a rather large domain. By definition, we can't represent a nonrealizable oriented matroid by a vector or hyperplane arrangement, and so the reader may be left with the sinking feeling that nonrealizable oriented matroids are unwieldy combinatorial gadgets to be dealt with only in an abstract or computational way.

In fact, nothing could be further from the truth. We conclude this chapter with two ways to describe a nonrealizable rank 3 oriented matroid pictorially.

The first is a description by points in a distorted affine plane. Not all oriented matroids are representable in this way, but this approach is powerful enough to prove an important result (Section 2.7.2).

The second approach is a description by oriented "topological equators" on the sphere. The discussion here is actually a preview of Chapter 4, where we'll prove that *every* oriented matroid has a description of this form. This marvelous result allows us to think of the whole subject of oriented matroids as concerning the combinatorics of arrangements of topological equators, rather than the combinatorics of big sets of sign vectors satisfying axioms.

2.7.1 Nonrealizable Oriented Matroids from Distortions in Affine Planes

Consider an arrangement $\mathcal{A} = \{p_i : i \in E\}$ of points in an affine plane, not all collinear. $\mathcal{A}$ determines a rank 3 (realizable) oriented matroid, and we can read off a chirotope χ for it as follows. For each $i, j, k \in E$, we consider a curve γ in the affine plane beginning at p_i, proceeding linearly to p_j, then proceeding linearly to p_k. Then

$$\chi(i,j,k) = \begin{cases} 0 & \text{if } p_i, p_j, p_k \text{ are collinear,} \\ + & \text{if } \gamma \text{ curves counterclockwise,} \\ - & \text{if } \gamma \text{ curves clockwise.} \end{cases}$$

For instance, for the arrangement of nine points shown in Figure 2.5, the chirotope χ_P satisfies $\chi_P(1,2,3) = 0$ and $\chi_P(1,2,j) = -$ for each $j > 3$.

2.7 Realizable vs. Nonrealizable Oriented Matroids

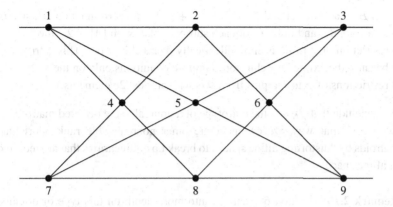

Figure 2.5 The Pappus point configuration.

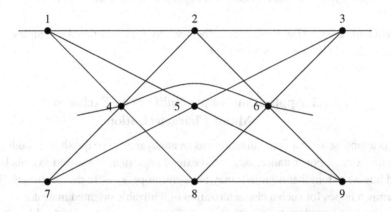

Figure 2.6 A non-Pappus pseudoconfiguration.

Figure 2.5 illustrates *Pappus's Theorem*, which says that for every point arrangement satisfying the collinearities indicated, the points 4, 5, and 6 must be collinear.

To get a nonrealizable oriented matroid, we will spit in the face of Pappus's Theorem. Namely, let's pretend that the points 4, 5, and 6 are not collinear, but rather point 5 lines below the line through points 4 and 6, as in Figure 2.6.

From this we can read off a new alternating function χ_{NP}, with $\chi_{NP}(4,5,6) = -$ and $\chi_{NP}(i,j,k) = \chi_P(i,j,k)$ whenever $\{i,j,k\} \neq \{4,5,6\}$. Two observations show that χ_{NP} is a chirotope:

- If we remove any one of the nine points from Figure 2.6, the resulting restriction of χ_{NP} is the chirotope of an affine point arrangement.

- Let E be a finite set and $\chi : E^r \to \{0, +, -\}$ a nonzero function. Then χ is a chirotope if and only if, for each subset A of E with $|A| \leq 2r$, the restriction of χ to A^r is either identically 0 or a chirotope. (This is true because the axioms for chirotopes impose conditions only on the restrictions of χ to r-tuples from sets with at most $2r$ elements.)

We conclude that χ_{NP} is the chirotope of a nonrealizable oriented matroid.

In a similar way, we can generate other nonrealizable rank 3 oriented matroids by "distorting affine space" to break up collinearities that are required in affine space.

Remark 2.76 Section 6.5 will go into more depth on this type of oriented matroid representation. In particular, the informal description used here will be nailed down to a precise definition of *pseudoconfigurations of points*.

Problem 2.77 Use Desargue's Theorem to find additional examples of nonrealizable rank 3 oriented matroids.

2.7.2 Application: Impossibility of an Excluded Minor Characterization

It is tempting to treat nonrealizability as an annoyance to be quashed by adding more axioms. For instance, one could extend our oriented matroid axioms by adding an axiom that no minor may have chirotope χ_{NP}. In this section we'll dampen hopes for such a characterization of realizable oriented matroids.

Every oriented matroid with a nonrealizable minor is nonrealizable. The problem of an *excluded minor characterization of realizability* asks: Is there a finite set N of isomorphism types of nonrealizable oriented matroids so that the realizable oriented matroids are exactly those with no minors in N?

For ordinary matroids, it was recently shown that for every finite field F there is a finite excluded minor characterization for realizability over F (Geelen et al. 2014). Explicit excluded minor characterizations are known for realizability of ordinary matroids over $GF(2)$ (Tutte 1958), $GF(3)$ (Björner 1995, Seymour 1979), and $GF(4)$ (Geelen et al. 2014). However, there is no such characterization for fields of characteristic 0: As Vamos so wistfully put it, *the missing axiom of matroid theory is lost forever* (Vámos 1978, see also Mayhew et al. 2014, 2018 – these papers actually take a broader perspective on the "missing axiom" than we discuss here).

For oriented matroids, there is not even a finite excluded minor characterization of realizable rank 3 oriented matroids. There are several proofs of this

2.7 Realizable vs. Nonrealizable Oriented Matroids

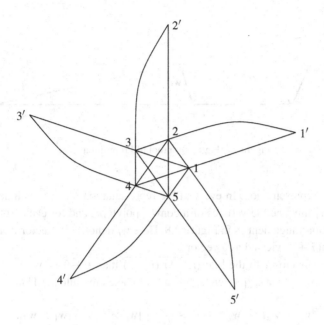

Figure 2.7 The nonrealizable oriented matroid B_5.

(cf. section 8.3 of Björner et al. 1999): We provide one from Goodman and Pollack (1980a).

For each $n \geq 5$ we will construct a nonrealizable rank 3 oriented matroid B_n on $2n$ elements such that each restriction to $2n - 1$ elements is realizable. One can see the idea immediately from B_5, shown in Figure 2.7. In general, the set of elements of B_n is $E_n = \{1, \ldots, n, 1', \ldots, n'\}$, and we draw these elements as intersections of distorted "lines" in an affine plane by making $\{1, \ldots, n\}$ the vertices of a convex n-gon P and, for each $i \in \{1, \ldots, n\}$, the point i' is the intersection of the "lines" $\overline{i, (i + 3)}$ and $\overline{(i + 1), (i + 2)}$. As in our figure, we take each i' to be on the opposite side of the line $\overline{i, (i + 1)}$ from the remaining vertices of P. From this we read off a function $\chi : (E_n)^3 \to \{0, +, -\}$, and we note that the restriction of χ to $(E_n \setminus \{j\})^3$ is the chirotope of a realizable oriented matroid for each j. Thus by our earlier observation χ is in fact the chirotope of an oriented matroid B_n.

Proposition 2.78 (Goodman and Pollack 1980a) B_n is nonrealizable for each $n \geq 5$.

Proof: Assume by way of contradiction there is a realization of B_n as an arrangement of vectors in $\mathbb{R}^3$. Then, since $(E_n)^+$ is a covector of B_n, B_n has

86 Oriented Matroids

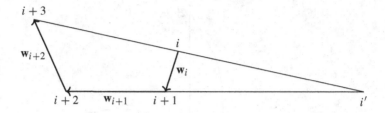

Figure 2.8 Figure for proof of Proposition 2.78.

an affine representation. In every affine representation, the points indexed by $\{1,\ldots,n\}$ must be the vertices of a convex polytope, and for each $i \in [n]$ we have a subarrangement as in Figure 2.8. Here $\mathbf{w}_i$ denotes the vector from point i to point $i+1$, viewed as a vector in $\mathbb{R}^3$.

Since i is closer to the line $\overline{(i+1),(i+2)}$ than $i+3$ is, we have $|\mathbf{w}_i \times \mathbf{w}_{i+1}| < |\mathbf{w}_{i+1} \times \mathbf{w}_{i+2}|$. (Here subscripts are taken modulo n.) Thus

$$|\mathbf{w}_1 \times \mathbf{w}_2| > |\mathbf{w}_2 \times \mathbf{w}_3| > \cdots > |\mathbf{w}_n \times \mathbf{w}_1| > |\mathbf{w}_1 \times \mathbf{w}_2|,$$

a contradiction. □

2.7.3 Nonrealizable Oriented Matroids from Pseudosphere Arrangements

Definition 2.79 Let T be homeomorphic to S^d. A subset S of T is a **pseudosphere** if there is a homeomorphism $H: T \to S^d$ sending S to the equator $\{\mathbf{x} \in S^d : x_{d+1} = 0\}$.

We also allow the **degenerate pseudosphere** T.

A **signed pseudosphere** in T is a triple $\mathcal{S} = (S, S^+, S^-)$, where S is a pseudosphere in T and

- if S is degenerate, then $S^+ = S^- = \emptyset$, and
- otherwise S^+, S^- are the two open balls in T bounded by S.

S^+ and S^- are called the **sides** of S.

(The terminology is standard but unfortunate, since a pseudosphere really is a sphere topologically. A better term might have been "pseudoequator.")

Definition 2.80 An **arrangement of signed pseudospheres** in T is a multiset $(\mathcal{S}_e : e \in E)$ of signed pseudospheres in T satisfying all of the following:

1. For every $A \subseteq E$, the set $S_A := \bigcap_{e \in A} S_e$ is a topological sphere. (Recall that the empty set is the sphere of dimension -1.)
2. For every $e \in E$ and $A \subseteq E \setminus \{e\}$, either $S_A \subseteq S_e$ or $S_A \cap S_e$ is a pseudosphere in S_A with sides $S_A \cap S_e^+$ and $S_A \cap S_e^-$.
3. Each intersection of a collection of closed sides is either a topological sphere or a topological open ball.

(Aside: A hard theorem (Mandel 1982) says that the last condition in Definition 2.80 is redundant.)

Definition 2.81 An arrangement $(S_e : e \in E)$ of signed pseudospheres in S^d is

- **centrally symmetric** if, for each $e \in E$, $S_e = -S_e$, and
- **essential** if $\bigcap_{e \in E} S_e = \emptyset$.

We can get a collection of sign vectors from an arrangement of pseudospheres in the same way we did for an arrangement of oriented equators in S^d in Section 1.3.4. Given an arrangement $\mathcal{A} = (S_e : e \in E)$ of signed pseudospheres in S^d and $x \in S^d$, let $\text{sign}_{\mathcal{A}}(x)$ be the element of $\{0, +, -\}^E$ with $\text{sign}_{\mathcal{A}}(x)(e) = 0$ if $x \in S_e$ and $\text{sign}_{\mathcal{A}}(x)(e) = \alpha \in \{+, -\}$ if $x \in S_e^{\alpha}$. Let $\mathcal{V}^*(\mathcal{A}) = \{\text{sign}_{\mathcal{A}}(x) : x \in S^d\} \cup \{0\}$. Thus $\mathcal{V}^*(\mathcal{A}) \setminus \{0\}$ indexes the cells in the cell decomposition of S^d given by $\mathcal{A}$.

Theorem 2.82 *Let $\mathcal{A}$ be a centrally symmetric, essential arrangement of signed pseudospheres in S^d. Then $\mathcal{V}^*(\mathcal{A})$ is the covector set of a rank $d + 1$ oriented matroid.*

Proof: Symmetry of $\mathcal{A}$ implies symmetry of $\mathcal{V}^*(\mathcal{A})$. To see both the Composition and Elimination Axioms: For every $A \subseteq E$ and $Z \in \{0, +, -\}^A$, define $C_Z = \bigcap_{e \in Z^0} S_e \cap \bigcap_{e \in \text{supp}(Z)} S_e^{Z(e)}$. In particular, if $Z \in \mathcal{V}^*(\mathcal{A}) \setminus \{0\}$ then C_Z is the open cell corresponding to Z, and if Z is a nonzero restriction of an element of $\mathcal{V}^*(\mathcal{A}) \setminus \{0\}$ then C_Z is an open ball in the sphere $\bigcap_{e \in Z^0} S_e$.

Consider $X, Y \in \mathcal{V}^*(\mathcal{A}) \setminus \{0\}$ with $X \neq -Y$. Let $A = E \setminus S(X, Y)$, and define $Z \in \{0, +, -\}^A$ by $Z(a) = \max(X(a), Y(a))$. Then C_Z is an open ball in the sphere $\bigcap_{e \in Z^0} S_e$, and C_X and C_Y are contained in the closure of C_Z.

Thus there is a path $\gamma : [0, 1] \to S^d$ with $\gamma(0) \in C_X$, $\gamma(1) \in C_Y$, and $\gamma((0, 1)) \subseteq C_Z$. From this we can see our axioms.

- For all $\epsilon > 0$ sufficiently small, $\text{sign}_{\mathcal{A}}(\gamma(\epsilon)) \geq X$ and $\text{supp}(\text{sign}_{\mathcal{A}}(\gamma(\epsilon))) = \text{supp}(X) \cup \text{supp}(Y)$. (We can see this by considering a small open ball around x and its intersection with each S_e.) Also, by the definition of Z we have $\text{sign}_{\mathcal{A}}(\gamma(\epsilon))(e) = Y(e)$ for each $e \in \text{supp}(Y) \setminus \text{supp}(X)$. Thus $\text{sign}_{\mathcal{A}}(\gamma(\epsilon)) = X \circ Y$.

- If $e \in S(X,Y)$, then since $\gamma(0)$ and $\gamma(1)$ are in different connected components of $S^d \backslash S_e$, there must be a $t \in (0,1)$ such that $\gamma(t) \in S_e$. Then $\text{sign}_A(\gamma(t))$ is an elimination of e between X and Y.

Thus $\mathcal{V}^*(\mathcal{A})$ is the covector set of an oriented matroid $\mathcal{M}$. To see the rank of $\mathcal{M}$, we prove by induction that, even without the condition that the arrangement is essential, if $A \subseteq E$ has rank r in $\mathcal{M}$ then S_A has dimension $d - r$. An important pair of observations for our induction: For every $A \subset E$,

- $\mathcal{V}^*(\mathcal{M} \backslash A)$ indexes the cells in the cell decomposition of S^d given by the arrangement $(\mathcal{S}_e : e \in E \backslash A)$, and
- $\mathcal{V}^*(\mathcal{M}/A)$ indexes the cells in the cell decomposition of S_A given by the arrangement $(\mathcal{S}_e \cap S_A : e \in E \backslash A)$.

If $r = 0$ then $S_A = S^d$. If $r = 1$ then $\mathcal{M}(A)$ is rank 1, hence $\mathcal{V}^*(\mathcal{M}(A))$ has the form $\{\mathbf{0}, \pm X\}$, and S_A is a pseudosphere in S^d.

For larger r, let a be a nonloop in A. Then $A \backslash a$ has rank $r - 1$ in $\mathcal{M}/a$. Notice $S_A = \bigcap_{b \in A \backslash a}(S_b \cap S_a)$, By our observation on contraction, we can view this as an intersection of pseudospheres in the sphere $S_a \cong S^{d-1}$, and so by the induction hypothesis S_A has dimension $(d - 1) - (r - 1)$.

Thus $d - \text{rank}(\mathcal{M})$ is the dimension of $\bigcap_{e \in E} S_e$. Since the arrangement is essential, this dimension is -1. □

Let's have another look at Pappus's Theorem, this time viewing it as a statement about planes in $\mathbb{R}^3$ with a particular intersection pattern. Figure 2.9 illustrates the spherical version of this theorem: The points a, b, and c in this figure must lie on a common equator. If we choose a positive side for each indicated equator, then we have an arrangement $\mathcal{A}_P$ of signed equators, and $\mathcal{V}^*(\mathcal{A}_P)$ is the covector set of a realizable oriented matroid.

Now consider Figure 2.10, with the nine pseudospheres signed however you like to give an arrangement $\mathcal{A}_{NP}$. But the oriented matroid with covector set $\mathcal{V}^*(\mathcal{A}_{NP})$ is not realizable: A realization would satisfy the hypotheses of Pappus's Theorem but not the conclusion.

2.7.4 Final Comments on Nonrealizable Oriented Matroids

"Most" oriented matroids are nonrealizable: For a fixed rank r and number n of elements, there are upper bounds on the number of realizable oriented matroids (cf. 8.7.5 in Björner et al. 1999) and lower bounds on the total number of oriented matroids (cf. 7.4.2 in Björner et al. 1999) that show that as n/r approaches infinity, the proportion of realizable oriented matroids approaches 0.

2.7 Realizable vs. Nonrealizable Oriented Matroids

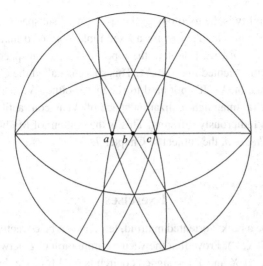

Figure 2.9 The spherical version of Pappus's Theorem.

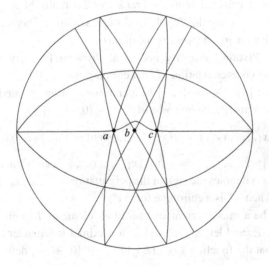

Figure 2.10 The non-Pappus oriented matroid.

To preview some coming topics: The first big theorem in oriented matroid theory, beyond the cryptomorphisms, is the Topological Representation Theorem (the subject of Chapter 4), which says that *every* oriented matroid can be represented by an arrangement of pseudospheres. Thus, from a topological viewpoint all oriented matroids are "close to" being realizable. Another question on realizability is the topology of the space of all subspace realizations of a fixed oriented matroid. That is, for a fixed rank r oriented matroid $\mathcal{M}$ with elements $\{1, 2, \ldots, n\}$, we'll look at the topology of the subspace of $G(r, \mathbb{R}^n)$ of all spaces with oriented matroid $\mathcal{M}$. This space is called the *Grassmannian realization space* of $\mathcal{M}$. It's not hard to see that if either $\mathcal{M}$ or $\mathcal{M}^*$ is rank 1 or 2 then the Grassmannian realization space of $\mathcal{M}$ is contractible. In rank 3 the situation is ludicrously different. This is the content of Mnëv's celebrated *Universality Theorem*, the subject of Chapter 7.

Exercises

2.1 Let $\mathcal{M}$ be a rank 2 oriented matroid, let $X \neq -Y$ be covectors of $\mathcal{M}$, and let $e \in S(X, Y)$. Prove that the vector elimination of e between X and Y is unique. If X and Y are signed cocircuits of $\mathcal{M}$, prove that this vector elimination is also the unique circuit elimination of e between X and Y.

2.2 Prove every oriented matroid of rank 1 or 2 is realizable.
(Once we prove the Topological Representation Theorem in Chapter 4 we'll have a much easier proof of this.)

2.3 Use the Pivoting Property and Dual Pivoting Property to prove the following characterization of dual chirotopes:
Let $\chi : [n]^r \to \{0, +, -\}$ be a chirotope for an oriented matroid $\mathcal{M}$ with elements $[n]$. Define $\chi^* : [n]^{n-r} \to \{0, +, -\}$ by

$$\chi^*(x_1, \ldots, x_{n-r}) = \chi(y_1, \ldots, y_r)\text{sign}(x_1, \ldots, x_{n-r}, y_1, \ldots, y_r),$$

where $\{y_1, \ldots, y_r\} = [n] \setminus \{x_1, \ldots, x_{n-r}\}$ and $\text{sign}(x_1, \ldots, x_{n-r}, y_1, \ldots, y_r)$ denotes the sign of the permutation $(x_1, \ldots, x_{n-r}, y_1, \ldots, y_r)$ of $[n]$. Then χ^* is a chirotope for $\mathcal{M}^*$.

2.4 Let $\mathcal{M}$ be a rank r oriented matroid on elements E with chirotope χ, let $A \subset E$, and let $\{a_1, \ldots, a_k\}$ be a maximal independent subset of A. Prove that the function $\chi_{/A} : (E \setminus A)^{r-k} \to \{0, +, -\}$ defined by

$$\chi_{/A}(x_1, \ldots, x_{r-k}) = \chi(x_1, \ldots, x_{r-k}, a_1, \ldots, a_k)$$

is a chirotope for $\mathcal{M}/A$.

2.5 Let $\mathcal{M}$ be a rank r oriented matroid and χ a chirotope for M. For each $x_2, \ldots, x_r \in E$, let $X_{x_2, \ldots, x_r} \in \{0, +, -\}^E$ be the sign vector with, for each $e \in E$,

$$X_{x_2, \ldots, x_r}(e) = \chi(e, x_2, \ldots, x_r).$$

Prove that $\mathcal{C}^*(\mathcal{M}) := \{\pm X_{x_2, \ldots, x_r} : x_2, \ldots, x_r \in E\} \setminus \mathbf{0}$.

(This characterization is due to Michael Dobbins.)

2.6 (Bachem and Wanka 1988) This exercise is to prove an oriented matroid generalization of Carathéodory's Theorem. (See Exercise 1.9.)

Let $\mathcal{M}$ be a rank r oriented matroid on elements E. Let $A \subset E$ and $f \in E \backslash A$. Prove that if $f \in \widehat{\text{conv}}(A)$ then there is a subset A' of A with at most r elements such that $f \in \widehat{\text{conv}}(A')$.

2.7 Let $\mathcal{C}$ be the signed circuit set of an oriented matroid on elements E and $\mathcal{V}^*$ the covector set. Let $\mathcal{C}^\uparrow$ be the set of all elements of $\{0, +, -\}^E$ that are greater than or equal to some element of $\mathcal{C}$, and let $\mathcal{V}^*_\downarrow$ be the set of all elements of $\{0, +, -\}^E$ that are less than or equal to some element of $\mathcal{V}^*$. Prove that $\mathcal{C}^\uparrow \dot\cup \mathcal{V}^*_\downarrow = \{0, +, -\}^E$.

3

Elementary Operations and Properties

Some aspects of vector arrangements and subspaces have immediate analogs for oriented matroids. We have seen a few already: deletion, contraction, loops, and coloops. This chapter introduces a few more and explores some poset properties.

3.1 Isomorphism

Proposition 3.1 *Let $\mathcal{M}$ be an oriented matroid on elements E and $\mathcal{N}$ an oriented matroid on elements F. Let $f: E \to F$ be a bijection. Also, let $\tilde{f}: \{0, +, -\}^E \to \{0, +, -\}^F$ be the map taking each X to $X \circ f$, and let $f^{\text{rank}(\mathcal{M})}: E^{\text{rank}(\mathcal{M})} \to F^{\text{rank}(\mathcal{M})}$ be the product function. The following are equivalent:*

1. *$\tilde{f}$ restricts to a bijection $\mathcal{V}(\mathcal{M}) \to \mathcal{V}(\mathcal{N})$.*
2. *$\tilde{f}$ restricts to a bijection $\mathcal{V}^*(\mathcal{M}) \to \mathcal{V}^*(\mathcal{N})$.*
3. *$\tilde{f}$ restricts to a bijection $\mathcal{C}(\mathcal{M}) \to \mathcal{C}(\mathcal{N})$.*
4. *$\tilde{f}$ restricts to a bijection $\mathcal{C}^*(\mathcal{M}) \to \mathcal{C}^*(\mathcal{N})$.*
5. *$\text{rank}(\mathcal{N}) = \text{rank}(\mathcal{M})$, and if $\chi_{\mathcal{M}}: E^{\text{rank}(\mathcal{M})} \to \{0, +, -\}$ is a chirotope for $\mathcal{M}$ then $\chi_{\mathcal{M}} \circ f^{\text{rank}(\mathcal{M})}$ is a chirotope for $\mathcal{N}$.*

The proof is straightforward.

Definition 3.2 We say $\mathcal{M}$ is **isomorphic** to $\mathcal{N}$ if there is a function f such that the equivalent conditions of Proposition 3.1 hold.

3.2 Reorientation and Cyclic/Acyclic Oriented Matroids

Definition 3.3 For any sign vector $X \in \{0, +, -\}^E$ and $A \subseteq E$, the *reorientation* of X at A is the sign vector X_{-A} defined by

$$X_{-A}(e) = \begin{cases} X(e) & \text{if } e \notin A, \\ -X(e) & \text{if } e \in A. \end{cases}$$

For any collection $\mathcal{F} \subseteq \{0, +, -\}^E$ of sign vectors and any $A \subseteq E$, the *reorientation* of $\mathcal{F}$ at A is $\mathcal{F}_{-A} := \{X_{-A} : X \in \mathcal{F}\}$.

Proposition 3.4 *Let $\mathcal{M}$ be an oriented matroid on elements E, and let $A \subseteq E$. There is an oriented matroid $\mathcal{M}_{-A}$ with vector set $\mathcal{V}(\mathcal{M})_{-A}$, covector set $\mathcal{V}^*(\mathcal{M})_{-A}$, and chirotope χ_{-A} related to a chirotope χ for $\mathcal{M}$ by*

$$\chi_{-A}(x_1, \ldots, x_r) = (-1)^{|A \cap \{x_1, \ldots, x_r\}|} \chi(x_1, \ldots, x_r).$$

Problem 3.5 Prove this from the cryptomorphisms of Chapter 2.

Definition 3.6 The oriented matroid $\mathcal{M}_{-A}$ is called the *reorientation* of $\mathcal{M}$ at A.

If $\mathcal{M}$ is realized by an arrangement ($\mathbf{v}_e : e \in E$) of vectors, then reorientation at A gives the oriented matroid realized by ($\epsilon_e \mathbf{v}_e : e \in E$), where $\epsilon_e = -1$ if $e \in A$ and $\epsilon_e = 1$ if $e \notin A$. If $\mathcal{M}$ is realized by a subspace W of $\mathbb{R}^n$, then $\mathcal{M}_A$ is realized by the subspace obtained by reflecting W across all of the coordinate hyperplanes corresponding to elements of A.

Definition 3.7 A sign vector is **positive** if it has some positive coordinates and no negative coordinates.

An oriented matroid is **acyclic** if it does not have a positive signed circuit.

An oriented matroid is **totally cyclic** if every element is in the support of a positive signed circuit.

Equivalently, an oriented matroid is acyclic if it does not have a positive vector, and because every vector of $\mathcal{M}$ is a composition of conformal signed circuits, an oriented matroid on elements E is totally cyclic if E^+ is a vector.

Orthogonality tells us that $\mathcal{M}$ is acyclic if and only if $\mathcal{M}^*$ is totally cyclic. An example of an oriented matroid which is neither acyclic nor totally cyclic is given by the vector arrangement $((1, 0), (-1, 0), (0, 1))$. A realizable oriented matroid is acyclic if and only if it can be realized by an affine point arrangement.

Definition 3.8 A **tope** of an oriented matroid $\mathcal{M}$ is a maximal covector of $\mathcal{M}$.

If $\mathcal{M}$ is an oriented matroid on elements E then there is a bijection between the topes of $\mathcal{M}$ and the reorientations of $\mathcal{M}$ that are acyclic. The bijection sends a tope $T = A^+ B^-$ to the reorientation $\mathcal{M}_{-B}$, which has as a tope $T_{-B} = E^+$. In particular, every oriented matroid with no loops can be reoriented to an acyclic oriented matroid. Properties of $\mathcal{M}$ are easily derived from properties of $\mathcal{M}_{-A}$, so we can often use reorientation to assume that the oriented matroid we're working with is acyclic.

Remark 3.9 In earlier literature on oriented matroids, from before the language of vectors and covectors became standard, the topes of $\mathcal{M}$ were sometimes called the *acyclic reorientations* of $\mathcal{M}$.

3.3 Parallel, Simple, Uniform

3.3.1 Parallel Elements

Proposition 3.10 *Let $\mathcal{M}$ be a rank r oriented matroid and e, f be distinct nonloops of $\mathcal{M}$. The following are equivalent:*

1. *There is an element of $\mathcal{C}(\mathcal{M})$ with support $\{e, f\}$.*
2. *$X(e) = X(f)$ for all $X \in \mathcal{V}^*(\mathcal{M})$ or $X(e) = -X(f)$ for all $X \in \mathcal{V}^*(\mathcal{M})$.*
3. *For all $X \in \mathcal{V}^*(\mathcal{M})$, $X(e) = 0$ if and only if $X(f) = 0$.*
4. *For all $X \in \mathcal{V}^*(\mathcal{M})$, $X(e) = 0$ implies $X(f) = 0$.*
5. *Either $\chi(e, x_2, \ldots, x_r) = \chi(f, x_2, \ldots, x_r)$ for all $x_2, \ldots, x_r \in E$ or $\chi(e, x_2, \ldots, x_r) = -\chi(f, x_2, \ldots, x_r)$ for all $x_2, \ldots, x_r \in E$.*
6. *Either $r = 1$ or $\chi(e, f, x_3, \ldots, x_r) = 0$ for all $x_3, \ldots, x_r \in E$.*

Problem 3.11 Prove this.

Definition 3.12 Let e and f be a pair for which the statements of Proposition 3.10 hold. If $X(e) = -X(f)$ for each signed circuit X with support $\{e, f\}$ we say e and f are **parallel**. If $X(e) = X(f)$ for each signed circuit X with support $\{e, f\}$ we say e and f are **antiparallel**.

The **parallelism class** of an element e is the set of elements parallel or antiparallel to e, together with e itself.

Some authors also use "parallel" for any pair e, f satisfying the statements of Proposition 3.10.

3.3.2 Simple Oriented Matroids

Definition 3.13 $\mathcal{M}$ is **simple** if it has no loops and each parallellism class has a single element.

Proposition 3.14 *Let $\mathcal{M}$ be an oriented matroid on E and let E' be obtained from E by deleting all loops of $\mathcal{M}$ and all but one element of each parallelism class of $\mathcal{M}$. Then the map $\mathcal{V}^*(\mathcal{M}) \to \mathcal{V}^*(\mathcal{M}(E'))$ taking each X to its restriction to E' is a poset isomorphism.*

This is easily proved by constructing the inverse map.

Because of this result, for many problems it suffices to consider only simple oriented matroids. For instance, in Chapter 4 we will prove the Topological Representation Theorem, which is a result about the isomorphisn type of the poset $\mathcal{V}^*(\mathcal{M})$, and we'll make the argument, well, simpler by restricting to simple oriented matroids.

If $\mathcal{M}$ is simple then every deletion $\mathcal{M}\backslash A$ is also simple. A contraction $\mathcal{M}/A$ may not be simple. For instance, if $\mathcal{M}$ has a signed circuit with support $\{a,b,c\}$ then b and c are parallel in $\mathcal{M}/\{a\}$.

3.3.3 Uniform Oriented Matroids

Definition 3.15 A rank r oriented matroid is **uniform** if every set of r elements is a basis.

In unoriented matroid theory, uniformity is a boring property – for every $r \leq |E|$ there is a unique uniform rank r matroid on elements E. But for $n \gg r$ there are typically many uniform rank r oriented matroids on n elements, and even many reorientation classes of rank r oriented matroids on n elements.

3.4 Direct Sum

Let E_1 and E_2 be disjoint sets. If $X_1 : E_1 \to \{0, +, -\}$ and $X_2 : E_2 \to \{0, +, -\}$ are functions, let $X_1 X_2 : E_1 \cup E_2 \to \{0, +, -\}$ be the function that restricts to X_1 and X_2.

Proposition 3.16 *Let $\mathcal{M}_1$ and $\mathcal{M}_2$ be oriented matroids on disjoint sets E_1 resp. E_2. Let r_1 and r_2 be the ranks of $\mathcal{M}_1$ resp. $\mathcal{M}_2$, and let χ_1 and χ_2 be chirotopes for $\mathcal{M}_1$ resp. $\mathcal{M}_2$. There is an oriented matroid $\mathcal{M}$ of rank $r_1 + r_2$ such that*

- $\mathcal{V}(\mathcal{M}) = \{XY : X \in \mathcal{V}(\mathcal{M}_1), Y \in \mathcal{V}(\mathcal{M}_2)\}$,
- $\mathcal{C}(\mathcal{M}) = \{X\mathbf{0} : X \in \mathcal{C}(\mathcal{M}_1)\} \cup \{\mathbf{0}X : X \in \mathcal{C}(\mathcal{M}_2)\}$,
- $\mathcal{V}^*(\mathcal{M}) = \{XY : X \in \mathcal{V}^*(\mathcal{M}_1), Y \in \mathcal{V}^*(\mathcal{M}_2)\}$,
- $\mathcal{C}^*(\mathcal{M}) = \{X\mathbf{0} : X \in \mathcal{C}^*(\mathcal{M}_1)\} \cup \{\mathbf{0}X : X \in \mathcal{C}^*(\mathcal{M}_2)\}$,
- $\mathcal{M}$ *has a chirotope χ satisfying:*

1. For every $x_1, \ldots, x_{r_1} \in E_1$ and $y_1, \ldots, y_{r_2} \in E_2$,

$$\chi(x_1, \ldots, x_{r_1}, y_1, \ldots, y_{r_2}) = \chi_1(x_1, \ldots, x_{r_1})\chi_2(y_1, \ldots, y_{r_2})$$

and

2. for every $z_1, \ldots, z_{r_1+r_2} \in E_1 \cup E_2$, if $|\{z_1, \ldots, z_{r_1+r_2}\} \cap E_1| \neq r_1$ then $\chi(z_1, \ldots, z_{r_1+r_2}) = 0$.

Problem 3.17 Prove this.

Definition 3.18 This oriented matroid is called the **direct sum** of $\mathcal{M}_1$ and $\mathcal{M}_2$ and is denoted $\mathcal{M}_1 \oplus \mathcal{M}_2$.

If $\mathcal{M}_1$ is the oriented matroid of a vector arrangement $(\mathbf{v}_e : e \in E_1)$ in a vector space V_1 and $\mathcal{M}_2$ is the oriented matroid of a vector arrangement $(\mathbf{v}_e : e \in E_2)$ in a vector space V_2 then $\mathcal{M}_1 \oplus \mathcal{M}_2$ is the oriented matroid of the arrangement $((\mathbf{v}_e, \mathbf{0}) : e \in E_1) \cup ((\mathbf{0}, \mathbf{v}_e) : e \in E_2)$ in $V_1 \oplus V_2$. If $\mathcal{M}_1$ is the oriented matroid of a subspace W_1 of $\mathbb{R}^{E_1}$ and $\mathcal{M}_2$ is the oriented matroid of a subspace W_2 of $\mathbb{R}^{E_2}$ then $\mathcal{M}_1 \oplus \mathcal{M}_2$ is the oriented matroid of $W_1 \oplus W_2 \subseteq \mathbb{R}^{E_1} \oplus \mathbb{R}^{E_2}$.

Definition 3.19 An oriented matroid is **connected** if it is not a direct sum.

An oriented matroid is connected if and only if its underlying matroid is connected. See, for instance, chapter 4 of Oxley (1992) for more on connectivity of matroids.

3.5 Extensions

Definition 3.20 Let $\mathcal{M}$ be an oriented matroid with set of elements E. An **extension** of $\mathcal{M}$ is an oriented matroid $\mathcal{M}'$ with set of elements $E' \supseteq E$ such that $\mathcal{M}'(E) = \mathcal{M}$. In other words, $\mathcal{M}'$ is an extension of $\mathcal{M}$ if and only if $\mathcal{M}$ is a deletion of $\mathcal{M}'$. The extension is **proper** if no element of $E' \backslash E$ is a loop or coloop.

A **single-element extension** of $\mathcal{M}$ is an extension of $\mathcal{M}$ with elements $E \cup \{f\}$. Such an extension is sometimes called an **extension by the element** f and denoted $\mathcal{M} \cup f$.

The notation $\mathcal{M} \cup f$ is ambiguous – there are typically many nonisomorphic extensions of $\mathcal{M}$ by f.

Example 3.21 If e is an element of $\mathcal{M}$ then we can take an extension $\mathcal{M} \cup f$ with f parallel to e, or an extension with f antiparallel to e: In either case

Section 3.3.1 tells us what the chirotope, signed circuits, and so on of the extension will be.

Given a realization $\mathcal{A}$ of $\mathcal{M}$, we can make many extensions of $\mathcal{M}$ just by adding one more element to $\mathcal{A}$. But coming up with the set of all realizations of $\mathcal{M}$ is not a simple matter. Just to inculcate proper fear and respect for the topic, we have the following observations about an $\mathcal{M}$ realized by an arrangement $\mathcal{A} = (p_e : e \in E)$ of points in the affine plane.

- Describing the set of oriented matroids of extensions $(p_e : e \in E \cup \{f\})$ of $\mathcal{A}$ in terms of $\mathcal{M}$ is not easy. If you don't believe this, draw a cloud of seven points in the plane and try it.
- In fact, $\mathcal{M}$ doesn't give us enough information to describe this set. For example, consider the oriented matroid $\mathcal{M}_{\text{hex}}$ on elements [6] realized by the vertices of a convex hexagon, with vertices numbered consecutively. Depending on what hexagon we choose to realize $\mathcal{M}_{\text{hex}}$, the diagonals $\overline{14}$, $\overline{25}$, and $\overline{36}$ may or may not have a common point. So some realizations of $\mathcal{M}_{\text{hex}}$ give us an extension $\mathcal{M}_{\text{hex}} \cup f$ with all three of $1^+4^+f^-, 2^+5^+f^-$, and $3^+6^+f^-$ as signed circuits, while some do not.
- $\mathcal{M}$ may have nonrealizable extensions.

The broad subject of single-element extensions is tricky. It will be discussed in more detail in Chapter 6, but that discussion will in some ways make the subject even scarier.

The dual of a proper extension is a proper *lifting*.

Definition 3.22 Let $\mathcal{M}$ be a rank r oriented matroid on elements E. A **lifting** of $\mathcal{M}$ is a rank $r + 1$ oriented matroid $\mathcal{M}'$ on elements $E \cup \{f\}$ such that $\mathcal{M}'/f = \mathcal{M}$. The lifting is **proper** if f is not a coloop of $\mathcal{M}'$.

Problem 3.23 Let $\mathcal{M}$ be an oriented matroid on elements E, and $\mathcal{M}'$ an oriented matroid on elements $E \cup \{f\}$. Show that $\mathcal{M}'$ is a proper extension of $\mathcal{M}$ if and only if $(\mathcal{M}')^*$ is a proper lifting of $\mathcal{M}$.

3.6 Poset Properties of $\mathcal{V}^*(\mathcal{M})$

Recall that $\{0, +, -\}$ is a poset with 0 the unique minimum and $+, -$ both maxima. The product $\{0, +, -\}^E$ is ordered componentwise, and $\mathcal{V}(\mathcal{M})$ and $\mathcal{V}^*(\mathcal{M})$ are subposets of $\{0, +, -\}^E$. In this section we'll phrase our results in terms of $\mathcal{V}^*(\mathcal{M})$ because it's the set that will usually be natural for us to look at. Since $\mathcal{V}(\mathcal{M}) = \mathcal{V}^*(\mathcal{M}^*)$, one can easily adapt these results to vector sets.

Notation 3.24 If a poset P has a unique minimum, we denote that element by $\hat{0}$. If P has a unique maximum, that element is denoted $\hat{1}$.

Definition 3.25 A poset is **bounded** it it has a $\hat{0}$ and a $\hat{1}$.

$\mathcal{V}^*(M)$ has a unique minimum $\mathbf{0}$. If $\mathcal{M}$ is rank 0 then $\mathcal{V}^*(M) = \{\mathbf{0}\}$. Otherwise, $\mathcal{V}^*(\mathcal{M})$ has no unique maximum, but it will sometimes be useful to look at the bounded poset $\mathcal{V}^*(\mathcal{M}) \cup \{\hat{1}\}$ obtained by adjoining a maximum element.

Definition 3.26 If x and y are elements of a poset P, a **least upper bound**, or **join**, for x and y is an element z of P such that $x \leq z$, $y \leq z$, and if z' is an element of P that is greater than or equal to both x and y then $z \leq z'$. If a join for x and y exists, it is denoted $x \vee y$. A **greatest lower bound**, or **meet**, for x and y is an element z of P such that $x \geq z$, $y \geq z$, and if z' is an element of P that is less than or equal to both x and y then $z \geq z'$. If a meet for x and y exists, it is denoted $x \wedge y$.

P is a **lattice** if every pair of elements has a meet and join.

In particular, every finite lattice is bounded.

Lemma 3.27 *A finite poset P is a lattice if and only if $\hat{0} \in P$ and every pair of elements has a join.*

Proof: One direction is by definition. To see the converse, consider $x, y \in P$. Certainly $P_{x,y} := \{z \in P : z \leq x \text{ and } z \leq y\}$ is nonempty, since it contains $\hat{0}$. Also, $P_{x,y}$ is finite, so the join of all elements of $P_{x,y}$ exists and is $x \wedge y$. □

Proposition 3.28 *For every oriented matroid $\mathcal{M}$, $\mathcal{V}(\mathcal{M}) \cup \{\hat{1}\}$ is a lattice.*

Proof: We use the previous lemma. Recall that $\mathbf{0} \in \mathcal{V}(\mathcal{M}) \cup \{\hat{1}\}$ is a $\hat{0}$. To see that every pair X, Y of elements has a join, consider two cases:

1. If $S(X, Y) = \emptyset$, then $X \circ Y = Y \circ X = X \vee Y$.
2. If $S(X, Y) \neq \emptyset$, then $X \vee Y = \hat{1}$.

□

Definition 3.29 $\mathcal{V}^*(\mathcal{M}) \cup \{\hat{1}\}$ is called the **face lattice** of $\mathcal{M}$.

Problem 3.30 Show that for every $X, Y \in \mathcal{V}(\mathcal{M})$ that $X \wedge Y$ is the composition of all signed circuits that are less than or equal to both X and Y. (By convention, the empty composition is $\mathbf{0}$.)

3.7 Flats

The notion of *flats* of an oriented matroid generalizes intersections of elements of a hyperplane arrangement and the notion of linear span of elements in a vector arrangement. This is an idea from unoriented matroid theory. We can think of flats as an unsigned analog to covectors. As we shall see, there is a forgetful map from $\mathcal{V}^*(\mathcal{M})$ to the poset $\mathcal{F}(\mathcal{M})$ of flats of $\mathcal{M}$. The underlying matroid of $\mathcal{M}$ is characterized by $\mathcal{F}(\mathcal{M})$.

Lemma 3.31 *1. If $X \in \mathcal{V}^*(\mathcal{M})$ then X is a tope if and only if X^0 is the set of loops of $\mathcal{M}$.*
2. If $Y \in \mathcal{V}^*(\mathcal{M})$ then $Y \backslash Y^0$ is a tope of $\mathcal{M}/Y^0$.

Proof: 1. If e is a nonloop and $X(e) = 0$ then there is a $Z \in \mathcal{V}^*(\mathcal{M})$ such that $Z(e) \neq 0$. Thus $X < X \circ Z$.

(2) follows from (1). □

The following generalizes Proposition 2.50.

Proposition 3.32 *Let $\mathcal{M}$ be an oriented matroid on elements E and let $F \subseteq E$. The following are equivalent:*

1. *$\mathcal{M}/F$ has no loops.*
2. *$F = X^0$ for some $X \in \mathcal{V}^*(\mathcal{M})$.*
3. *$\mathrm{rank}(F) < \mathrm{rank}(F \cup \{e\})$ for every $e \in E \backslash F$.*

Proof: (1)$\Rightarrow$(2): Let Y be a tope of $\mathcal{M}/F$. By Lemma 3.31 $Y^0 = \emptyset$, and $X := YF^0 \in \mathcal{V}^*(\mathcal{M})$.

(2)$\Rightarrow$(3): Let I be a maximal independent subset of F, and assume by way of contradiction that $I \cup \{e\}$ is dependent for some $e \in E \backslash F$. Thus there is a $Y \in \mathcal{C}(\mathcal{M})$ such that $e \in \mathrm{supp}(Y) \subseteq I \cup \{e\}$. But then $X \cdot Y = \{X(e)Y(e)\} \neq \{0\}$, contradicting orthogonality of X and Y.

(3)$\Rightarrow$(1): Assume by way of contradiction that $\mathcal{M}/F$ has a loop e. Then there is a $Y \in \mathcal{V}(\mathcal{M})$ with $e \in \mathrm{supp}(Y) \subseteq F \cup \{e\}$. Since Y is a composition of signed circuits, there is a $Y' \in \mathcal{C}(\mathcal{M})$ with $e \in \mathrm{supp}(Y') \subseteq F \cup \{e\}$. Let $I = \mathrm{supp}(Y') \backslash e$, and let G be a maximal independent subset of F containing I. Then G is also a maximal independent subset of $F \cup \{e\}$ containing I, and so $\mathrm{rank}(F) = \mathrm{rank}(F \cup \{e\})$. □

Definition 3.33 A set F as in Proposition 3.32 is called a **flat** of $\mathcal{M}$.

Let $\mathcal{F}(\mathcal{M})$ be the poset of flats of $\mathcal{M}$, ordered by inclusion.

Definition 3.34 A **chain** in a poset P is a totally ordered subset.

Proposition 3.35 *The order-reversing map*
$$\mathcal{V}^*(\mathcal{M}) \to \mathcal{F}(\mathcal{M})$$
$$X \to X^0$$
takes each maximal chain in $\mathcal{V}^*(\mathcal{M})$ bijectively to a maximal chain in $\mathcal{F}(\mathcal{M})$. Further, every maximal chain in $\mathcal{F}(\mathcal{M})$ is the image of a maximal chain in $\mathcal{V}^*(\mathcal{M})$.

Proof: Certainly if $\mathbf{0} = X_0 < X_1 < \cdots < X_k$ is a maximal chain in $\mathcal{V}^*(\mathcal{M})$ then $\mathbf{0}^0 \supset X_1^0 \supset \cdots \supset X_k^0$ is a chain in $\mathcal{F}(\mathcal{M})$. Also $E = \mathbf{0}^0$ is the unique maximum of $\mathcal{F}(\mathcal{M})$, and X_k is a tope, so X_k^0, which is the set of loops of $\mathcal{M}$, is the unique minimum of $\mathcal{F}(\mathcal{M})$. To see that $\mathbf{0}^0 \supset X_1^0 \supset \cdots \supset X_k^0$ is a maximal chain in $\mathcal{F}(\mathcal{M})$, it remains to see that, for each i, there is no flat F such that $X_i^0 \supset F \supset X_{i+1}^0$.

Assume by way of contradiction there is an F with $X_i^0 \supset F \supset X_{i+1}^0$. Then $F = Y^0$ for some $Y \in \mathcal{V}^*(\mathcal{M})$. For every such Y, we have $Y^0 = (X_i \circ Y)^0$ and $X_i \circ Y > X_i$, and so the set $S := \{Y \in \mathcal{V}^*(\mathcal{M}) : X_i < Y \text{ and } Y^0 \supset X_{i+1}^0\}$ is nonempty. Choose an element Y_0 of this set with $S(Y_0, X_{i+1})$ minimal. $S(Y_0, X_{i+1}) \neq \emptyset$ since otherwise we have $X_i < Y_0 < X_{i_1}$, contradicting maximality of our chain in $\mathcal{V}^*(\mathcal{M})$. So eliminating some element e between Y_0 and X_{i+1} gives us a $Z \in \mathcal{V}^*(\mathcal{M})$ with $S(Z, X_{i+1}) \subset S(Y_0, X_{i+1})$ and $Z^0 \supseteq X_{i+1}^0 \cup \{e\}$.

To get our contradiction, we show that $Z > X_i$. Since $X_i < Y_0$ and $X_i < X_{i+1}$, for each $f \in \text{supp}(X_i)$ we have $Y_0(f) = X_{i+1}(f) = X_i(f)$, and so $Z(f) = X_i(f)$. Thus $Z \geq X_i$. Additionally, there is some $g \in Y_0^0 \setminus X_{i+1}^0$, and for this g we have $Z(g) = X_{i+1}(g) \neq 0$ and $X_i(g) \leq Y(g) = 0$. Thus $Z \neq X_i$.

We conclude that $Z \in S$, contradicting our assumption on Y_0.

To see the second statement: Given a maximal chain $F_0 \supset \cdots \supset F_k$ in $\mathcal{F}(\mathcal{M})$, Proposition 3.32 says that for each i there is a $Y_i \in \mathcal{V}^*(\mathcal{M})$ such that $Y_i^0 = F_i$. For each i, let $X_i = Y_1 \circ \cdots \circ Y_i$. Then $X_0 < \cdots < X_k$ maps to $F_0 \supset \cdots \supset F_k$. □

Example 3.36 The left-hand side of Figure 3.1 shows an affine picture of a rank 3 hyperplane arrangement. If we choose a positive side for each hyperplane then we get an oriented matroid $\mathcal{M}$ whose poset $\mathcal{F}(\mathcal{M})$ of flats is shown on the right-hand side of Figure 3.1.

The intersection of two flats is a flat, since $X^0 \cap Y^0 = (X \circ Y)^0$. Thus for each $A \subseteq E$, the intersection $\bar{A}$ of all flats containing A is the smallest flat

3.7 Flats

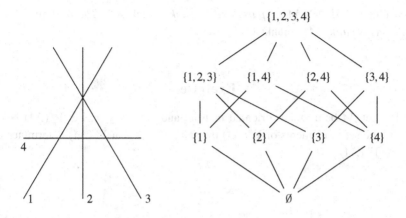

Figure 3.1 An arrangement and its poset of flats.

containing A. The function $\mathcal{P}(E) \to \mathcal{P}(E)$ taking each set A to $\bar{A}$ is a closure operator.

Example 3.37 If $(\mathcal{H}_e : e \in E)$ is a signed hyperplane arrangement and $\mathcal{M}$ is its oriented matroid, then for each $A \subset E$ the closure $\bar{A}$ is $\{e : \bigcap_{a \in A} H_a^0 \subseteq H_e^0\}$.

Example 3.38 If $(\mathbf{v}_e : e \in E)$ is a vector arrangement and $\mathcal{M}$ its oriented matroid, then for each $A \subset E$ the closure $\bar{A}$ is the set of all e such that $\mathbf{v}_e$ is in the linear span of $\{\mathbf{v}_a : a \in A\}$.

Definition 3.39 The **length** of a finite chain is one less than the number of elements of the chain. P is **pure** if every maximal chain has the same length. In this case, we call this length the **rank** of P. A chain with smallest element p and largest element q is **saturated** if it is maximal in the interval $[p, q]$.

If x is an element of a pure poset P then the **rank** of x in P is the maximal length of a chain in P with maximal element x, and the **corank** of x in P is the maximal length of a chain in P with minimal element x.

Proposition 3.40 $\mathcal{F}(\mathcal{M})$ is pure, and its rank is $\text{rank}(\mathcal{M})$. For each $F \in \mathcal{F}(\mathcal{M})$, its rank in $\mathcal{F}(\mathcal{M})$ coincides with its rank in the sense of Definition 2.48.

Proof: Consider a maximal chain $F_0 \subset F_1 \subset \cdots \subset F_k = E$ in $\mathcal{F}(\mathcal{M})$. We have that F_0 is the set of loops of $\mathcal{M}$. For each $i > 0$ choose $e_i \in F_i \setminus F_{i+1}$. Then $\{e_1, \ldots, e_i\}$ is a maximal independent subset of F_i. In particular, $k = \text{rank}(\mathcal{M})$, and the two notions of rank for F_i coincide. □

Corollary 3.41 $\mathcal{V}^*(\mathcal{M})$ *is pure, and its rank is* rank$(\mathcal{M})$. *The rank of* $X \in \mathcal{V}^*(\mathcal{M})$ *is* $rank(\mathcal{M}) - \text{rank}(X^0)$.

Exercise

3.1 Let $\mathcal{M}$ be a rank r oriented matroid and $1 \leq k \leq r$. Let $\mathcal{V}_k^*(\mathcal{M})$ be the set of elements of $\mathcal{V}^*(\mathcal{M})$ of rank k. Show that $\mathcal{V}_k^*(\mathcal{M})$ determines $\mathcal{V}^*(\mathcal{M})$.

4

The Topological Representation Theorem

In Section 2.7.3 we saw that every centrally symmetric, essential arrangement $\mathcal{A}$ of signed pseudospheres gives the covector set $\mathcal{V}^*(\mathcal{A})$ of an oriented matroid in the same way that an arrangement of signed hyperplanes gives the covector set of a realizable oriented matroid. In this chapter we prove that *every* oriented matroid arises from a centrally symmetric, essential arrangement of signed pseudospheres. (We also show that every arrangement of signed pseudospheres can be "symmetrized," so the assumption that the arrangement is centrally symmetric is unnecessary.)

Theorem 4.1 (Topological Representation Theorem) (Folkman and Lawrence 1978) *Let E be a finite set. For $\mathcal{W} \subseteq \{0, +, -\}^E$, the following are equivalent:*

1. $\mathcal{W} = \mathcal{V}^*(\mathcal{M})$ *for some oriented matroid $\mathcal{M}$ of rank r with elements E.*
2. $\mathcal{W} = \mathcal{V}^*(\mathcal{A})$ *for some essential arrangement of signed pseudospheres $\mathcal{A} = (\mathcal{S}_e : e \in E)$ in S^{r-1}.*
3. $\mathcal{W} = \mathcal{V}^*(\mathcal{A})$ *for some essential, centrally symmetric arrangement of signed pseudospheres $\mathcal{A} = (\mathcal{S}_e : e \in E)$ in S^{r-1}.*

In particular, the poset of nonzero covectors of a rank r oriented matroid is isomorphic to the poset of cells in a very nice cell decomposition of S^{r-1}.

The Topological Representation Theorem was proved by Lawrence in his PhD thesis. The article of Folkman and Lawrence (1978) in which it first appeared established many foundations of oriented matroid theory. Folkman's contribution was posthumous: Preliminary notes he had made on a signed version of matroid theory fed into the ideas Lawrence developed. The proof here follows the proof given in Björner et al. (1999), which in turn draws from proofs in Mandel (1982) and Lawrence (1984).

To briefly outline the proof, which spans most of this chapter: We need to prove that various subposets of $\mathcal{V}^*(\mathcal{M}) \setminus \{0\}$ are isomorphic to face posets of various cell decompositions of spheres or balls. The key idea for this is *shelling*. Loosely put, a shelling of a regular cell complex is an ordering $\sigma_1, \sigma_2, \ldots, \sigma_k$ of its maximal cells with certain properties that, together with some properties of the poset P of cells in the cell complex, imply that each of the spaces $\sigma_1, \sigma_1 \cup \sigma_2, \ldots, \bigcup_{i=1}^{k-1} \sigma_i$ is a ball and that $\bigcup_{i=1}^{k} \sigma_i$ is a ball or sphere. The properties required of a shelling can be demonstrated by combinatorial properties of P: The ordering $\sigma_1, \sigma_2, \ldots, \sigma_k$ should be a *recursive coatom ordering* of $P \cup \{\hat{0}, \hat{1}\}$, and $P \cup \{\hat{0}, \hat{1}\}$ should be *thin* or *subthin*.

In Section 4.1 we'll explain this framework in more detail and show that every poset satisfying these properties is the face poset of a regular cell decomposition of a ball or sphere. Section 4.2 will nail down the subposets of $\mathcal{V}^*(\mathcal{M})$ of interest to us. They are analogous to intersections of closed half-spaces given by a hyperplane arrangement, and we call them *convex*. Section 4.3 will prove thinness and subthinness results.

Sections 4.4 through 4.6 will give our recursive coatom orderings. A coatom ordering of $\mathcal{V}^*(\mathcal{M}) \cup \{\hat{1}\}$ is an ordering of the topes. From looking at a picture of an arrangement of pseudospheres in S^2, it seems that a likely route to a shelling $X_1, \ldots, X_N$ of the resulting cell complex is to choose one tope to be X_1, then to list the remaining topes in order of increasing distance from X_1, where the distance from X_1 to Y is measured by the number of pseudospheres one must cross in a path from X_1 to Y. This is the idea of our recursive coatom ordering. We express the idea of distance from X_1 in terms of the *tope poset* (Section 4.4).

The recursive part of our recursive coatom ordering requires us to find, for each covector X, a recursive coatom ordering on $\{Y \in \mathcal{V}^*(\mathcal{M}) : Y \leq X\}$. We get such an ordering by generalizing Brugesser and Mani's classic shelling of the faces of a convex polytope (Section 4.5).

4.1 Shellings and Recursive Coatom Orderings

In this section we step back from oriented matroids to study some topological and combinatorial tools.

4.1.1 Preliminaries on Posets

We'll be interested in subposets of the face lattice $\mathcal{V}^*(\mathcal{M}) \cup \{\hat{1}\}$, introduced in Section 3.6.

Definition 4.2 If $x < y$ are elements of a poset P, the **closed interval** $[x, y]$ is $\{z \in P : x \leq z \leq y\}$.

4.1 Shellings and Recursive Coatom Orderings

Definition 4.3 A poset is **bounded** if it has both a unique minimum and a unique maximum. A finite poset P is **graded** if it is pure and bounded.

By Corollary 3.41, $\mathcal{V}^*(\mathcal{M}) \cup \{\hat{1}\}$ is graded. Every interval $[x, y]$ in a graded poset is graded.

Definition 4.4 If $x, y \in P$, we say x **covers** y, denoted $x \gtrdot y$, if $x > y$ and there is no z such that $x > z > y$.

If P is bounded, the elements covering $\hat{0}$ are called **atoms**. The elements covered by $\hat{1}$ are called **coatoms**.

Example 4.5 Atoms of $\mathcal{V}^*(\mathcal{M}) \cup \{\hat{1}\}$ are signed cocircuits. Coatoms of $\mathcal{V}^*(\mathcal{M}) \cup \{\hat{1}\}$ are topes.

Definition 4.6 An **order ideal** of a poset P is a $I \subseteq P$ such that whenever $x \in I$ and $y < x$, we have $y \in I$. An **upper order ideal**, or **filter**, is a $I \subseteq P$ such that whenever $x \in I$ and $y > x$, we have $y \in I$.

Definition 4.7 A **total order** on a set S is a partial order on S such that every two elements x, y are comparable (i.e., $x < y$ or $y < x$ or $x = y$). A **linear extension** of a poset P with partial order $\leq$ is a total order $<^*$ on the underlying set of P such that $x <^* y$ whenever $x < y$.

Problem 4.8 Prove that if I is an order ideal in a finite poset P then P has a linear extension in which the elements of I come first.

Definition 4.9 Let P be a poset of finite rank. The *order complex* of P, denoted ΔP, is the abstract simplicial complex consisting of all chains in P.

For an introduction to simplicial complexes, see, for instance, Munkres (1984) or Björner (1995). We'll distinguish between an *abstract simplicial complex*, which is a collection of finite sets, closed under taking subsets, and a *geometric simplicial complex*, which is a collection of geometric simplices, closed under taking faces. A geometric realization of an abstract simplicial complex A will be denoted $\|A\|$: Each element σ of A has a corresponding element in $\|A\|$, denoted $\|\sigma\|$.

A poset map $f: P \to Q$ induces a simplicial map $\Delta f: \Delta P \to \Delta Q$. The resulting map of geometric simplicial complexes is denoted $\|\Delta f\|$.

4.1.2 Preliminaries on Topological Balls

Let $B^d \subset \mathbb{R}^d$ denote the standard d-ball and S^d the standard d-sphere.

For S a subset of a topological space T, we let $\overline{S}$ denote the closure of S.

Lemma 4.10 *If B_1 and B_2 are two topological d-balls and $h: \partial B_1 \to \partial B_2$ is a homeomorphism then h extends to a homeomorphism $B_1 \to B_2$.*

Proof: Fix homeomorphisms $g_i: B_i \to B^d$. Let $\hat{g}_i: \partial B_i \to \partial B^d$ be the restriction of g_i. Then $\hat{g}_2 \circ h \circ \hat{g}_1^{-1}$ is a self-homeomorphism of ∂B^d. Extend this homeomorphism radially to a self-homeomorphism H of B^d. Then $g_2^{-1} \circ H \circ g_1: B_1 \to B_2$ is the desired homeomorphism. □

Lemma 4.11 *Let $d \geq 1$. For a $(d-1)$-sphere $S \subseteq S^d$, the following are equivalent.*

1. *There is a self-homeomorphism H of S^d sending S to the equator $\{\mathbf{x} \in S^d : x_{d+1} = 0\}$.*
2. *The closure of each connected component of $S^d \setminus S$ is homeomorphic to a d-ball.*

Recall (Definition 2.79) that such an S is a *pseudosphere*. Of course, $(d-1)$-spheres in S^d that are *not* pseudospheres are strange beasts. One example of such a monster is the Alexander Horned Sphere – see, for instance, Hatcher (2002) for details.

Proof: That (1) implies (2) is clear. To see the converse, let B_1 and B_2 be the closures of the connected components. Then $\partial B_1 = \partial B_2 = S$. Since S is a $(d-1)$-sphere, there is a homeomorphism h from S to the equator of S^d. Let U and L be the upper resp. lower hemispheres of S^d. Then, by Lemma 4.10, h extends to homeomorphisms $B_1 \to U$ and $B_2 \to L$. These homeomorphisms glue to give the desired H. □

Lemma 4.12 *Let B_1 and B_2 be topological d-balls such that $B_1 \cap B_2 = \partial B_1 \cap \partial B_2$ is a $(d-1)$-ball and for $i \in \{1,2\}$ the closure of $\partial B_i \setminus (B_1 \cap B_2)$ is a $(d-1)$-ball. Then $B_1 \cup B_2$ is a topological d-ball.*

Proof: Fix a homeomorphism from $\partial(B_1 \cap B_2)$ to the equator in S^{d-1}. By Lemma 4.10 this homeomorphism extends to a homeomorphism $B_1 \cap B_2 \to \{x \in B^d : x_d = 0\}$ and to homeomorphisms from $\overline{\partial B_1 \setminus (B_1 \cap B_2)}$ to the upper hemisphere of S^{d-1} and from $\overline{\partial B_2 \setminus (B_1 \cap B_2)}$ to the lower hemisphere. Thus we have homeomorphisms from ∂B_1 to the boundary of the top half of B^d and from ∂B_2 to the boundary of the bottom half of B^d, and these homeomorphisms agree on $B_1 \cap B_2$. Apply Lemma 4.10 again to get homeomorphisms from B_1 and B_2 to the top and bottom hemispheres of B^d, respectively. These homeomorphisms glue to give a homeomorphism $B_1 \cup B_2 \to B^d$. □

A similar proof shows the following.

4.1 Shellings and Recursive Coatom Orderings

Lemma 4.13 *Let B_1 and B_2 be d-balls such that $B_1 \cap B_2 = \partial B_1 = \partial B_2$. Then $B_1 \cup B_2 \cong S^d$.*

4.1.3 Regular Cell Complexes

Definition 4.14 A **regular cell** in a Hausdorff space X is a subset of X which is homeomorphic to a closed (nonempty) ball. A **finite regular cell complex** is a finite set R of regular cells such that

1. For every two cells σ and τ in R, either $\sigma \cap \tau = \emptyset$ or $\sigma \cap \tau$ is a union of elements of R.
2. For every two cells σ and τ in R, if $\sigma \neq \sigma \cap \tau$ then $\sigma \cap \tau \subseteq \partial \sigma$.
3. For every $\sigma \in R$, $\partial \sigma$ is a union of elements of R.

A regular cell complex is **pure** if every maximal cell has the same dimension.

Definition 4.15 A **regular cell decomposition** of a space X is a regular cell complex R such that $\bigcup_{\sigma \in R} \sigma = X$. We call X the **underlying space** of R.

A finite regular cell complex is a poset, ordered by inclusion. This poset is pure if and only if the cell complex is pure.

Definition 4.16 The **augmented face poset** of a regular cell complex R is the poset $R \cup \{\hat{0}, \hat{1}\}$.

Problem 4.17 Let R be a regular cell complex such that every nonempty intersection of cells is a single cell. Prove that $R \cup \{\hat{0}, \hat{1}\}$ is a lattice.

Regular cell complexes are combinatorially nice because the combinatorics of the poset of cells determines the topology of the underlying space, as we now show.

Proposition 4.18 *Let R be a regular cell complex, viewed as a poset. There is a homeomorphism from $\|\Delta R\|$ to the underlying space of R, in which the preimage of a cell σ is $\|\Delta R_{\leq \sigma}\|$.*

If R is a simplicial complex, then the homeomorphism of Proposition 4.18 can be taken to be a simplicial isomorphism from $\|\Delta R\|$ to the barycentric subdivision of R. The barycenter of a general regular cell isn't defined, and so the barycentric subdivision is not defined, but a homeomorphism as in Proposition 4.18 gives a notion of barycentric subdivision of a regular cell complex.

Proof: For each k let $R^{(k)}$ be the k-skeleton of R (the set of all cells of R of dimension at most k). We recursively construct homeomorphisms h_k from $\|\Delta R^{(k)}\|$ to the underlying space of $R^{(k)}$. When $k = 0$ the map just sends each 0-dimensional cell to itself. For larger k, assume we have h_{k-1}, sending each subcomplex $\|\Delta R_{\leq \sigma}\|$ to σ, and consider a k-cell τ. Then h_{k-1} maps $\|\Delta R_{<\tau}\|$ homeomorphically to the boundary of τ. In particular, $\|\Delta R_{<\tau}\|$ is a $(k-1)$-sphere. The maximal elements of $\Delta R_{\leq \tau}$ are obtained from the maximal elements of $\Delta R_{<\tau}$ by adjoining τ as an element, and so $\|\Delta R_{\leq\tau}\|$ is a cone on $\|\Delta R_{<\tau}\|$. Thus $\|\Delta R_{\leq\tau}\|$ is a k-ball, and our homeomorphism from its boundary to the boundary of τ extends to a homeomorphism from $\|\Delta R_{\leq\tau}\|$ to τ, by Lemma 4.10. Extending in this way for each k-cell τ defines h_k. □

Corollary 4.19 *Let R and R' be regular cell complexes and $f: R \to R'$ be a poset isomorphism. Then there is a homeomorphism of the underlying spaces taking each $\sigma \in R$ to $f(\sigma)$.*

Proof: f induces an isomorphism $\Delta R \to \Delta R'$, and hence a homeomorphism $\|\Delta R\| \to \|\Delta R'\|$. □

4.1.4 From Posets to Balls and Spheres

A shelling of a pure regular cell complex is an ordering of its maximal cells satisfying a recursive property. The combinatorialization of shelling consists of an ordering of the coatoms of a graded poset satisfying a recursive definition. The recursion, together with a condition on intervals of rank 2, then guarantees

1. that the lattice is the face lattice of a regular cell complex, and
2. that the underlying space of this cell complex is a topological ball or sphere,[1] with the coatom order as a shelling order.

Definition 4.20 Let P be a pure d-dimensional regular cell complex. An order $\sigma_1, \sigma_2, \ldots, \sigma_m$ of its maximal cells is a **shelling** if either $d = 0$ or $d \geq 1$ and the following hold:

1. For every $i > 1$ the subcomplex $\partial \sigma_i \cap \bigcup_{j=1}^{i-1} \partial \sigma_j$ is pure and $(d-1)$-dimensional.

---

[1] Actually, the recursion guarantees that the lattice is the face poset of a piecewise-linear (PL) cell complex whose underlying space is a piecewise-linear ball or sphere. See Chapter 9 for more details.

4.1 Shellings and Recursive Coatom Orderings

2. For every $i > 1$ the subcomplex $\partial \sigma_i$ has a shelling in which the $(d-1)$-cells of $\partial \sigma_i \cap \bigcup_{j=1}^{i-1} \partial \sigma_j$ come first.
3. $\partial \sigma_1$ has a shelling.

If P has a shelling, we say P is **shellable**.

Example 4.21 Consider a regular cell complex whose maximal cells are three line segments that share a common endpoint. Every order of the line segment is a shelling. This tells us that a shelling alone is not enough to ensure that the underlying space is a ball or sphere.

Example 4.22 Consider a cell complex whose maximal cells are the four sides of a square. Then an ordering $\sigma_1, \sigma_2, \sigma_3, \sigma_4$ of these four sides is a shelling if and only if σ_2 is not the side opposite σ_1.

Example 4.23 Consider the cell complex in Figure 4.1 and the indicated numbering $\tau_1, \ldots, \tau_5$ of its maximal cells. It is straightforward to check that this is a shelling, by checking that for each $i > 1$ the boundary of τ_i has a shelling as in Example 4.22. But it's important to note that this order on the faces of τ_i need not be induced by the order on the squares. The boundary of the square τ_5 has a shelling in which the intersections of τ_5 with $\tau_2, \tau_3,$ and τ_4 come first. But the order of these three edges in the shelling is not the ordering indicated by the subscripts on these adjacent squares.

This example illustrates a common phenomenon: Often the shelling orders on intersections $\partial \sigma_i \cap \bigcup_{j=1}^{i-1} \partial \sigma_j$ do not come in a straightforward way from the shelling order on the maximal cells. Rather, the various shelling orders all come from a more subtle global picture.

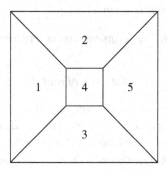

Figure 4.1 A recursive coatom ordering of the face poset of a cube.

110   The Topological Representation Theorem

We now introduce the primary tool of this chapter: a purely combinatorial formulation of shelling.

Notation 4.24   If P is a poset with $\hat{0}$ and $x \in P$, then $[x]$ denotes the interval $[\hat{0}, x]$ in P.
If P is bounded then the set of coatoms of P is denoted $\text{coat}(P)$.

Remark 4.25   We will only use the notation $[x]$ in this chapter: Elsewhere we will use the easier-to-remember notation $P_{\leq x}$. In our current discussion, the notation $[x]$ will save us from complicated subscripts.

Definition 4.26   (Definition 4.7.17 in Björner et al. 1999) Let P be a graded poset. An ordering $x_1, \ldots, x_m$ of its coatoms is a **recursive coatom ordering** if either $\text{rank}(P) \leq 2$ or $\text{rank}(P) > 2$ and for each $1 \leq i \leq m$, there is a distinguished subset $Q_i \subseteq \text{coat}[x_i]$ such that

1. $[x_i] \cap (\bigcup_{j<i}[x_j]) = \bigcup_{y \in Q_i}[y]$, and
2. $[x_i]$ has a recursive coatom ordering in which the elements of Q_i come first.

Notice in this definition that $Q_1 = \emptyset$, and if $i > 1$ then $\hat{0} \in [x_i] \cap (\bigcup_{j<i}[x_j])$, so $Q_i \neq \emptyset$.

Definition 4.27   A poset P is **thin** if every interval $[x, y]$ of rank 2 has cardinality four. P is **subthin** if

1. every interval $[x, y]$ of rank 2 with $y \neq \hat{1}$ has cardinality four, and
2. every interval $[x, \hat{1}]$ of rank 2 has cardinality three or four, with at least one such interval of cardinality three.

Problem 4.28   Verify that if P is the augmented face poset of a regular cell decomposition of a manifold then P is thin. Also, show that if P is the augmented face poset of a regular cell decomposition of a manifold with nonempty boundary then P is subthin.

The key result connecting recursive coatom orderings and shellings is the following.

Theorem 4.29   Let P be a graded poset of rank $d + 2$. Then

1. P is the face lattice of a shellable regular cell decomposition of a d-sphere if and only if P is thin and admits a recursive coatom ordering;
2. P is the face lattice of a shellable regular cell decomposition of a d-ball if and only if P is subthin and admits a recursive coatom ordering.

4.1 Shellings and Recursive Coatom Orderings 111

Further, the recursive coatom ordering is a shelling order.

Lemma 4.30 *Let P be thin or subthin and let $x_1, \ldots, x_m$ be a recursive coatom ordering of P. For each $1 < i \leq m$ let $P_i = (\bigcup_{j<i}[x_j]) \cup \{\hat{1}\}$. The following hold for each such i.*

1. *P_i is subthin.*
2. *$([x_i] \cap \bigcup_{j<i}[x_j]) \cup \{\hat{1}\}$ is thin or subthin.*
3. *The ordering $x_1, \ldots, x_{i-1}$ is a recursive coatom ordering on P_i.*

Proof: 1. Certainly P_i is thin or subthin. Also, since $[x_i] \cap \bigcup_{j<i}[x_j]$ contains some element y of Q_i, the interval $[y, \hat{1}]$ in P is $\{y, \hat{1}, x_i, x_j\}$ for some $j < i$. Thus the interval $[y, \hat{1}]$ in P_i is just $\{y, \hat{1}, x_j\}$.

2. Since $[x_i] \cap \bigcup_{j<i}[x_j]$ is an order ideal in P, the only question is on intervals $[z, \hat{1}]$ in P_i of rank 2. (Thus we're looking at certain elements z of $[x_i]$ such that the interval $[z, x_i]$ in P has rank 2.) Denote the interval in P_i as $[z, \hat{1}]_i$ to distinguish it from the interval $[z, \hat{1}]$ in P.

Since $[x_i] \cap \bigcup_{j<i}[x_j] = \bigcup_{y \in Q_i}[y]$, we know $[z, \hat{1}]_i \supseteq \{z, \hat{1}, y\}$ for some $y \in Q_i$. Each element of $[z, \hat{1}]_i \setminus \{\hat{1}\}$ is in the interval $[z, x_i]$ in P, so $|[z, \hat{1}]_i| \leq 4$. Thus $[x_i] \cap \bigcup_{j<i}[x_j]$ is thin or subthin.

3. This is clear. □

Proof of Theorem 4.29: The forward direction is immediate for both parts. To see the converse, let $R = P \setminus \{\hat{0}, \hat{1}\}$. We will construct a collection of sets $\{\sigma_x : x \in R\}$ and then show that it's the regular cell complex desired.

Fix a geometric realization $\|\Delta R\|$ of ΔR. For each chain γ in R let τ_γ denote its geometric realization in $\|\Delta R\|$. For each $x \in R$ let

$$\sigma_x = \bigcup_{\substack{\gamma \in \Delta R: \\ \max(\gamma) \leq x}} \tau_\gamma.$$

For each $x, y \in R$ we have $x < y$ if and only if $\sigma_x \subset \sigma_y$, and so the poset $\{\sigma_x : x \in R\}$, ordered by inclusion, is isomorphic to R. Also, each maximal element of $\{\sigma_x : x \in R\}$ is a union of chains with $d + 1$ elements, hence has dimension d.

The proof proceeds by induction on d. The minimal case is immediate. Above the minimal case, we first show that $\{\sigma_x : x \in R\}$ is a regular cell complex. For each $x \in R$, $[x]$ is the augmented face poset of $R_{<x}$. We have a realization of $\Delta R_{<x}$ as a subcomplex of $\|\Delta R\|$. Using this realization, the cells σ_y we define for $y \in R_{<x}$ coincide with the cells we defined by considering

y as an element of R. Because $[x]$ is graded, thin, and has a recursive coatom ordering, the induction hypothesis tells us that $\{\sigma_y : y \in R_{<x}\}$ is a regular cell decomposition of a sphere. The set of simplices making up σ_x consists of the 0-simplex τ_x and pairs $\tau_y, \tau_{y \cup \{x\}}$ for all nonempty $\gamma \in \Delta R_{<x}$. Thus σ_x is a cone on the sphere $\bigcup_{y<x} \sigma_y$ with apex τ_x.

This verifies two of the conditions for $\{\sigma_x : x \in R\}$ to be a regular cell complex: Each cell is a ball, and each cell's boundary is a union of cells in the complex. To see the remaining two conditions, note that for each $x, y \in R$,

$$\sigma_x \cap \sigma_y = \bigcup_{\substack{\gamma \in \Delta R: \\ \max(\gamma) \leq x \\ \max(\gamma) \leq y}} \tau_\gamma = \bigcup_{\substack{z \in R: \\ z \leq x \\ z \leq y}} \sigma_z.$$

So indeed $\sigma_x \cap \sigma_y$ is a union of elements of our complex, and if $\sigma_x \nsubseteq \sigma_y$ then

$$\sigma_x \cap \sigma_y \subseteq \bigcup_{z<x} \sigma_z = \partial \sigma_x.$$

To see that the underlying space of $\{\sigma_x : x \in R\}$ is a ball or sphere, we do an additional induction, this time on $|\operatorname{coat}(P)|$. If P has only one coatom then P is subthin and our underlying space is a single cell. Above this case, let $x_1, \ldots, x_m$ be a recursive coatom ordering for P. Lemma 4.30 tells us that P_m is subthin and has a recursive coatom ordering, so the induction hypothesis tells us that $\{\sigma_y : y \leq x_i \text{ for some } i < m\}$ is a regular cell decomposition of a d-ball B_m. Let $S_m = [x_m] \cap \bigcup_{j<m}[x_j] = \bigcup_{y \in Q_m}[y]$. Then

$$\sigma_{x_m} \cap B_m = \bigcup_{\substack{y: \\ y<x_m}} \sigma_y \cap \bigcup_{\substack{y, j: \\ j<m \\ y \leq x_j}} \sigma_y$$

$$= \bigcup_{y \in S_m} \sigma_y.$$

Lemma 4.30 tells us that $S_m \cup \{\hat{1}\}$ is thin or subthin and admits a recursive coatom ordering, and so by the induction hypothesis $\sigma_{x_m} \cap B_m$ is a ball or sphere. Since each maximal cell in $\sigma_{x_m} \cap B_m$ is σ_y for some $y \in Q_m \subseteq \operatorname{coat}([x_m])$, this ball or sphere has dimension $d - 1$. Certainly $\sigma_{x_m} \cap B_m$ is contained in the boundary of σ_m. To see that it is also contained in the boundary of B_m, we show that every maximal cell in $\sigma_{x_m} \cap B_m$ is in the boundary of B_m. Since $S_m = \bigcup_{y \in Q_m}[y]$, the maximal cells in $\sigma_{x_m} \cap B_m$ are σ_y with $y \in Q_m$. For such a y, we have $y < x_m$ and $y < x_j$ for some $j < m$. Because

$|[y, \hat{1}]| \leq 4$, this j is unique. Thus σ_y is contained in exactly one maximal cell σ_{x_j} of B_m, and so σ_y is in the boundary of B_m.

We conclude that $\bigcup_{x \in R} \sigma_x = \sigma_m \cup B_m$ consists of two balls of dimension d intersecting in their boundary in a ball or sphere of dimension d, and so $B_m \cup \sigma_m$ is a ball or sphere of dimension d. The result of Problem 4.28 tells us that this union is a ball if P is subthin and a sphere if P is thin. □

From Corollary 4.19 we have the following.

Corollary 4.31 *Let R be a regular cell complex.*

1. *If the augmented face poset of R is graded of rank $d + 2$, thin, and has a recursive coatom ordering, then the underlying space of R is a d-sphere.*
2. *If the augmented face poset of R is graded of rank $d + 2$, subthin, and has a recursive coatom ordering, then the underlying space of R is a d-ball.*

4.2 Convexity

Consider an arrangement of signed hyperplanes in $\mathbb{R}^n$ and the resulting partitions of the unit ball and the unit sphere in $\mathbb{R}^n$. Every nonempty intersection of closed half-spaces from the arrangement intersects the unit ball in a convex set, which is topologically a ball, and intersects the unit sphere in a spherically convex set, which is topologically either a ball or a sphere. We will define a closed convex set in $\mathcal{V}^*(\mathcal{M})$ to be analogous to an intersection of closed half-spaces with the unit ball. As part of our proof of the Topological Representation Theorem, we will prove analogs to the above topological descriptions.

Definition 4.32 Let $\mathcal{M}$ be an oriented matroid on elements E. Let $A \subseteq E$, and let $Y \in \{0, +, -\}^A$.

Define $C(Y) := \{X \in \mathcal{V}^*(M) : \forall a \in A \ X(a) \leq Y(a)\}$. We call $C(Y)$ the **closed convex set** in $\mathcal{V}^*(\mathcal{M})$ determined by Y.

If $\mathcal{V}^*(\mathcal{M})$ is the sign vector set associated to an arrangement of pseudospheres $(\mathcal{S}_e : e \in E)$ then $C(Y)$ is the set of sign vectors associated to $\bigcap_{a \in A} (S_a \cup S_a^{\epsilon})$.

Example 4.33 For each $A \subseteq E$, $C(A^0) \cong \mathcal{V}^*(\mathcal{M}/A)$.

Example 4.34 If $Y \in \mathcal{V}^*(\mathcal{M})$ then $C(Y)$ is the interval $[Y] = \{X \in \mathcal{V}^*(\mathcal{M}) : X \leq Y\}$ in $\mathcal{V}^*(\mathcal{M})$.

Example 4.35 If $A = \emptyset$ and Y is the empty vector then $C(Y) = \mathcal{V}^*(\mathcal{M})$.

Our proof of the Topological Representation Theorem will prove that each nonempty $C(Y)\setminus\{0\}$ is isomorphic to a regular cell decomposition of a ball or sphere. The order complex $\Delta C(Y)$ is a cone on $\Delta(C(Y)\setminus\{0\})$, hence is a topological ball.

Proposition 4.36 *All maximal elements of $C(Y)$ have the same support. If e is a nonloop of $\mathcal{M}$ and $\epsilon \in \{+, -\}$ then a maximal element of $C(e^\epsilon)$ is a tope of $\mathcal{M}$.*

Proof: $C(Y)$ is closed under composition. Thus if X_1 and X_2 are maximal elements of $C(Y)$ then $X_1 \circ X_2 = X_1$ and $X_2 \circ X_1 = X_2$, and so $X_1^0 \subseteq X_2^0$ and $X_2^0 \subseteq X_1^0$.

To see the second statement: Since e is a nonloop, by the Symmetry Axiom there is at least one $X \in \mathcal{V}^*(\mathcal{M})$ with $X(e) = \epsilon$. Every tope greater than or equal to X is also in $C(e^\epsilon)$. □

Corollary 4.37 *If $A \subseteq E$ and $Y \in \{0, +, -\}^A$ then $C(Y) \cup \{\hat{1}\}$ is graded. The rank of $C(Y)$ is the rank in $\mathcal{V}^*(\mathcal{M})$ of a maximal element of $C(Y)$.*

Proof: Since all maximal elements of $C(Y)$ have the same support, and $C(Y)$ is an order ideal in $\mathcal{V}^*(\mathcal{M})$, the result follows from Corollary 3.41. □

Corollary 4.38 *If e is a nonloop of $\mathcal{M}$ and $\epsilon \in \{+, -\}$ then $C(e^\epsilon) \cup \{\hat{1}\}$ is graded, and its rank is $\operatorname{rank}(\mathcal{M}) + 1$.*

4.3 Intervals in the Face Lattice

The set of all topes of $\mathcal{M}$ is denoted $\mathcal{T}(\mathcal{M})$.

The next proposition tells us that issues about intervals in the face lattice boil down to two types of intervals: those of the form $[0, \hat{1}]$ and those of the form $[T]$ with T a tope.

Proposition 4.39 *1. Let X be an element of $\mathcal{V}^*(\mathcal{M})$. Then $[X, \hat{1}] \cong \mathcal{V}^*(\mathcal{M}(X^0)) \cup \{\hat{1}\}$.*
2. Let $[X, Y]$ be an interval in $\mathcal{V}^*(\mathcal{M})$. Then there exists an oriented matroid $\mathcal{M}'$ and tope Y' of $\mathcal{T}(\mathcal{M}')$ such that $[X, Y] \cong \mathcal{V}^*(\mathcal{M}')_{\leq Y'}$.

Proof: 1. Consider the poset map $[X, \hat{1}] \to \mathcal{V}^*(\mathcal{M}(X^0)) \cup \{\hat{1}\}$ sending $\hat{1}$ to $\hat{1}$ and sending every other Z to the restriction of Z to X^0.

4.3 Intervals in the Face Lattice

To see this map is surjective, consider $Y \in \mathcal{V}^*(\mathcal{M}(X^0))$. Y is the restriction of some $Y' \in \mathcal{V}^*(\mathcal{M})$. For every such Y' we have $X \circ Y' \in [X, \hat{1}]$, and $X \circ Y'$ maps to Y.

To see this map is injective: Consider $Y, Z \in [X, \hat{1}]$ such that Y and Z have the same restriction to X^0. Then, since $Y \geq X$ and $Z \geq X$ implies Y and Z have the same restriction to $\mathrm{supp}(X)$, we have $Y = Z$.

2. If $Y \in \mathcal{T}(\mathcal{M})$ then by the same argument as (1) the desired $\mathcal{M}'$ is $\mathcal{M}(X^0)$ and the desired Y' is $Y \setminus (E \setminus X^0)$. Otherwise, take $\mathcal{M}' = \mathcal{M}(X^0)/Y^0$ and $Y' = Y \setminus (Y^0 \cup (E \setminus X^0))$. □

Corollary 4.40 $\mathcal{V}^*(\mathcal{M}) \cup \{\hat{1}\}$ *is thin for every* $\mathcal{M}$.

Proof: If every maximal interval in $\mathcal{V}^*(\mathcal{M}) \cup \{\hat{1}\}$ has rank 2 or 3, then $\mathcal{M}$ is rank 1 or 2, hence is realizable. By examining realizations in these cases we see that all intervals of rank 2 in $\mathcal{V}^*(\mathcal{M}) \cup \{\hat{1}\}$ with $\mathcal{M}$ of rank 1 or 2 has four elements. Thus by Proposition 4.39 we see that for all $\mathcal{M}$, intervals of rank 2 in $\mathcal{V}^*(\mathcal{M}) \cup \{\hat{1}\}$ have four elements. □

Corollary 4.41 *For every* $A \subseteq E$ *and* $Y \in \{0, +, -\}^A$, *the set* $C(Y) \cup \{\hat{1}\}$ *is either* $\{\hat{0}, \hat{1}\}$ *or is thin or subthin. If e is a nonloop and $\epsilon \in \{+, -\}$ then $C(e^\epsilon) \cup \{\hat{1}\}$ is subthin, and the intervals $[X, \hat{1}]$ in $C(e^\epsilon) \cup \{\hat{1}\}$ of rank 2 with only three elements are exactly those with $X(e) = 0$.*

Proof: $C(Y)$ is an order ideal in $\mathcal{V}^*(\mathcal{M})$. Thus each interval of rank 2 in $C(Y)$ is an interval of rank 2 in $\mathcal{V}^*(\mathcal{M})$, so has four elements.

Now consider intervals $[X, \hat{1}]$ in $C(Y) \cup \{\hat{1}\}$ of rank 2. Let B be the complement of the support of a maximal element of $C(Y)$. Then $C'(Y) := \{Z \setminus B : Z \in C(Y)\} \cong C(Y)$ by Proposition 4.36, and $C'(Y)$ is a closed convex set in $\mathcal{M}/B$ whose maximal elements are topes. Thus each interval $[X, \hat{1}]$ of rank 2 in $C'(Y) \cup \{\hat{1}\}$ is contained in an interval of rank 2 in $\mathcal{V}^*(\mathcal{M}/B) \cup \{\hat{1}\}$, hence has three or four elements.

For the second statement: Since e is a nonloop, there is some $Z \in \mathcal{V}^*(\mathcal{M}) \setminus \{0\}$ with $Z(e) = \epsilon$. For each $X \in C(e^\epsilon)$, we have a chain $\{X, X \circ Z, \hat{1}\}$ in $C(e^\epsilon)$, and every extension of this to a maximal chain in the interval $[X, \hat{1}]$ in $\mathcal{V}^*(\mathcal{M}) \cup \{\hat{1}\}$ is also a maximal chain in $C(e^\epsilon) \cup \{\hat{1}\}$. In particular, if the interval $[X, \hat{1}]$ in $C(e^\epsilon) \cup \{\hat{1}\}$ has rank 2, then the interval $[X, \hat{1}]$ in $\mathcal{V}^*(\mathcal{M}) \cup \{\hat{1}\}$ has rank 2. For such an interval, we have the following.

1. If $X(e) = \epsilon$, then this interval is an interval in $\mathcal{V}^*(\mathcal{M}) \cup \{\hat{1}\}$, so has order four.
2. If $X(e) = 0$, then the elements of the interval $[X, \hat{1}]$ in $\mathcal{V}^*(\mathcal{M}) \cup \{\hat{1}\}$ are $\{X, \hat{1}, X \circ Z, X \circ -Z\}$. The interval $[X, \hat{1}]$ in $C(e^\epsilon) \cup \{\hat{1}\}$ is $\{X, X \circ Z, \hat{1}\}$.

Let F be the flat of $\mathcal{M}$ whose elements are the loops of $\mathcal{M}$ together with all elements of the parallelism class of e. F is rank 1 in $\mathcal{M}$, so by Proposition 3.35 applied to F, there is an X of corank 2 in $\mathcal{V}^*(\mathcal{M}) \cup \{\hat{1}\}$ with $X(e) = 0$. Thus the interval $[X, \hat{1}]$ in $C(e^{\epsilon}) \cup \{\hat{1}\}$ has rank 2 and falls into our second case above. Thus there is at least one interval $[X, \hat{1}]$ of rank 2 in $C(e^{\epsilon}) \cup \{\hat{1}\}$ with exactly three elements. □

4.4 Tope Posets

At the center of the proof of the Topological Representation Theorem is a recursive coatom ordering on $\mathcal{V}^*(\mathcal{M}) \cup \{1\}$ – that is, a very nice ordering of $\mathcal{T}(\mathcal{M})$. As this section will show, every choice of a distinguished tope B determines a partial order on $\mathcal{T}(\mathcal{M})$, in which B is $\hat{0}$ and $-B$ is $\hat{1}$. Recall (Problem 4.8) that every finite poset has a linear extension. Section 4.6 will show that every linear extension of the partial order on $\mathcal{T}(\mathcal{M})$ induced by B is a recursive coatom ordering.

Definition 4.42 If $B \in \mathcal{T}(\mathcal{M})$, define the **tope poset** $\mathcal{T}(\mathcal{M}, B)$ to be the poset with elements $\mathcal{T}(\mathcal{M})$ and partial order

$$T_1 \leqslant_B T_2 \text{ if and only if } S(B, T_1) \subseteq S(B, T_2).$$

Notation 4.43 For every $X \leqslant_B Y$, denote the interval $\{Z : X \leqslant_B Z \leqslant_B Y\}$ in $\mathcal{T}(\mathcal{M}, B)$ by $[X, Y]_B$.

For the remainder of our proof of the Topological Representation Theorem, it will be convenient to assume our oriented matroids to be simple (Definition 3.13). Recall that, for every oriented matroid $\mathcal{M}$, by deleting all loops and all but one element of each parallelism class we get a simple oriented matroid $\mathcal{M}\backslash A$ with $\mathcal{V}^*(\mathcal{M}) \cong \mathcal{V}^*(\mathcal{M}\backslash A)$ by the isomorphism $X \to X\backslash A$. If $\mathcal{A}' := (\mathcal{S}_e : e \in E\backslash A)$ is an arrangement of signed pseudospheres with $\mathcal{V}^*(\mathcal{A}') = \mathcal{M}^*(\mathcal{M}\backslash A)$, then we can extend $\mathcal{A}'$ to an arrangement $\mathcal{A} := (\mathcal{S}_e : e \in E)$ with $\mathcal{V}^*(\mathcal{A}) = \mathcal{M}^*(\mathcal{M})$ by defining $\mathcal{S}_e$ to be the degenerate pseudosphere for each loop e of $\mathcal{M}$, defining $\mathcal{S}_e = \mathcal{S}_f$ for each $e \in A$ parallel to $f \notin A$, and defining $\mathcal{S}_e = (S_f, S_f^-, S_f^+)$ for each $e \in A$ antiparallel to $f \notin A$. So to prove the Topological Representation Theorem it's enough to prove it for simple oriented matroids.

Proposition 4.44 *Let $\mathcal{M}$ be simple and $X \in \mathcal{V}^*(\mathcal{M})$.*

1. *$X \in \mathcal{T}(\mathcal{M})$ if and only if $X^0 = \emptyset$.*
2. *X has corank 1 in $\mathcal{V}^*(\mathcal{M})$ if and only if $|X^0| = 1$.*

3. If $X, Y \in \mathcal{T}(\mathcal{M})$ then there is a Z such that $X \succ Z \prec Y$ if and only if $|S(X, Y)| = 1$.

Proof: Lemma 3.31 applied to a simple $\mathcal{M}$ says that $X^0 = \emptyset$ if and only if X is a tope. Thus if $|X^0| = 1$ then X has corank 1.

If X^0 has distinct elements e and f then, since e is not parallel or antiparallel to f, there is some Z such that $Z(e) = 0$ and $Z(f) \neq 0$. Let T be a tope. Then $X < X \circ Z < X \circ Z \circ T$, and so X has corank of at least 2.

The forward direction is clear from (2). Conversely, if $S(X, Y) = \{e\}$ then the elimination Z of e between X and Y satisfies $Z < X$, $Z < Y$, and $Z^0 = \{e\}$, so $X \succ Z \prec Y$. □

Proposition 4.45 *Let $\mathcal{M}$ be simple and $X, Y \in \mathcal{T}(\mathcal{M})$. If $|S(X, Y)| > 1$ then there is a $Z \in \mathcal{T}(\mathcal{M})$ such that $\emptyset \neq S(X, Z) \subset S(X, Y)$.*

Proof: Let $e, f \in S(X, Y)$, and let Z' be an elimination of e between X and Y. For each $g \notin S(X, Y)$, we have $Z'(g) = X(g) = Y(g) \neq 0$, and so for each $Z'' \geq Z'$ we have $S(X, Z'') \subseteq S(X, Y)$. We then have three cases:

1. If $Z'(f) = X(f)$, let $Z = Z' \circ Y$. Then $e \in S(X, Z) \subseteq S(X, Y) \setminus \{f\}$.
2. If $Z'(f) = Y(f)$, let $Z = Z' \circ X$. Then $f \in S(X, Z) \subseteq S(X, Y) \setminus \{e\}$.
3. If $Z'(f) = 0$, since e and f are not parallel, not antiparallel, and not loops, there is a $U \in \mathcal{V}^*(\mathcal{M})$ such that $U(e) = 0$ and $U(f) = X(f)$. Let $Z = Z' \circ U \circ Y$. Then $e \in S(X, Z) \subseteq S(X, Y) \setminus \{f\}$.

□

Proposition 4.46 *Let $\mathcal{M}$ be simple and $X, Y, B \in \mathcal{T}(\mathcal{M})$. The following are equivalent:*

1. X covers Y in $\mathcal{T}(\mathcal{M}, B)$.
2. $S(X, B) = S(Y, B) \cup \{e\}$ for some e.
3. $Y \prec_B X$ and there is a $Z \in \mathcal{V}^*(\mathcal{M})$ such that Z is covered by both X and Y in $\mathcal{V}^*(\mathcal{M})$.

Proof: That (2) implies (1) is clear. That (3) and (2) are equivalent follows from the second part of Proposition 4.44, along with the observation that $Y \prec_B X$ if and only if $S(X, B) = S(Y, B) \dot\cup S(X, Y)$.

To see (1) implies (2): If $|S(X, Y)|$ were greater than 1, then by Proposition 4.45 we could find a tope Z such that $\emptyset \neq S(Y, Z) \subset S(X, Y)$, and hence

$$S(Y, B) \subset S(Y, B) \cup S(Y, Z) \subset S(Y, B) \cup S(X, Y)$$
$$S(Y, B) \subset S(Z, B) \qquad\qquad \subset S(X, B).$$

□

Corollary 4.47 *Let $\mathcal{M}$ be simple and $B \in \mathcal{T}(\mathcal{M})$. $\mathcal{T}(\mathcal{M}, B)$ is a graded poset. For each $T \in \mathcal{T}(\mathcal{M})$, the rank of T in $\mathcal{T}(\mathcal{M}, B)$ is $|S(T, B)|$.*

For each maximal chain $B = T_0 \leqslant_B T_1 \leqslant_B \cdots \leqslant_B T_n = -B$ in $\mathcal{T}(\mathcal{M}, B)$ and each $i \in [n]$, we have $S(T_i, B) = S(T_{i-1}, B) \cup \{e_i\}$ for some e_i. The sequence $e_1, \ldots, e_n$ is a total ordering of the elements of $\mathcal{M}$.

Definition 4.48 We call the sequence $e_1, \ldots, e_n$ of Corollary 4.47 the **sequence induced by the chain** $T_0 \leqslant_B T_1 \leqslant_B \cdots \leqslant_B T_n$.

4.5 A Recursive Coatom Ordering on the Face Poset of a Covector

In this section we'll apply our earlier discussion on recursive coatom orderings of general graded posets P to posets $\mathcal{V}^*(\mathcal{M})_{\leq X}$ with $X \in \mathcal{V}^*(\mathcal{M})$. As in the earlier discussion, we will denote $\mathcal{V}^*(\mathcal{M})_{\leq X}$ by $[X]$.

Definition 4.49 A **facet** of a convex polytope is a maximal proper face.

Let $\mathcal{M}$ be an oriented matroid and $X \in \mathcal{V}^*(\mathcal{M}) \setminus \{0\}$. A **facet** of X is a coatom of $[X]$.

The most charming of all shellings is Brugesser and Mani's shelling of the face poset of a convex polytpe (Bruggesser and Mani 1971). The idea is as follows. Think of a convex polytope as a planet and put a spaceship on one facet F_1. Send the spaceship out in a straight line. This line should be in general position with respect to the facets of the polytope. As the spaceship takes off, the only facet its occupants can see is F_1. As the spaceship gets further from the polytope, the remaining facets come into view one at a time. The astronauts record the order $F_1, F_2, \ldots$ in which the facets come into view. Once the ship gets to infinity, it can see all the facets $F_1, F_2, \ldots, F_a$ on one side. Then the ship returns to the polytope, approaching from infinity on the other side. One by one, the facets on the other side of the polytope disappear from view. The order in which they disappear is recorded as $F_{a+1}, \ldots, F_b$. The resulting order on the facets is a recursive coatom order. (See Figure 4.2.)

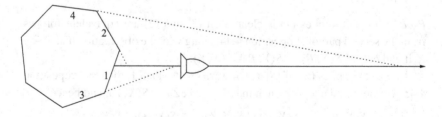

Figure 4.2 Shelling the faces of Planet Polytope.

4.5 A Recursive Coatom Ordering on the Face Poset of a Covector

To see the recursive aspect of the ordering, consider a facet F_j. We'll assume it appeared to the astronauts before they went to infinity. Let Q_j be the set of all facets of F_j of the form $F_j \cap F_i$ with $i < j$. Consider the polytope F_j and lower-dimensional astronauts in the hyperplane H containing F_j, positioned at the point p where our original rocket ship crossed H. When these astronauts look toward F_j, the facets of F_j they can see are the facets $F_j \cap F_i$ for which F_i is visible to higher-dimensional astronauts at p. (For either type of astronaut, the facets that are visible are the ones that span a hyperplane separating p from P.) These are exactly the elements of Q_j. A generic choice of a line intersecting both F_j and p describes a trajectory in H for a lower-dimensional rocket ship to follow to get an ordering of the facets of F_j in which the elements of Q_j come first.

This section will establish an analogous recursive coatom ordering on each interval $[0, B]$ in $V^*(\mathcal{M})$ with B a tope. The analog to a direction for the spaceship to travel is a maximal chain in $\mathcal{T}(\mathcal{M}, B)$. The analog to a facet coming into view – say, a facet F with $F(e) = 0$ – is the sign on e changing as we move up the chain. (See Figure 4.3.) Since we are assuming $\mathcal{M}$ to be simple, the sign changes on one element at a time.

Specifically, given a simple $\mathcal{M}$ and $B \in \mathcal{T}(\mathcal{M})$, fix a maximal chain $\mu := \{B = R_0 \prec_B R_1 \prec_B \cdots \prec_B R_n = -B\}$ in $\mathcal{T}(\mathcal{M}, B)$. Corollary 4.47 says that μ induces an order on the elements of $\mathcal{M}$, and Proposition 4.44 says that for each coatom Y of $[B]$ there is a unique element e such that $Y(e) = 0$. Thus μ

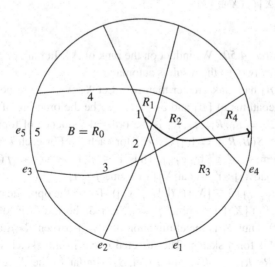

Figure 4.3 Shelling the faces of a tope R.

induces an order $\prec_\mu$ on coat[B], in which $X \prec_\mu Y$ if X^0 comes before Y^0 in the ordering of elements induced by μ.

Theorem 4.50 (Mandel 1982) *This is a recursive coatom ordering of* [B].

Section 4.6 will find a recursive coatom ordering on all of $\mathcal{V}^*(\mathcal{M}) \cup \{\hat{1}\}$. Theorem 4.50 is the key to verifying the recursiveness of the ordering there.

Lemma 4.51 *Let X and Y be elements of $\mathcal{V}^*(\mathcal{M})$ such that* supp(Y) $\subseteq$ supp(X) *and* $Y \nleq X$. *Then there is some* $Z \in$ coat[X] $\cap (X \boxplus Y)$.

In the realizable case, this lemma says there is a facet Z of X whose realization is in the interior of the convex hull of the realizations of X and Y. Metaphorically, if an astronaut in the space spanned by planet X at a point Y not on the surface of X is looking towards X then Z is in her line of sight.

Proof: Assume not, and choose a counterexample with $|S(X, Y)|$ minimal. Choose $f \in S(X, Y)$ and eliminate f between X and Y to get Y' with $S(X, Y') \subset S(X, Y)$ and $Y'(f) = 0$. We then have two cases:

1. If $Y' < X$ then let Z be a coatom of $[Y', X]$. Then for each $g \in S(X, Y)$ we have $Z(g) \in X(g) \boxplus Y(g)$ vacuously, and for each $g \notin S(X, Y)$ we have $X(g) \geq Y(g)$, and so $X(g) \boxplus Y(g) = \{X(g)\}$ and $0 \neq X(g) = Y'(g) = Z(g)$.
2. Otherwise, by the minimality assumption we get Z satisfying the conditions for X and Y'. But for each g we have $X(g) \boxplus Y'(g) \subseteq X(g) \boxplus Y(g)$, and so $Z \in$ coat[X] $\cap (X \boxplus Y)$.

□

Proof of Theorem 4.50: We induct on the rank of $\mathcal{M}$. In rank one or two, [B] has rank one or two, so the result is automatic.

Assume [B] has rank greater than two. Let $X_1, X_2, \ldots, X_t$ be the ordered sequence of coatoms of [B] and $e_1, e_2, \ldots, e_n$ be the ordering of elements of $\mathcal{M}$. (Recall $B = R_0, R_1, \ldots, R_n$ is the ordered sequence of topes spawning this setup. So $S(B, R_j) = \{e_1, \ldots, e_j\}$ for each j.) For each $i \in [t]$ denote the single element of X_i^0 by $e_{f(i)}$. Thus $1 \leq f(1) < \cdots < f(t) \leq n$. Let $Q_i = \{Y \in \text{coat}[X_i] : Y \in \text{coat}[X_j] \text{ for some } j < i\}$.

Clearly $\bigcup_{Y \in Q_i} [Y] \subseteq [X_i] \cap (\bigcup_{j<i} [X_j])$. To see the opposite containment, let $Z \in [X_i] \cap [X_j]$ for some $j < i$. Let S_i be the facet of $R_{f(i)}$ with $S_i(e_{f(i)}) = 0$. Thus S_i is the elimination of $e_{f(i)}$ between $R_{f(i)}$ and $R_{f(i)-1}$. (See Figure 4.4 for a sketch. The directed curve in the sketch indicates the chain $B = R_0, R_1, \ldots, R_{f(i)}$ in $\mathcal{T}(\mathcal{M}, B)$, similar to the directed curve in Figure 4.3.) Let $Y = Z \circ S_i$. We wish to apply Lemma 4.51 to X_i and Y. We have supp(Y) = $E - \{e_{f(i)}\}$ = supp(X_i) and

4.5 A Recursive Coatom Ordering on the Face Poset of a Covector

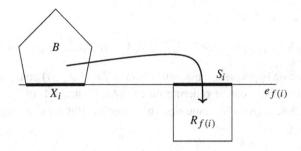

Figure 4.4 Figure for the proof of Theorem 4.50.

$$S(Y, X_i) = Z^0 \cap S(S_i, X_i)$$
$$= Z^0 \cap S(R_{f(i)-1}, X_i)$$
$$= Z^0 \cap S(R_{f(i)-1}, B)$$
$$= Z^0 \cap \{e_1, \ldots, e_{f(i)-1}\}.$$

Since $Z(e_{f(j)}) = 0$ and $f(j) < f(i)$, we see $S(Y, X_i) \neq \emptyset$, so $Y \not\leq X_i$.

Applying Lemma 4.51 to X_i and Y gives $\hat{Z} \in \text{coat}[X_i]$ such that $\hat{Z}(e) = X_i(e)$ for every $e \notin S(Y, X_i)$. In particular, $Z \leq \hat{Z}$ and $e_{f(i)} \in \hat{Z}^0 \subseteq \{e_1, \ldots, e_{f(i)}\}$. Also, by thinness of $\mathcal{V}^*(\mathcal{M}) \cup \{\hat{1}\}$ (Corollary 4.40), $\hat{Z}$ must be covered by some X_k with $k \neq i$, and since $f(k) \in \hat{Z}^0$ we see $k < i$. Thus, $\hat{Z} \in \text{coat}[X_k]$ for some $k < i$. This proves the first condition for a recursive coatom ordering.

To see the second condition, let X_i be a coatom of $[B]$. We wish to apply the induction hypothesis to see that $[X_i]$ has a recursive coatom ordering in which the elements of Q_i come first. $X_i \setminus e_{f(i)}$ is a tope of $\mathcal{M}/\{e_{f(i)}\}$, and the interval $[X_i \setminus e_{f(i)}]$ in $\mathcal{V}^*(\mathcal{M}/\{e_{f(i)}\})$ is isomorphic to the interval $[X_i]$ in $\mathcal{V}^*(\mathcal{M})$. A small complication is that $\mathcal{M}/\{e_{f(i)}\}$ might not be simple. We will be a little bit careful in finding a simple deletion. Notice that no triple $\{f(i), f(j), f(k)\}$ is dependent in $\mathcal{M}$, since no sign vector with support $\{f(i), f(j), f(k)\}$ is orthogonal to all three of X_i, X_j, and X_k. Thus each parallelism class in $\mathcal{M}/\{e_{f(i)}\}$ contains at most one element of the form $e_{f(j)}$. So, let $\widetilde{\mathcal{M}}$ be obtained from $\mathcal{M}/\{e_{f(i)}\}$ by deleting all but one element of each parallelism class, with the condition that we do not delete elements of the form $e_{f(j)}$. Let A be the set of deleted elements together with $e_{f(i)}$. Then the map $Z \to Z \setminus A$ is a poset isomorphism from $\{Y \in \mathcal{V}^*(\mathcal{M}) : Y(e_{f(i)}) = 0\}$ to $\mathcal{V}^*(\widetilde{\mathcal{M}})$. It remains to show that the interval $[X_i \setminus A]$ in $\mathcal{V}^*(\widetilde{\mathcal{M}})$ has a recursive coatom order in which the elements of $Q_i \setminus A = \{Y \setminus A : Y \in Q_i\}$ come first. Because we were careful in our choice of A, the elements of $Q_i \setminus A$ are the coatoms Y of $[X_i \setminus A]$ such that $Y(e_{f(j)}) = 0$ for some $j < i$.

$S_i \backslash A$ is a tope of $\widetilde{\mathcal{M}}$, and

$$S(X_i \backslash A, S_i \backslash A) = S(B, R_{f(i)-1}) \backslash A = \{e_1, \ldots, e_{f(i)-1}\} \backslash A.$$

By the induction hypothesis, a maximal chain in $\mathcal{T}(\widetilde{\mathcal{M}}, X_i \backslash A)$ containing $S_i \backslash A$ induces a recursive coatom ordering on $[X_i \backslash A]$. Coatoms Y of $[X_i \backslash A]$ with $Y^0 \subseteq S(X_i \backslash A, S_i \backslash A)$ will come first in this order. But these are exactly the elements of $Q_i \backslash A$. □

4.6 Recursive Coatom Orderings on the Face Lattice and on Convex Sets

Let $\mathcal{M}$ be a simple oriented matroid and $C(Y)$ a nonempty closed convex set in $\mathcal{M}$ containing at least one tope. This section will construct a recursive coatom ordering on $\mathcal{V}^*(\mathcal{M}) \cup \{\hat{1}\}$ that restricts to a recursive coatom ordering of $C(Y) \cup \{\hat{1}\}$. (For a set $C(Y)$ not containing a tope, the same argument in an appropriate contraction of $\mathcal{M}$ gives a recursive coatom ordering of $C(Y) \cup \{\hat{1}\}$.)

Theorem 4.52 (Lawrence 1984) *Let $\mathcal{M}$ be a simple oriented matroid and B a tope of $\mathcal{M}$. Every linear extension of the partial order $\leqslant_B$ is a recursive coatom ordering of $\mathcal{V}^*(\mathcal{M}) \cup \{\hat{1}\}$.*

Proof: Let $B = R_0 \prec R_1 \prec \cdots \prec R_m = -B$ be the linear extension of $\leqslant_B$. For each $i \in [m]$ let

$$Q_i = \{X \in \text{coat}[R_i] : X^0 \subseteq S(B, R_i)\}.$$

We first show that for each $i \in \{0, \ldots, m\}$,

$$Q_i = \text{coat}[R_i] \cap \bigcup_{j<i} \text{coat}[R_j]. \tag{4.1}$$

Let $X \in \text{coat}[R_i]$. Then the interval $[X, \hat{1}]$ has rank 2, so thinness of $\mathcal{V}^*(\mathcal{M}) \cup \{\hat{1}\}$ tells us that X is covered by R_i and exactly one other tope R_j. Since $X \circ B$ and $X \circ -B$ are distinct topes, we see $\{R_i, R_j\} = \{X \circ B, X \circ -B\}$. Also, $S(B, X \circ -B) = S(B, X \circ B) \dot\cup X^0$. Thus

$$X \in Q_i \Leftrightarrow R_i = X \circ -B \text{ and } R_j = X \circ B$$
$$\Leftrightarrow R_j \prec_B R_i$$
$$\Leftrightarrow j < i.$$

We now check the two requirements for a recursive coatom ordering.

4.6 Recursive Coatom Orderings on the Face Lattice & on Convex Sets 123

To see $[R_i] \cap \bigcup_{j<i}[R_j] = \bigcup_{X \in Q_i}[X]$ by (4.1) we have,

$$\bigcup_{X \in Q_i} [X] \subseteq [R_i] \cap \bigcup_{j<i}[R_j].$$

Conversely, if $Y \leq R_i$ and $Y \leq R_j$ with $j < i$, then consider the interval $[Y \circ B, Y \circ -B]_B$ in $\mathcal{T}(\mathcal{M}, B)$. A tope X is in this interval if and only if

$$S(Y, B) = S(Y \circ B, B) \subseteq S(X, B) \subseteq S(Y \circ -B, B) = S(Y, B) \cup Y^0.$$

Thus X is in this interval if and only if

- for each $e \in S(Y, B)$, we have $X(e) = -B(e) = Y(e)$, and
- for each $e \in \operatorname{supp}(Y) \setminus S(Y, B)$, we have $X(e) = B(e) = Y(e)$.

Thus $[Y \circ B, Y \circ -B]_B = \{X \in \mathcal{T}(\mathcal{M}) : X \geq Y\}$. In particular, R_i and R_j are both in this interval, and since $j < i$ we know R_i is not the smallest element of this interval. Thus there is some $k < i$ such that R_i covers R_k in this interval. By Proposition 4.46 this implies there is a $Z \in \operatorname{coat}[R_i] \cap \operatorname{coat}[R_k]$, and $Z \geq Y$.

Finally, we need to see that for every tope R_i, the interval $[R_i]$ has a recursive coatom ordering in which the elements of Q_i come first.

In Section 4.5 we saw how to get a recursive ordering on $[R_i]$ from a maximal chain in $\mathcal{T}(\mathcal{M}, R_i)$. Take such a maximal chain that contains B. Then the coatoms of R_i that come first in the resulting recursive coatom ordering correspond to elements of $S(B, R_i)$, i.e., are the elements of Q_i. □

Corollary 4.53 *Let $\mathcal{M}$ be an oriented matroid on elements E, let A be a subset of E, and let $Y \in \{0, +, -\}^A$. If $C(Y) \setminus \{0\} \neq \emptyset$ and $C(Y) \neq \mathcal{V}^*(\mathcal{M})$ then $C(Y) \setminus \{0\}$ is the face poset of a shellable ball.*

Proof: All maximal elements of $C(Y)$ have the same support S (Proposition 4.36), and $C(Y) \setminus (E \setminus S)$ is a convex set in $\mathcal{V}^*(\mathcal{M}/(E \setminus S))$ containing a tope. Also, $C(Y) \setminus (E \setminus S)$ contains a tope of $\mathcal{V}^*(\mathcal{M}/(E \setminus S))$. Thus it suffices to prove (1) for convex sets containing a tope.

If $C(Y)$ contains a tope B, then $\mathcal{T}(\mathcal{M}) \cap C(Y)$ is an ideal in $\mathcal{T}(\mathcal{M}, B)$. Thus (Problem 4.8) there is a linear extension of the partial order $\leq_B$ in which the elements of $\mathcal{T}(\mathcal{M}) \cap C(Y)$ come first. So by the third part of Lemma 4.30, the recursive coatom ordering of Theorem 4.52 restricts to a recursive coatom ordering on $C(Y) \cup \{\hat{1}\}$. By Corollary 4.41 we know $C(Y)$ is subthin, and so by Theorem 4.29.2 we have our conclusion. □

4.7 Covector Sets Give Pseudosphere Arrangements

Now we can prove the hard direction of the Topological Representation Theorem.

Theorem 4.54 *For each rank r oriented matroid $\mathcal{M}$ there is an essential arrangement $\mathcal{A}$ of signed pseudospheres in S^{r-1} such that $\mathcal{V}^*(\mathcal{A}) = \mathcal{V}^*(\mathcal{M})$.*

Proof: Corollary 4.40 and Theorem 4.52 tell us that $\mathcal{V}^*(\mathcal{M}) \cup \{\hat{1}\}$ is thin and has a recursive coatom ordering, hence is the augmented face poset of a regular cell decomposition R of the sphere. Since the rank of the poset $\mathcal{V}^*(\mathcal{M})$ is r, this sphere has dimension $r - 1$. Fix such a cell decomposition and identify each cell with its corresponding element of $\mathcal{V}^*(\mathcal{M})$.

For each $e \in E$, let S_e denote the union of all open cells X with $X(e) = 0$, and for each $\epsilon \in \{+, -\}$, let S_e^ϵ denote the union of all open cells X with $X(e) = \epsilon$. Let $\mathcal{S}_e$ be the triple (S_e, S_e^+, S_e^-). If e is a loop, then $\mathcal{S}_e$ is the degenerate signed pseudosphere. Otherwise, since $\{X \in \mathcal{V}^*(\mathcal{M}) : X(e) = 0\} \cong \mathcal{V}^*(\mathcal{M}/e)$, we have that S_e is a sphere of dimension $r - 2$. Also, $S_e^+ \cup S_e$ is the underlying space of $C(e^+) \backslash \{0\}$ and $S_e^- \cup S_e$ is the underlying space of $C(e^-) \backslash \{0\}$. Theorem 4.52 says that each of $C(e^+) \cup \{\hat{1}\}$ and $C(e^-) \cup \{\hat{1}\}$ has a recursive coatom ordering, and Corollaries 4.38 and 4.41 say that each of these posets is subthin of rank r, and that the intervals $[X, \hat{1}]$ of size 3 are exactly those with $X(e) = 0$. Thus $S_e^+ \cup S_e$ and $S_e^- \cup S_e$ are closed balls with boundary S_e, and so $\mathcal{S}_e$ is a signed pseudosphere.

Similarly we see that for each $A \subseteq E$ that $S_A := \bigcap_{e \in A} S_e$ is a topological sphere, and that if $S_A \not\subseteq S_f$ then $S_A \cap S_f$ is a pseudosphere in S_A with sides $S_A \cap S_f^+$ and $S_A \cap S_f^-$. The intersection of a collection of closed sides is the underlying space of a subcomplex $C(Y) \backslash \{0\}$, hence is thin or subthin by Corollary 4.41, and has a recursive coatom ordering by Theorem 4.52.

Thus $\mathcal{A} := (\mathcal{S}_e : e \in E)$ is an arrangement of pseudospheres. For each cell X in the cell complex R and each point x in the interior of X, we have $\text{sign}_\mathcal{A}(x) = X$. Thus $\mathcal{V}^*(\mathcal{A}) \backslash \{0\}$ is the poset of cells in R, and so $\mathcal{V}^*(\mathcal{A}) = \mathcal{V}^*(\mathcal{M})$. Since every cell corresponds to a nonzero element of $\mathcal{V}^*(\mathcal{M})$, this arrangement is essential. □

4.7.1 Centrally Symmetric Arrangements

The preceding arguments on thinness and recursive coatom orderings show that every rank r oriented matroid can be represented by an essential arrangement of signed pseudospheres in S^{r-1}. We still need to verify that this arrangement can be taken to be centrally symmetric.

4.7 Covector Sets Give Pseudosphere Arrangements

Proposition 4.55 (cf. 5.2.2 in Björner et al. 1999) *If $\mathcal{A} = (\mathcal{S}_e : e \in E)$ is an arrangement of signed pseudospheres in S^{r-1} representing an oriented matroid $\mathcal{M}$, then there is a homeomorphism $S^{r-1} \to S^{r-1}$ taking $\mathcal{A}$ to a centrally symmetric arrangement.*

Proof: Step 1 is to find a fixed-point-free involution $\alpha \colon S^{r-1} \to S^{r-1}$ fixing each S_e. The topological representation of $\mathcal{M}$ gives a homeomorphism $g \colon \Delta(\mathcal{V}^*(\mathcal{M})\backslash\{0\}) \to S^{r-1}$ taking each $X \in \mathcal{V}^*(\mathcal{M})\backslash\{0\}$ to a point in the interior of the corresponding cell. Also, there is a fixed-point-free involution $i \colon \mathcal{V}^*(\mathcal{M})\backslash\{0\} \to \mathcal{V}^*(\mathcal{M})\backslash\{0\}$ sending each X to $-X$. Since i is a poset map, it induces a topological map $\|\Delta i\| \colon \|\Delta \mathcal{V}^*(\mathcal{M})\backslash\{0\}\| \to \|\Delta \mathcal{V}^*(\mathcal{M})\backslash\{0\}\|$. The composition $g^{-1} \circ \|\Delta i\| \circ g$ is our α. Note that α reverses the two sides of each S_e.

Step 2 is to find, for every such α, a homeomorphism $h \colon S^{r-1} \to S^{r-1}$ such that $h\alpha(x) = -h(x)$ for each x. We get this by induction on r. If $r = 1$ then h is the identity.

For larger r, let $f \in E$. Then there is a self-homeomorphism h_1 of S^{r-1} taking S_f to the equator S^{r-2}. Let $\alpha_f = h_1 \circ \alpha \circ h_1^{-1}|_{S^{r-2}}$. Then α_f is a fixed-point-free involution of S^{r-2}, so by the induction hypothesis, there is a self-homeomorphism h_f of S^{r-2} such that $h_f \alpha_f(x) = -h_f(x)$ for each x.

Recall every self-homeomorphism of the sphere ∂B^{r-1} extends to all of B^{r-1}, so h_f extends to a self-homeomorphism h_2^+ of the top hemisphere S_+ of S^{r-1}. Define a self-homeomorphism h_2^- of the bottom hemisphere S_- by $h_2^-(x) = -h_2^+(h_1 \circ \alpha \circ h_1^{-1}(x))$. (Since α reverses the two sides of S_f, this is well defined.) Together h_2^+ and h_2^- define a self-homeomorphism h_2 of S^{r-1}.

Now define $h := h_2 \circ h_1$. To verify that $h \circ \alpha(x) = -h(x)$ for all x:
If $x \in S^{r-2}$ then

$$\begin{aligned} h\alpha(x) &= h_2 h_1 \alpha(x) \\ &= h_2 h_1 (h_1^{-1} \alpha_f h_1)(x) \\ &= h_2 \alpha_f h_1(x) \\ &= -h_2 h_1(x) \\ &= -h(x). \end{aligned}$$

If $x \in S_+$ then

$$\begin{aligned} h\alpha(x) &= h_2 h_1 \alpha(x) \\ &= h_2^- h_1 \alpha(x) \\ &= (-h_2^+ h_1 \alpha h_1^{-1}) h_1 \alpha(x) \\ &= -h_2^+ h_1(x) \\ &= -h_2 h_1(x) \\ &= -h(x). \end{aligned}$$

If $x \in S_-$ then

$$\begin{aligned} h\alpha(x) &= h_2^+ h_1 \alpha(x) \\ &= (h_2^+ h_1 \alpha h_1^{-1}) h_1(x) \\ &= -h_2^- h_1(x) \\ &= -h_2 h_1(x) \\ &= -h(x). \end{aligned}$$

Step 3 is to notice that applying h to $\mathcal{A}$ gives the desired centrally symmetric arrangement: For each $S_e \in \mathcal{A}$ and each $x \in S_e$, we have $-h(x) = h\alpha(x)$. Since α fixes S_e, we have that $-h(x)$ is in $h(S_e)$. □

4.8 Topological Interpretations of Oriented Matroid Concepts

With the Topological Representation Theorem in hand, we may, when we choose, mostly ignore the first three chapters of this book and study oriented matroids by thinking about wiggly equators on spheres. Let's go through ideas introduced in previous chapters and interpret them in terms of topological representations. Fix a rank r oriented matroid $\mathcal{M}$ with topological representation $(S_e : e \in E)$ in S^{r-1}.

An element e is a *loop* if S_e is the degenerate hyperplane. Two nonloops e, f are *parallel* if $S_e = S_f$ and are *antiparallel* if $S_e = S_f$ and $S_e^+ = -S_f^-$.

We obtain a *reorientation* of $\mathcal{M}$ by replacing some pseudospheres (S_e, S_e^+, S_e^-) with (S_e, S_e^-, S_e^+).

We can extend our discussion of flats in Section 3.7 to discuss intersections of pseudospheres. Recall that for $A \subseteq E$ we define $S_A = \bigcap_{a \in A} S_a$, and that $\mathcal{F}(\mathcal{M})$ is the poset of flats of $\mathcal{M}$.

Proposition 4.56 *Let $\mathcal{M}$ be rank r and $(S_e : e \in E)$ a topological representation of $\mathcal{M}$ in S^{r-1}.*

1. *There is a bijection β from $\mathcal{F}(\mathcal{M})$ to the set of all intersections S_A of pseudospheres in our topological representation, taking a flat F to S_F.*
   In particular, if $A \subseteq E$ and F is the smallest flat containing A then $S_A = S_F$.
2. *If $A \subseteq E$ then $S_A \cong S^{r-\text{rank}(A)-1}$.*

Example 4.57 Let $\mathcal{M}$ be the oriented matroid with topological representation given in Figure 4.5, with the pseudospheres signed however you wish. Part 2 of Proposition 4.56 tells us that a 3-element subset A of $\{1, 2, 3, 4\}$ is a basis if

4.8 Topological Interpretations of Oriented Matroid Concepts 127

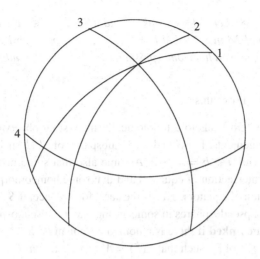

Figure 4.5 Figure for Example 4.57.

and only if $S_A \cong S^{-1}$. (By definition, $S^{-1} = \emptyset$.) So from the figure we see that the only three-element set that is not a basis is $\{1,2,3\}$. For each 2-element set A we see $S_A \cong S^0$, so A is independent. Part 1 of Proposition 4.56 tells us that $\{1,2,3\}$ is the smallest flat containing $\{1,2\}$, and likewise is the smallest flat containing $\{1,3\}$ or $\{2,3\}$.

Proof: 1. This follows from Proposition 3.35. For each $A \subseteq E$ S_A is the union of all cells indexed by nonzero covectors X such that $X(A) = 0$. By Proposition 3.35 we see that if F is the smallest flat containing A and X is a covector with $X(A) = 0$, then $X(F) = 0$. Thus β is surjective. Further, if F is a flat and $g \in E \backslash F$ then from Proposition 3.35 we see there is a covector X such that $X(F) = 0$ and $X(g) \neq 0$. Thus $S_F \neq S_G$ for each $G \not\subseteq F$, and so β is injective.

2. We induct on rank(A), with the rank 0 case being clear. For higher rank, let A have rank k and $A' \subset A$ be a maximal subset of rank $k - 1$. Then for each $a \in A \backslash A'$ we have that a is not in the smallest flat containing A', so $S_{A' \cup \{a\}} \subsetneq S_{A'}$. By the definition of pseudosphere arrangement, we see that $S_{A' \cup \{a\}} = S_{A'} \cap S_a$ is codimension 1 in $S_{A'}$, and since rank$(A' \cup \{a\}) = $ rank(A), we have that $A' \cup \{a\}$ and A have the same closure, so $S_{A' \cup \{a\}} = S_A$. □

From Proposition 4.56 we see that an element f of $\mathcal{M}$ is a *coloop* if $S_{E \backslash \{f\}} \cong S^0$.

The next result generalizes the fact that a single nondegenerate pseudosphere in S^{r-1} is mapped by some self-homeomorphism of S^{r-1} to an equator.

Proposition 4.58 Let $\mathcal{M}$ be rank r and $(\mathcal{S}_e : e \in E)$ a topological representation of $\mathcal{M}$ in S^{r-1}. Let $I = \{e_1, \ldots, e_k\} \subseteq E$. Then I is independent if and only if there is a homeomorphism $h \colon S^{r-1} \to S^{r-1}$ sending each $\mathcal{S}_{e_i}$ to $\mathbf{e}_i^\perp \cap S^{r-1}$.

Problem 4.59 Prove this.

Proposition 4.58 leads to a topologically interesting observation on pseudosphere arrangements. Let S and T be subspaces of S^{r-1} such that $S \cong S^a$ and $T \cong S^b$ with $a + b = r - 2$. Assume also that some homeomorphism takes S to an intersection of equators and that some homeomorphism takes T to an intersection of equators. This is the case, for instance, if S and T are each intersections of pseudospheres in some arrangement of pseudospheres. We say that S and T are **linked** if there is a homeomorphism $h \colon S^{r-1} \to S^{r-1}$ and a linear subspace V of $\mathbb{R}^r$ such that $h(S) = V \cap S^{r-1}$ and $h(T) = V^\perp \cap S^{r-1}$. Equivalently, S and T are linked if they are disjoint and $S^{r-1} \setminus S$ retracts to T.

Corollary 4.60 Let $\mathcal{M}$ be rank r and $(\mathcal{S}_e : e \in E)$ a topological representation of $\mathcal{M}$ in S^{r-1}. Let $A, B \subset E$ such that S_A and S_B are disjoint and $\dim(S_A) + \dim(S_B) = r - 2$. Then S_A and S_B are linked.

Proof: Let $\hat{A}$ and $\hat{B}$ be maximal independent subsets of A and B. From Proposition 4.56 we have that $|\hat{A}| + |\hat{B}| = r$, and since $S_{\hat{A} \cup \hat{B}} = S_{\hat{A}} \cup S_{\hat{B}} = S_A \cap S_B = \emptyset$, we have $\operatorname{rank}(\hat{A} \cup \hat{B}) = r$. Thus $\hat{A} \cup \hat{B}$ is a basis. Applying Proposition 4.58 then gives our result. □

Problem 4.61 This problem will show that a topological representation of a direct sum $\mathcal{M} \oplus \mathcal{M}'$ can be obtained from topological representations of $\mathcal{M}$ and $\mathcal{M}'$ by taking joins.

Let E and E' be disjoint finite sets. Let $\mathcal{M}$ and $\mathcal{M}'$ be oriented matroids on E and E' respectively, with topological representations $(\mathcal{S}_e : e \in E)$ in S^{r-1} and $(\mathcal{S}_e : e \in E')$ in $S^{r'-1}$. For each signed pseudosphere $\mathcal{S}_e = (S_e, S_e^+, S_e^-)$ in S^{r-1}, let $\mathcal{S}_e * S^{r'-1}$ denote the triple $(S_e * S^{r'-1}, S_e^+ * S^{r'-1}, S_e^- * S^{r'-1})$ in the sphere $S^{r-1} * S^{r'-1}$. Similarly for each signed pseudosphere $\mathcal{S}_e = (S_e, S_e^+, S_e^-)$ in $S^{r'-1}$, let $S^{r-1} * \mathcal{S}_e$ denote the triple $(S^{r-1} * S_e, S^{r-1} * S_e^+, S^{r-1} * S_e^- * S)$ in the sphere $S^{r-1} * S^{r'-1}$.

Show that the union of the arrangements $(\mathcal{S}_e * S^{r'-1} : e \in E)$ and $(S^{r-1} * \mathcal{S}_e : e \in E')$ is an arrangement of signed pseudospheres, and further that it is a topological representation of $\mathcal{M} \oplus \mathcal{M}'$.

A special case of 4.61 is when $\mathcal{M}'$ is rank 1 with one element, so that $\mathcal{M} \oplus \mathcal{M}'$ is the extension of $\mathcal{M}$ by a coloop. In this case we get a topological representation of $\mathcal{M} \oplus \mathcal{M}'$ as a suspension of a topological representation of $\mathcal{M}$.

4.8 Topological Interpretations of Oriented Matroid Concepts

If $A \subseteq E$ then the *deletion* $\mathcal{M}\backslash A$ has topological representation ($\mathcal{S}_e : e \in E\backslash A$). If $\mathcal{M}\backslash A$ has rank $r' < r$ then this arrangement isn't essential: $S_{E\backslash A} \cong S^{r-r'-1}$. In this case every basis B' of $\mathcal{M}\backslash A$ extends to a basis $B' \cup B$ of $\mathcal{M}$, and S_B is a sphere of dimension $r' - 1$ linked with $S_{E\backslash A}$. The arrangement $(\mathcal{S}_e \cap S_B : e \in E\backslash A)$ in S_B is a topological realization of $\mathcal{M}\backslash A$.

If $A \subseteq E$ then the *contraction* $\mathcal{M}/A$ has topological representation $((\mathcal{S}_e \cap S_A, S_e^+ \cap S_A, S_E^- \cap S_A) : e \in E\backslash A)$ in S_A. In particular, rank 2 contractions $\mathcal{M}/A$ of $\mathcal{M}$ are represented in $\mathcal{M}$ by 1-dimensional spheres S_A, which we call **pseudocircles**.

Next we consider how to read off *vectors*. Let $X = A^+B^- \in \{0, +, -\}^E$. Recall that $\mathcal{V} = (\mathcal{V}^*)^{\perp}$, and that $\mathcal{V}^* = -\mathcal{V}^*$. Thus $X \in \mathcal{V}$ if and only if

$$\{Y \in \mathcal{V}^* : X \cdot Y \subseteq \{0, +\}\} = \{Y \in \mathcal{V}^* : X \cdot Y = \{0\}\}.$$

In terms of the topological representation, this says that $X \in \mathcal{V}$ if and only if

$$\bigcap_{a \in A} \overline{S_a^+} \cap \bigcap_{b \in B} \overline{S_b^-} = S_{A \cup B}.$$

Example 4.62 Consider the oriented matroid $\mathcal{M}$ with topological representation shown in Figure 4.6. Neither $a^+b^+c^+e^+$ nor $a^+b^+c^+e^-$ is in $\mathcal{V}(\mathcal{M})$, because $S_{\{a,b,c,e\}} = \emptyset$ and both $S_a^+ \cap S_b^+ \cap S_c^+ \cap S_e^+$ and $S_a^+ \cap S_b^+ \cap S_c^+ \cap S_e^-$ are nonempty. On the other hand, $a^+b^+c^+d^-$ is in $\mathcal{V}(\mathcal{M})$ because $S_a^+ \cap S_b^+ \cap S_c^+ \cap S_d^- = \emptyset$. The reader is encouraged to draw an additional signed pseudosphere to add an element f so that $S_d \cap S_e \subset S_f$ and then find the signed circuit supported on $\{d, e, f\}$.

Example 4.63 Let $\mathcal{M}$ be an oriented matroid on elements E, and let $A \subseteq E$. Recall from Section 2.6 the convex hull $\widehat{\mathrm{conv}}(A)$ of A (not to be confused with the convex hull of a set of covectors!) by

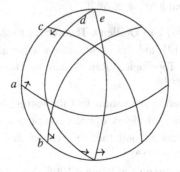

Figure 4.6 Figure for Example 4.62.

$\widehat{\text{conv}}(A) = \{e \in E : e \in A \text{ or } \hat{A}^+ e^- \in \mathcal{V}(\mathcal{M}) \text{ for some } \hat{A} \subseteq A\}$.

In a topological representation $(\mathcal{S}_e : e \in E)$ of $\mathcal{M}$, we have $e \in \widehat{\text{conv}}(A)$ if and only if $\bigcap_{a \in A} (S_a \cup S_a^+) \subseteq S_e \cup S_e^+$.

The Dual Pivoting Property allows us to read off *chirotope* information from a topological representation. A 0-cell in our cell decomposition corresponds to a signed cocircuit Y. For each $H = \{h_2, \ldots, h_r\} \subset E$ such that S_H consists of the cells corresponding to $\pm Y$ and every $e, f \in E$, we have

- $\chi(e, h_2, \ldots, h_r) \chi(f, h_2, \ldots, h_r) = +$ if our 0-cell is in $S_e^+ \cap S_f^+$ or in $S_e^- \cap S_f^-$, and
- $\chi(e, h_2, \ldots, h_r) \chi(f, h_2, \ldots, h_r) = -$ if our 0-cell is in $S_e^+ \cap S_f^-$ or in $S_e^- \cap S_f^+$.

4.9 Some Homotopy Results, and an Application

This section will apply the Topological Representation Theorem to get a topological proof of a combinatorial result. We include this only to amuse the more topologically inclined reader, and we assume some knowledge of poset topology, including Quillen's Theorem A (Theorem 5.26), the Euler characteristic, and the Möbius inversion for posets (cf. Stanley 1997). The third part of Problem 4.64 requires knowledge of good covers and nerves, but this part isn't necessary to the discussion. See Section 5.3.1 for some of the relevant background.

Problem 4.64 This problem concerns the spaces $\mathcal{V}_\downarrow$ and $\mathcal{C}^\uparrow$ introduced in Exercise 2.7. Consider a rank r oriented matroid with signed circuit set $\mathcal{C}$, vector set $\mathcal{V}$, and covector set $\mathcal{V}^*$. We'll give two proofs that $\|\Delta(\mathcal{V}_\downarrow^* \backslash \{0\})\| \simeq S^{r-1}$ and one proof that $\|\Delta \mathcal{C}^\uparrow\| \simeq S^{n-r-1}$.

1. (Björner et al. 1999) Use Quillen's Theorem A to show that the inclusion $\mathcal{V}^* \backslash \{0\} \hookrightarrow \mathcal{V}_\downarrow^* \backslash \{0\}$ induces a homotopy equivalence of order complexes. Apply the Topological Representation Theorem to conclude that $\|\Delta(\mathcal{V}_\downarrow^* \backslash \{0\})\| \simeq S^{r-1}$.
2. Use Quillen's Theorem A to show that the inclusion $\mathcal{V} \backslash \{0\} \hookrightarrow (\mathcal{V} \backslash \{0\})^\uparrow$ induces a homotopy equivalence of order complexes. Apply the Topological Representation Theorem and the observation that $\mathcal{C}^\uparrow = (\mathcal{V} \backslash \{0\})^\uparrow$ to conclude that $\|\Delta \mathcal{C}^\uparrow\| \simeq S^{n-r-1}$.
3. (Edelman 1984, Anderson and Wenger 1996) Assume $\mathcal{M}$ has no loops. Let $(\mathcal{S}_e : e \in E)$ be a topological representation of $\mathcal{M}$ in S^{r-1}. Show that the

4.9 Some Homotopy Results, and an Application

set $\bigcup_{e \in E} \{S_e^+, S_e^-\}$ is a good cover of S^{r-1} by open sets. It follows that its nerve is homotopy equivalent to S^{r-1} (cf. corollary 4G.3 in Hatcher 2002). Find an isomorphism between this nerve and $\mathcal{V}_\downarrow^* \backslash \{0\}$. (We're using the convention that the empty set is an element of every simplicial complex – in particular, the empty set is an element of every nerve.)

The third part of Problem 4.64 explicitly found an isomorphism from $\mathcal{V}_\downarrow^*$ to a simplicial complex, but even without this we can see $\mathcal{V}_\downarrow^*$ as a simplicial complex on the vertex set $\bigcup_{e \in E} \{e_+, e_-\}$ by identifying each sign vector $A^+ B^-$ with $\{a_+ : a \in A\} \cup \{b_- : b \in B\}$. So the first part of the problem shows that this simplicial complex is homotopy equivalent to S^{r-1}. Throughout the following we will view $\mathcal{V}_\downarrow^*$ as a simplicial complex.

Corollary 4.65 $\sum_{X \in \mathcal{V}_\downarrow^*} (-1)^{|X|} = (-1)^r$.

Proof: The reduced Euler characteristic of $\mathcal{V}_\downarrow^* \backslash \{0\}$ is $\sum_{X \in \mathcal{V}_\downarrow^*} (-1)^{|X|-1} - 1 = \sum_{X \in \mathcal{V}_\downarrow^*} (-1)^{|X|-1}$, and the reduced Euler characteristic of S^{r-1} is $(-1)^{r-1}$. So the result follows from the first (or third) part of Problem 4.64. □

We can apply Corollary 4.65 to count the number of topes of an oriented matroid in terms of its underlying matroid. This proof of Theorem 4.67 is due to Edelman. This is not the original proof or the most commonly known technique for proving this (see Remark 4.68), but it's an elegant application of the Topological Representation Theorem.

For convenience, we'll assume our oriented matroid $\mathcal{M}$ has no loops. In the following definition we talk about the characteristic polynomial of $\mathcal{M}$, but the reader familiar with unoriented matroids will recognize it as the characteristic polynomial of the underlying matroid.

Definition 4.66 Let $\mathcal{M}$ be a rank r oriented matroid on elements E. The **characteristic polynomial** of $\mathcal{M}$ is

$$p(\mathcal{M}, \lambda) = \sum_{S \subseteq E} (-1)^{r - \text{rank}(S)} \lambda^{|S|}.$$

Theorem 4.67 (Edelman 1984, Las Vergnas 1975b, Winder 1966, Zaslavsky 1975) *Let $\mathcal{M}$ be a rank r oriented matroid with no loops. The number of topes of $\mathcal{M}$ is $(-1)^r p(\mathcal{M}, -1)$.*

Proof: For $S \subseteq E$, let $\alpha(S, \mathcal{M}) = |\{X \in \mathcal{V}^*(\mathcal{M})_\downarrow : \text{supp}(X) = S\}|$. Since $\mathcal{M}$ has no loops, $\alpha(E, \mathcal{M})$ is the number of topes of $\mathcal{M}$. We can restate Corollary 4.65 as

$$\sum_{S \subseteq E}(-1)^{|S|}\alpha(S,\mathcal{M}) = (-1)^r. \qquad (4.2)$$

If $S \subseteq T \subseteq E$, then $\alpha(S,\mathcal{M}) = \alpha(S,\mathcal{M}(T))$. So applying Equation (4.2) to $\mathcal{M}(T)$, we see that for each $T \subseteq E$,

$$\sum_{S \subseteq T}(-1)^{|S|}\alpha(S,\mathcal{M}) = (-1)^{\text{rank}(T)}.$$

By Möbius inversion,

$$(-1)^{|E|}\alpha(E,\mathcal{M}) = \sum_{S \subseteq T} \mu(S,E)(-1)^{\text{rank}(S)},$$

where $\mu(S,E) = (-1)^{|E|-|S|}$ is the Möbius function of the power set of E. Thus

$$\alpha(E,\mathcal{M}) = (-1)^{|E|}\sum_{S \subseteq T}(-1)^{|E|-|S|}(-1)^{\text{rank}(S)}.$$

Multiplying both sides by $(-1)^{|E|+r}$ gives our result. □

For much more on enumerative issues on $\mathcal{V}^*(\mathcal{M})$, see section 4.6 in Björner et al. (1999). Another application of the third part of Problem 4.64 can be found in Anderson and Wenger (1996).

Remark 4.68 Variations on Theorem 4.67 were originally proved independently by Las Vergnas (Las Vergnas 1975b, 1980), whose statement was in terms of the Tutte polynomial, and Zaslavsky (1975), whose statement was for hyperplane arrangements but whose proof carries through for oriented matroids. (Oriented matroids were only beginning to emerge as a topic at the time of Zaslavsky's original proof.) The proofs in Las Vergnas (1980) and Zaslavsky (1975) are both purely combinatorial, using induction and deletion/contraction. Zaslavsky also gave a later proof that used some topology (Zaslavsky 1977). Going way back, Winder (1966) proved essentially the same result for hyperplane arrangements in 1966, using entirely different methods.

4.10 Historical Note

At this point, we have plenty of reasons to favor covectors over other characterizations of oriented matroids:

- the directness of the description $\mathcal{V}^*(V) = \{\text{sign}(\mathbf{x}) : \mathbf{x} \in V\}$ for a subspace V of $\mathbb{R}^n$,
- the simplicity of the duality $\mathcal{V}^* = \mathcal{V}^\perp$, and
- the Topological Representation Theorem.

But the theory of oriented matroids grew out of matroid theory, where vectors and covectors have little role. Recall that the unsigned circuits and unsigned cocircuits of the matroid underlying $\mathcal{M}$ are just the supports of the signed circuits and signed cocircuits of $\mathcal{M}$. In principle, we could also define unsigned vectors and unsigned covectors of the underlying matroid in the same way: The "unsigned vectors" would just be the unions of unsigned circuits, and similarly for "unsigned covectors." But these objects are relatively uninteresting from the perspective of matroid theory. For instance, a person whose interest in matroids arises from an interest in vector arrangements $(\mathbf{v}_e : e \in E)$ in $\mathbb{F}_2^n$ will interpret unsigned circuits as indicating the nonzero coefficients in minimal linear dependencies $\sum_{e \in E} a_e \mathbf{v}_e = \mathbf{0}$, but "unsigned vectors" have no similar interpretation.

For this reason, vectors and covectors came relatively late to the party. Amusingly, covectors do not appear in the paper in which the Topological Representation Theorem was first proved (Folkman and Lawrence 1978), although that paper does note that the cells in a topological representation correspond to conformal compositions of cocircuits. Vectors are discussed near the end of the foundational paper (Bland and Las Vergnas 1979) as "signed spans," without axioms. The vector axioms first appeared in the PhD theses of Mandel (1982) and Fukuda (1982), contemporaneous students of Edmonds, and recognition of the central role of covectors seems to have arisen from the collaboration of these three.

4.11 Final Notes

1. Recall from Section 1.4 the vision of oriented matroids as "combinatorial models for real subspaces of $\mathbb{R}^n$": Given a rank r subspace V of $\mathbb{R}^n$, the coordinate hyperplanes intersect V in an arrangement of signed hyperplanes, and the oriented matroid for this arrangement is viewed as a combinatorial model for V. Notice that this arrangement intersects the unit sphere in V in an arrangement of equators that is a topological representation of V. This realizable picture motivated a general philosophy: A rank r oriented matroid $\mathcal{M}$ with elements $\{1, 2, \ldots, n\}$ is a combinatorial analog to a rank r subspace of $\mathbb{R}^n$, and a topological representation of $\mathcal{M}$ is an analog to the unit sphere in this subspace.

2. Now that the Topological Representation Theorem has given us a nice generalization of hyperplane arrangements representing realizable oriented matroids, one might ask for a similar generalization of vector arrangements.

Is there a good notion of "pseudoconfiguration of vectors" so that every oriented matroid can be represented by such a beast?

We'll return to this question in Chapter 6, where we'll show the answer to be "No."

3. In Section 4.5 we saw that, in each topological representation of $\mathcal{M}$, the boundary of a cell can be shelled in much the same way that the boundary of a convex polytope can be shelled. This raises a question: Does each cell have the combinatorial type of a convex polytope? We'll see in Section 8.4 that the answer is "No."

Exercises

4.1 Let $S \subseteq \mathcal{T}(\mathcal{M})$. Show that S is the set of topes in a convex set if and only if S is an order ideal in $\mathcal{T}(\mathcal{M}, B)$ for each $B \in S$.

4.2 Recall the definition of the convex hull of a set of elements of $\mathcal{M}$ from Section 2.6. We say $A \subseteq E$ is *convex* if $\widehat{\text{conv}}(A) = A$. If M is acyclic (i.e., E^+ is a tope of $\mathcal{M}$) prove that the separation set $S(E^+, T)$ is convex for each tope T.

4.3 1. Use the Topological Representation Theorem to prove that every rank 2 oriented matroid is realizable.

   2. Let $\mathcal{M}$ be a simple rank 3 oriented matroid. Consider the cell decomposition of S^2 resulting from a topological representation of $\mathcal{M}$. Use Euler's formula $V - E + F = 2$ to deduce that $|X^0| = 2$ for some $X \in \mathcal{C}^*(\mathcal{M})$.

   3. Use this to deduce the *Sylvester–Gallai Theorem*: For every finite set of points in the affine plane, either there is a line containing all of the points or there is a line containing exactly two of the points.

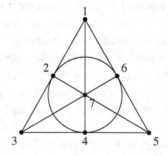

Figure 4.7 The Fano plane

4.4 The *Fano plane* is a rank 3 (unoriented) matroid illustrated in Figure 4.7. Its set of elements is [7], and the unsigned circuits of size less than 4 are the triples of collinear points in Figure 4.7, as well as $\{2, 4, 6\}$. (So we treat the circle in the figure as a "line.")

Show that the Fano matroid is not *orientable*, that is, there is no oriented matroid whose underlying matroid is the Fano Plane, by showing that there is no arrangement of signed pseudospheres in S^2 whose associated rank 2 flats are the rank 2 flats of the Fano plane.

5

Strong Maps and Weak Maps

This chapter will discuss two notions of morphism between oriented matroids. Our development in each case will be reminiscent of our development of oriented matroids themselves. For each notion of morphism,

1. we begin with a geometric motivation for a general combinatorial definition;
2. we find that our morphisms are not necessarily "realizable" (even when the oriented matroids involved are realizable!);
3. finally, we see that our morphisms are always representable topologically, as topological maps between topological representations of oriented matroids.

Either kind of morphism induces a poset structure on the poset of all oriented matroids on a fixed set E of elements. Posets ordered by weak maps lead to interesting topology that will be the subject of Chapter 10.

5.1 Strong Maps

If $W_1 \subseteq W_2$ are linear subspaces of $\mathbb{R}^E$ then by definition $\mathcal{V}^*(W_1) \subseteq \mathcal{V}^*(W_2)$. Exercise 1.5 expressed the same idea another way: If $\mathcal{M}$ is the oriented matroid associated to $M \in \text{Mat}(n,m)$ and $\mathcal{N}$ is the oriented matroid associated to AM for some matrix A, then $\mathcal{V}^*(\mathcal{N}) \subseteq \mathcal{V}^*(\mathcal{M})$.

These examples motivate the following definition.

Definition 5.1 Let $\mathcal{M}$ and $\mathcal{N}$ be oriented matroids with the same elements. We say there is a **strong map** from $\mathcal{M}$ to $\mathcal{N}$, denoted $\mathcal{M} \to \mathcal{N}$, if $\mathcal{V}^*(\mathcal{N}) \subseteq \mathcal{V}^*(\mathcal{M})$. We also say $\mathcal{N}$ is a **strong map image** of $\mathcal{M}$.

5.1 Strong Maps

Example 5.2 For every $X \in \mathcal{V}^*(\mathcal{M})$, the rank 1 oriented matroid with covector set $\{\pm X, \mathbf{0}\}$ is a strong map image of $\mathcal{M}$.

Example 5.3 If $\mathcal{M}$ is an oriented matroid on elements E and $\mathcal{M} \cup A$ is an extension then $\mathcal{M} \to (\mathcal{M} \cup A)/A$. Section 5.4.1 will discuss whether every strong map arises in this way.

Proposition 5.4 (Las Vergnas 1975a, Björner et al. 1999, Jarra and Lorscheid 2024) *Let $\mathcal{M}$ and $\mathcal{N}$ be oriented matroids on elements E with chirotopes $\chi_\mathcal{M}$ and $\chi_\mathcal{N}$. Let $r = \text{rank}(\mathcal{M})$ and $s = \text{rank}(\mathcal{N})$. The following are equivalent.*

1. $\mathcal{M} \to \mathcal{N}$.
2. $\mathcal{C}^*(\mathcal{N}) \subseteq \mathcal{V}^*(\mathcal{M})$.
3. $\mathcal{V}(\mathcal{M}) \subseteq \mathcal{V}(\mathcal{N})$.
4. $\mathcal{C}(\mathcal{M}) \subseteq \mathcal{V}(\mathcal{N})$.
5. $\mathcal{C}^*(\mathcal{N}) \perp \mathcal{C}(\mathcal{M})$.
6. *For each $x_2, \ldots, x_r, y_0, \ldots, y_s \in E$,*

$$0 \in \boxplus_{k=0}^{s} (-1)^k \chi_\mathcal{N}(y_0, \ldots, \widehat{y_k}, \ldots, y_s) \chi_\mathcal{M}(y_k, x_2, \ldots, x_r). \tag{5.1}$$

Proof: The proof of the first five equivalences are straightforward.

Assume $\mathcal{C}^*(\mathcal{N}) \perp \mathcal{C}(\mathcal{M})$, and consider $x_2, \ldots, x_r, y_0, \ldots, y_s \in E$. If $\{x_2, \ldots, x_r\}$ is dependent in $\mathcal{M}$ or if $\{y_0, \ldots, y_s\}$ does not contain a basis of $\mathcal{N}$ then clearly all terms in the summation (5.1) are 0. Otherwise, there is a cocircuit X of $\mathcal{M}$ such that $\{x_2, \ldots, x_r\}$ is a maximal independent subset of X^0, and there is a circuit Y of $\mathcal{N}$ whose support is contained in $\{y_0, \ldots, y_s\}$. By our hypothesis $X \perp Y$, so

$$0 \in \boxplus_{e \in E} X(e)Y(e). \tag{5.2}$$

Case 1: Assume every term in the summation (5.2) is 0. Then for each $y_k \in \text{supp}(Y)$ we have that $y_k \in X^0$. Since $\{x_2, \ldots, x_r\}$ is a maximal independent subset of X^0, we have that $\chi_\mathcal{M}(y_k, x_2, \ldots, x_r) = 0$. Also, for each $y_k \notin \text{supp}(Y)$ we have that $\chi_\mathcal{N}(y_0, \ldots, \widehat{y_k}, \ldots, y_s) = 0$. Thus every term in the summation (5.1) is 0.

Case 2: The summation (5.2) has a nonzero term, and so it must have at least two nonzero terms that are negatives of each other. Consider $k < l$ such that $X(y_k)Y(y_k) = -X(y_l)Y(y_l) \neq 0$, and so $X(y_k)X(y_l) = -Y(y_k)Y(y_l)$. By the Pivoting Property and Dual Pivoting Property (Theorem 2.54),

$$X(y_k)X(y_l) = \chi_\mathcal{M}(y_k, x_2, \ldots, x_r) \chi_\mathcal{M}(y_l, x_2, \ldots, x_r),$$

and

$$-Y(y_k)Y(y_l) = \chi_{\mathcal{N}}(y_k, y_0, \ldots, \widehat{y_k}, \ldots, \widehat{y_l}, \ldots, y_s)\chi_{\mathcal{N}}(y_l, y_0, \ldots, \widehat{y_k}, \ldots, \widehat{y_l}, \ldots, y_s)$$
$$= (-1)^{k-1}\chi_{\mathcal{N}}(y_0, \ldots, \widehat{y_l}, \ldots, y_s)(-1)^l \chi_{\mathcal{N}}(y_0, \ldots, \widehat{y_k}, \ldots, y_s).$$

Setting $X(y_k)X(y_l) = -Y(y_k)Y(y_l)$ and rearranging terms, we get two nonzero terms of the summation (5.1) with opposite signs.

The proof of the converse largely reverses these steps. □

Corollary 5.5 *If* $\mathcal{M} \to \mathcal{N}$ *then we have the following:*

1. $\mathcal{N}^* \to \mathcal{M}^*$.
2. *For each set A of elements,* $\mathcal{M}/A \to \mathcal{N}/A$.
3. *For each set A of elements,* $\mathcal{M}\backslash A \to \mathcal{N}\backslash A$.

Problem 5.6 Prove that if $\mathcal{M} \to \mathcal{N}$ then $\text{rank}(\mathcal{M}) \geq \text{rank}(\mathcal{N})$, with equality if and only if $\mathcal{M} = \mathcal{N}$.

Proposition 5.7 *The relation* $\mathcal{M} \geq_s \mathcal{N}$ *if and only if* $\mathcal{M} \to \mathcal{N}$ *is a partial order on oriented matroids.*

5.1.1 Realizations of Strong Maps

The following proposition gives a few interpretations of the motivation for the strong map definition.

Proposition 5.8 *Let* $\mathcal{M}$ *and* $\mathcal{N}$ *be oriented matroids on elements* $[n]$. *The following are equivalent. Each implies* $\mathcal{M} \to \mathcal{N}$.

1. *There are matrices M and A such that M is a realization of* $\mathcal{M}$ *and AM is a realization of* $\mathcal{N}$.
2. *There is a vector arrangement* $(\mathbf{v}_i : i \in [n])$ *in a vector space W and a linear map* $f : W \to W'$, *for some vector space* W', *such that* $(\mathbf{v}_i : i \in [n])$ *is a realization of* $\mathcal{M}$ *and* $(f(\mathbf{v}_i) : i \in [n])$ *is a realization of* $\mathcal{N}$.
3. *There are linear subspaces* $W' \subseteq W$ *of* $\mathbb{R}^m$ *such that W is a realization of* $\mathcal{M}$ *and* W' *is a realization of* $\mathcal{N}$.
4. *There is a signed hyperplane arrangement* $(\mathcal{H}_i : i \in [n])$ *in a vector space V and a linear subspace* V' *of V such that* $(\mathcal{H}_i : i \in [n])$ *is a realization of* $\mathcal{M}$ *and* $(\mathcal{H}_i \cap V' : i \in [n])$ *is a realization of* $\mathcal{N}$.

Proof: The equivalence of (1) and (2) is clear.

To see (1)⇔(3): Given M and A, let $W = \text{row}(M)$ and $W' = \text{row}(AM)$. Conversely, given W and W', choose a basis for W' and extend it to a basis for W. Make the elements of this basis the rows of a matrix M, and let A be a

5.1 Strong Maps

matrix (e.g., with a block of 0s and a block $\mathbf{I}$) so that the rows of AM constitute the basis of W'.

To see (3)$\Rightarrow$(4): Given W and W', as we saw in Chapter 1, the signed hyperplane arrangement $(\mathbf{e}_i^\perp \cap W : i \in [n])$ is a realization of $\mathcal{M}$ by a signed hyperplane arrangement, and $(\mathbf{e}_i^\perp \cap W' : i \in [n])$ is a realization of $\mathcal{N}$.

To see (4)$\Rightarrow$(2): Given signed hyperplane arrangements $(\mathcal{H}_i : i \in [n])$ in V realizing $\mathcal{M}$ and $(\mathcal{H}_i \cap V' : i \in [n])$ realizing $\mathcal{N}$, choose vectors $\mathbf{v}_i$ so that $\mathcal{H}_i = \mathbf{v}_i^\perp$ for each i. Thus $(\mathbf{v}_i : i \in [n])$ realizes $\mathcal{M}$. Let $f: V \to V'$ be orthogonal projection. Then $f(\mathbf{v}_i)^\perp = \mathcal{H}_i \cap V'$, so $(f(\mathbf{v}_i) : i \in [n])$ realizes $\mathcal{N}$. □

Definition 5.9 The pairs (M, AM), $((\mathbf{v}_i : i \in [n]), (f(\mathbf{v}_i) : i \in [n]))$, (W, W'), and $((\mathcal{H}_i : i \in [n]), (\mathcal{H}_i \cap V' : i \in [n]))$ of Proposition 5.8 are **matrix realizations**, **vector realizations**, **subspace realizations**, and **hyperplane realizations** of $\mathcal{M} \to \mathcal{N}$. We call each of these a **realization** of $\mathcal{M} \to \mathcal{N}$.

Example 5.10 Let $\mathcal{M}$ be realized by an arrangement $\mathcal{A}$ of vectors in a vector space V. Let $\mathbf{w} \in V$, and let $q: V \to V/\mathbb{R}\mathbf{w}$ be the quotient map. Then $(q(\mathbf{v}) : \mathbf{v} \in \mathcal{A})$ is a realization of a strong map image $\mathcal{N}$ of $\mathcal{M}$.

This is particularly useful when $\mathcal{A}$ defines a set of points in an affine subspace $\mathbb{A}$ of V and $\mathbb{R}\mathbf{w} \cap \mathbb{A} = \{\mathbf{w}'\}$ is nonempty. If we have a picture of the arrangement in $\mathbb{A}$, then we get a picture of the arrangement $(q(\mathbf{v}) : \mathbf{v} \in \mathcal{A})$ by declaring $\mathbf{w}'$ to be the origin.

In other words, given an affine point arrangement, we can produce a quotient arrangement and strong map image by choosing a point in the affine space to be the origin.

Definition 5.11 Let $\mathcal{M}$ and $\mathcal{N}$ be realizable oriented matroids such that $\mathcal{M} \to \mathcal{N}$.

- We say the strong map is **weakly realizable** if it has a realization.
- We say the strong map is $\mathcal{M}$-**realizable** if, for each realization of $\mathcal{M}$, there is a realization of $\mathcal{N}$ such that the pair is a realization of $\mathcal{M} \to \mathcal{N}$.
- We say the strong map is $\mathcal{N}$-**realizable** if, for each realization of $\mathcal{N}$, there is a realization of $\mathcal{M}$ such that the pair is a realization of $\mathcal{M} \to \mathcal{N}$.

A strong map between realizable oriented matroids may not be weakly realizable. We can see an example by thinking once again about Pappus's Theorem (Section 2.7.1). Let $\mathcal{M}$ be the rank 3 oriented matroid realized by the affine signed hyperplane arrangement on the top in Figure 5.1 (not including the dotted curve), and let $\mathcal{N}$ be the rank 2 oriented matroid realized by the signed hyperplane arrangement on the bottom. Then $\mathcal{M} \to \mathcal{N}$ – in fact, $\mathcal{V}^*(\mathcal{N})$

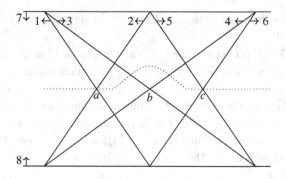

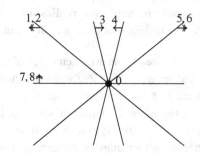

Figure 5.1 A strong map of realizable oriented matroids need not be weakly realizable.

is the subset of $V_*^*(\mathcal{M})$ corresponding to points along the dotted curve in the top picture, together with their negatives, the intersection of hyperplanes 7 and 8 at infinity, and **0**. But Pappus's Theorem tells us that in every realization of $\mathcal{M}$ by an oriented hyperplane arrangement, the intersections indicated by the points a, b, and c are coplanar. For a rank 2 strong map image $\mathcal{N}$ of $\mathcal{M}$ given by a line L added to the top arrangement in Figure 5.1, L intersects a if and only if 1 and 2 are parallel in $\mathcal{N}$, L intersects b if and only if 3 and 4 are parallel in $\mathcal{N}$, and L intersects c if and only if 5 and 6 are parallel in $\mathcal{N}$. Thus for every weakly realizable strong map $\mathcal{M} \to \mathcal{N}'$, if 1 and 2 are parallel in $\mathcal{N}'$ and 5 and 6 are parallel in $\mathcal{N}'$ then 3 and 4 are parallel in $\mathcal{N}'$. In particular, $\mathcal{M} \to \mathcal{N}$ is not weakly realizable.

A strong map $\mathcal{M} \to \mathcal{N}$ that is weakly realizable might not be $\mathcal{M}$-realizable and might not be $\mathcal{N}$-realizable.

To see an example of a weakly realizable strong map that is not $\mathcal{M}$-realizable, let $\mathcal{M}_{\text{hex}}$ be the rank 3 oriented matroid on six elements that

5.1 Strong Maps

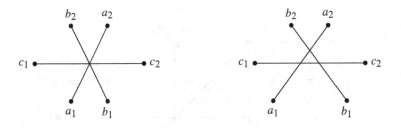

Figure 5.2 Two realizations with different realizable strong maps.

is realized by either of the affine point arrangements in Figure 5.2. Let **w** be the vector in $\mathbb{R}^3$ from **0** to the intersection of the three line segments shown in the left arrangement. Let $\mathcal{N}$ be the rank 2 oriented matroid given by the image of the left arrangement under the quotient map $\mathbb{R}^3 \to \mathbb{R}^3/\mathbb{R}\mathbf{w}$ (as in Example 5.10). Thus $\mathcal{M}_{\text{hex}} \to \mathcal{N}$, and in $\mathcal{N}$ we have parallelism classes $\{a_1, a_2\}$, $\{b_1, b_2\}$, and $\{c_1, c_2\}$. No rank 2 linear projection of the arrangement on the right will realize an oriented matroid with these parallelism classes.

To see an example of a weakly realizable strong map that is not $\mathcal{N}$-realizable, let $\mathcal{M}$ be the rank 3 oriented matroid on seven elements $\{a_1, a_2, b_1, b_2, c_1, c_2, \infty\}$ that is realized by the affine line arrangement given by solid lines in the top of Figure 5.3, together with an additional element at infinity. Let $\mathcal{N}$ be the strong map image given by the dotted line. At the bottom of Figure 5.3 is a realization of $\mathcal{N}$, but this realization can't arise from a realization of $\mathcal{M}$. To see this, consider a realization of $\mathcal{M}$ by hyperplanes, treat the hyperplane H_∞^0 as the hyperplane at infinity, and consider the resulting affine line arrangement in a plane A parallel to H_∞^0. In this line arrangement the lines corresponding to a_1 and a_2 must be parallel, because $\{a_1, a_2, \infty\}$ is dependent in $\mathcal{M}$, and likewise the lines corresponding to b_1 and b_2 must be parallel. If V is a line in A giving a strong map image containing the cocircuits of $\mathcal{N}$ with supports $\{b_1, b_2, c_1, c_2\}$ and $\{a_1, a_2, b_1, b_2, \infty\}$, then V is parallel to the lines corresponding to a_1 and a_2 and contains the intersection P of the lines corresponding to c_1 and c_2. In addition, basic Euclidean gemonetry tells us that the lines corresponding to b_1 and b_2 will intersect V at points equidistant from P. Consider the implications of this for the realization of $\mathcal{N}$ in the plane spanned by V. An affine line in this plane parallel to the line labeled ∞ will intersect the lines labeled b_1, c_1/c_2, and b_2 at points B_1, C, and B_2, and our parallelism argument says that B_1 and B_2 must be equidistant from C. Thus the hyperplane arrangement arising from V can't be as shown at the bottom of Figure 5.3.

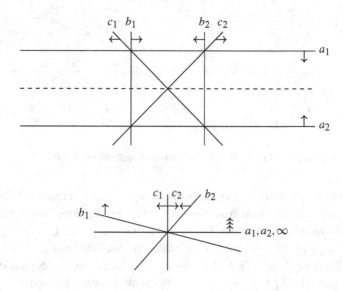

Figure 5.3 A weakly realizable strong map that is not $\mathcal{N}$-realizable.

Problem 5.12 Give an explanation in terms of hyperplane arrangements of the example illustrated by Figure 5.2, and give an explanation in terms of vector arrangements of the example illustrated by Figure 5.3.

5.2 Weak Maps

Consider a vector arrangement in $\mathbb{R}^n$ with oriented matroid $\mathcal{N}$, and consider the effect on $\mathcal{N}$ of moving the vectors very slightly. The movement is slight enough that, while some vectors that were in a common hyperplane initially are no longer so after the movement, every two vectors that were initially separated by a hyperplane remain separated. This is the idea behind *weak maps*.

Definition 5.13 Let $\mathcal{M}$ and $\mathcal{N}$ be oriented matroids with the same elements. We say there is a **weak map** from $\mathcal{M}$ to $\mathcal{N}$, denoted $\mathcal{M} \rightsquigarrow \mathcal{N}$, if for every $Y_2 \in \mathcal{V}^*(\mathcal{N})$ there is some $Y_1 \in \mathcal{V}^*(\mathcal{M})$ with $Y_2 \leq Y_1$. We also say $\mathcal{N}$ is a **weak map image** of $\mathcal{M}$.

A weak map $\mathcal{M} \rightsquigarrow \mathcal{N}$ is **rank-preserving** if $\text{rank}(\mathcal{M}) = \text{rank}(\mathcal{N})$.

Example 5.14 Figure 5.4 shows oriented matroids $\mathcal{M}_3 \rightsquigarrow \mathcal{M}_2 \rightsquigarrow \mathcal{M}_1 \rightsquigarrow \mathcal{M}_0$ given by vector arrangements. (In $\mathcal{M}_1$ element 3 is a loop, and in $\mathcal{M}_0$ both 2 and 3 are loops.)

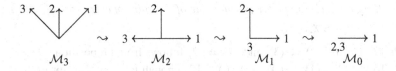

Figure 5.4 Three weak maps.

Example 5.15 If $\mathcal{M}$ and $\mathcal{N}$ are oriented matroids on the same set of elements, A is the set of nonloops of $\mathcal{N}$, and $\mathcal{M}(A) = \mathcal{N}(A)$, then $\mathcal{M} \rightsquigarrow \mathcal{N}$. In other words, one way to make a weak map image of $\mathcal{M}$ is to make some nonloops into loops.

The relation $\mathcal{M} \geq_w \mathcal{N}$ if and only if $\mathcal{M} \rightsquigarrow \mathcal{N}$ is a partial order, and if $\mathcal{M} \geq_s \mathcal{N}$ then $\mathcal{M} \geq_w \mathcal{N}$. The poset of all rank r oriented matroids on elements $[n]$, ordered by weak maps, is called the *MacPhersonian* MacP(r, n). (Israel Gel'fand proposed the name in honor of Robert MacPherson.) Its connection to the Grassmannian $G(r, \mathbb{R}^n)$ will be explored in Chapter 10.

As with strong maps, the "map" terminology is a holdover from unoriented matroid theory, and if we were making terminology from scratch, we would probably call $\geq_w$ the "weak partial order."

Proposition 5.16 Let $\mathcal{M} \rightsquigarrow \mathcal{N}$.

1. $\mathcal{M} \backslash e \rightsquigarrow \mathcal{N} \backslash e$.
2. If e is not a loop of $\mathcal{N}$ then $\mathcal{M}/e \rightsquigarrow \mathcal{N}/e$.

Proof: The first statement is clear. To see the second statement, let $Y_2 \in \mathcal{V}^*(\mathcal{N}/e)$. Then $Y_2 e^0 \in \mathcal{V}^*(\mathcal{N})$, and since e is not a loop of $\mathcal{N}$ there is a $Z \in \mathcal{V}^*(\mathcal{N})$ such that $Z(e) = +$. Consider the two covectors $(Y_2 e^0) \circ Z$ and $(Y_2 e^0) \circ (-Z)$ of $\mathcal{N}$. Since $\mathcal{M} \rightsquigarrow \mathcal{N}$ there are covectors X_+, X_- of $\mathcal{M}$ such that $X_+ \geq Y_2 \circ Z$ and $X_- \geq Y_2 \circ (-Z)$. Eliminating e between X_+ and X_- gives a $Y_1 e^0 \in \mathcal{V}^*(\mathcal{M})$ with $Y_1 \geq Y_2$. □

Proposition 5.17 (Björner et al. 1999) *Let $\mathcal{M}$ and $\mathcal{N}$ be oriented matroids with elements E. The following are equivalent:*

1. $\mathcal{M} \rightsquigarrow \mathcal{N}$, that is, for every $Y_2 \in \mathcal{V}^*(\mathcal{N})$ there is some $Y_1 \in \mathcal{V}^*(\mathcal{M})$ with $Y_2 \leq Y_1$.
2. For every $X_1 \in \mathcal{V}(\mathcal{M}) \backslash \{\mathbf{0}\}$ there is some $X_2 \in \mathcal{V}(\mathcal{N}) \backslash \{\mathbf{0}\}$ with $X_2 \leq X_1$.
3. For every $X_1 \in \mathcal{C}(\mathcal{M})$ there is some $X_2 \in \mathcal{C}(\mathcal{N})$ with $X_2 \leq X_1$.

If rank$(\mathcal{M})$ = rank$(\mathcal{N})$, *then each of the above statements is equivalent to:*

4. If $\chi_\mathcal{M}$ is a chirotope for $\mathcal{M}$ then one of the chirotopes $\chi_\mathcal{N}$ for $\mathcal{N}$ satisfies $\chi_\mathcal{N} \leq \chi_\mathcal{M}$.

Proof: Clearly (2) $\Rightarrow$ (3), and (3) $\Rightarrow$ (2) follows from Proposition 2.19.

To see (2)$\Rightarrow$(1): Let $Y_2 \in \mathcal{V}^*(\mathcal{N})$. It's enough to show that, for each $j \in \mathrm{supp}(Y_2)$, there is a $Y_{1,j} \in \mathcal{V}^*(\mathcal{M})$ such that $Y_{1,j}(j) = Y_2(j)$ and $Y_{1,j}(e) \leq Y_2(e)$ for all $e \in \mathrm{supp}(Y_2)$; if these exist then we can compose them to get the desired Y_1. So assume by way of contradiction that some $Y_{1,j}$ does not exist. Applying the Farkas Property (Definition 2.28) to $\mathcal{M}$ with the sign vector $Y_2 = A^+ B^- (C \dot\cup \emptyset)^0$, the nonexistence of $Y_{1,j}$ implies the existence of some $X_1 \in \mathcal{V}(\mathcal{M})$ such that $X_1(j) \neq 0$ and $X_1 \leq Y_2$. Thus by (2) there is an $X_2 \in \mathcal{V}(\mathcal{N})\setminus\{0\}$ with $X_2 \leq X_1 \leq Y_2$, and so $X_2 \not\perp Y_2$, a contradiction.

(1) $\Rightarrow$ (2) also follows from the Farkas Property. Let $X_1 = A^+ B^- C^0 \in \mathcal{V}(\mathcal{M})\setminus\{0\}$. We need to show there is a $j \in A \cup B$ and $X_2 \in \mathcal{V}(\mathcal{N})$ such that $X_2 \leq X_1$ and $X_2(j) \neq 0$. Assume by way of contradiction that for each j there is no such X_2. Applying Farkas to $\mathcal{N}$ with the sign vector $X_1 = A^+ B^- (C \dot\cup \emptyset)^0$, we see that for each $j \in A \cup B$ there is a $Y_{2,j} \in \mathcal{V}^*(\mathcal{N})$ such that $Y_{2,j}(e) \leq X_1(e)$ for each $e \in A \cup B$ and $Y_{2,j}(j) = X_1(j)$. A composition of all the $Y_{2,j}$, in whatever order is a $Y_2 \in \mathcal{V}^*(\mathcal{N})$ such that $Y_2(e) = X_1(e)$ for every $e \in A \cup B$. By (1) there is a $Y_1 \in \mathcal{V}^*(\mathcal{M})$ such that $Y_1 \geq Y_2 \geq X_1$. But then $X_1 \cdot Y_1 = \{+\}$, contradicting orthogonality of $X_1 \in \mathcal{V}(\mathcal{M})$ and $Y_1 \in \mathcal{V}^*(\mathcal{M})$.

Thus we have equivalence of the first three properties. Now assume $\mathrm{rank}(\mathcal{M}) = \mathrm{rank}(\mathcal{N})$.

To see (3)$\Rightarrow$(4): From (3) we see that every basis of $\mathcal{N}$ is a basis of $\mathcal{M}$. Choose a basis B_0 of $\mathcal{N}$, and then choose chirotopes $\chi_\mathcal{M}$ for $\mathcal{M}$ and $\chi_\mathcal{N}$ for $\mathcal{N}$ to coincide on B_0. Now consider another basis B of $\mathcal{N}$. By the Basis Exchange Axiom, there is a sequence $B_0, B_1, \ldots, B_k =: B$ of bases of $\mathcal{N}$ such that successive terms differ by a single element. That is, for each i there is a set H and elements e, f so that $B_i = H \cup \{e\}$ and $B_{i+1} = H \cup \{f\}$.

These bases of $\mathcal{N}$ are also bases of $\mathcal{M}$, and so there is a circuit $X_1 \in \mathcal{C}(\mathcal{M})$ such that $\{e, f\} \subseteq \mathrm{supp}(X) \subseteq \{e, f\} \cup H$. By (3) there is an $X_2 \in \mathcal{C}(\mathcal{N})$ such that $X_2 \leq X_1$, but since B_i and B_{i+1} are bases of $\mathcal{N}$ we know $\{e, f\} \subseteq \mathrm{supp}(X_2)$.

The Pivoting Property says, for each ordering $\{x_2, \ldots, x_r\}$ of H,

$$\chi_\mathcal{M}(e, x_2, \ldots, x_r) = -X_1(e) X_1(f) \chi_\mathcal{M}(f, x_2, \ldots, x_r),$$

and

$$\chi_\mathcal{N}(e, x_2, \ldots, x_r) = -X_2(e) X_2(f) \chi_\mathcal{N}(f, x_2, \ldots, x_r).$$

But X_1 and X_2 coincide on $\{e, f\}$. Thus, since $\chi_\mathcal{M}$ and $\chi_\mathcal{N}$ coincide on B_0, by following basis exchanges we see that they coincide on all bases of $\mathcal{N}$.

To see (4)⇒(3), we induct on the number of elements. The base case is when $\mathcal{M} = \mathcal{N}$ is the coordinate oriented matroid on E.

Above the minimal case, let $X_1 \in \mathcal{C}(\mathcal{M})$.

Case 1: If there is a $e \in E \backslash \text{supp}(X_1)$ then we apply the induction hypothesis to $\mathcal{M} \backslash e$ and $\mathcal{N} \backslash e$ to get $X_2 \in \mathcal{C}(\mathcal{N} \backslash e)$ such that $X_2 \leq X_1 \backslash e$. Then $X_2 e^0 \in \mathcal{C}(\mathcal{N})$ and $X_2 e^0 \leq X_1$.

Case 2: Assume $\text{supp}(X_1) = E$. If $\mathcal{N}$ has a loop e, then $e^{X_1(e)} \in \mathcal{C}(\mathcal{N})$ and $e^{X_1(e)} \leq X_1$. Otherwise let $X_2 \in \mathcal{C}(\mathcal{N})$ and $e \in \text{supp}(X_2)$. By Symmetry we may assume $X_2(e) = X_1(e)$. Since $\mathcal{M}$ is rank $|E| - 1$, so is $\mathcal{N}$, and so for each $f \in \text{supp}(X_2) \backslash e$ we have that $E \backslash e$ and $E \backslash f$ are bases of $\mathcal{N}$. Let $E \backslash \{e, f\} = \{x_2, \ldots, x_n\}$. By the Pivoting Property we see that

$$X_2(e)X_2(f) = -\chi_\mathcal{N}(e, x_2, \ldots, x_n)\chi_\mathcal{N}(f, x_2, \ldots, x_n) \neq 0$$
$$= -\chi_\mathcal{M}(e, x_2, \ldots, x_n)\chi_\mathcal{M}(f, x_2, \ldots, x_n)$$
$$= X_1(e)X_2(f),$$

and so $X_2(f) = X_1(f)$ for all $f \in \text{supp}(X_2)$. □

Corollary 5.18 *If* $\mathcal{M} \rightsquigarrow \mathcal{N}$ *then* $\text{rank}(\mathcal{M}) \geq \text{rank}(\mathcal{N})$.
If $\mathcal{M} \rightsquigarrow \mathcal{N}$ *and* $\mathcal{M}$ *and* $\mathcal{N}$ *have the same rank then* $\mathcal{M}^* \rightsquigarrow \mathcal{N}^*$.

The proof of the last statement follows from the last part of Proposition 5.17 and Exercise 2.3.

5.2.1 Realizations of Weak Maps

Definition 5.19 Let $\mathcal{M}$ and $\mathcal{N}$ be realizable oriented matroids such that $\mathcal{M} \rightsquigarrow \mathcal{N}$.

- We say the weak map is **weakly realizable** if there is a realization $(\mathbf{w}_e : e \in E)$ of $\mathcal{N}$ in $\mathbb{R}^{\text{rank}(\mathcal{M})}$ such that, for each $\epsilon > 0$, there is a realization $(\mathbf{v}_e : e \in E)$ of $\mathcal{M}$ in $\mathbb{R}^{\text{rank}(\mathcal{M})}$ such that $\|\mathbf{v}_e - \mathbf{w}_e\| < \epsilon$ for each e.
- We say the weak map is **strongly realizable** if, for each realization $(\mathbf{w}_e : e \in E)$ of $\mathcal{N}$ in $\mathbb{R}^{\text{rank}(\mathcal{M})}$ and each $\epsilon > 0$, there is a realization $(\mathbf{v}_e : e \in E)$ of $\mathcal{M}$ in $\mathbb{R}^{\text{rank}(\mathcal{M})}$ such that $\|\mathbf{v}_e - \mathbf{w}_e\| < \epsilon$ for each e.

A weak map between realizable oriented matroids is not necessarily weakly realizable! That is, if $\mathcal{M} \rightsquigarrow \mathcal{N}$ are two realizable oriented matroids, it's not necessarily true that there is a realization of $\mathcal{N}$ that can be perturbed

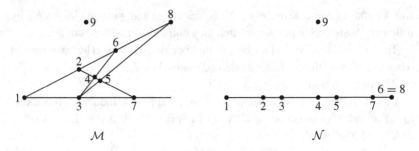

Figure 5.5 A weak map that is not weakly realizable.

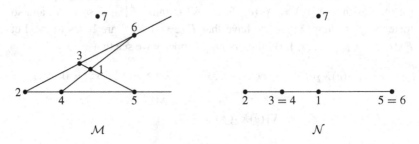

Figure 5.6 A weakly realizable but not strongly realizable weak map.

an arbitrarily small amount to get a realization of $\mathcal{M}$. In fact, it's not even true when $\mathcal{M}$ and $\mathcal{N}$ have the same rank. For instance, consider the rank 3 oriented matroids $\mathcal{M}$ and $\mathcal{N}$ shown in Figure 5.5. Certainly $\mathcal{M} \rightsquigarrow \mathcal{N}$. In every sufficiently small perturbation of a realization of $\mathcal{N}$ the distance between points 4 and 5 is greater than the distance between points 6 and 8, and hence the perturbed arrangement cannot realize $\mathcal{M}$.

A *weakly realizable weak map* $\mathcal{M} \to \mathcal{N}$ *between realizable oriented matroids may not be strongly realizable.* Again, this can happen even when both oriented matroids are of the same rank. Consider the two rank 3 oriented matroids represented by affine point arrangements in Figure 5.6. A realization of $\mathcal{N}$ as an affine point arrangement $(p_1, \ldots, p_7)$ can be perturbed to a realization of $\mathcal{M}$ by an arbitrarily small perturbation if and only if p_1 is the midpoint of the line segment $\overline{p_4 p_5}$.

5.2.2 Rank-Preserving Weak Maps and Realization Spaces

Our observations on realizability of weak maps offer some insight on realization spaces of oriented matroids. Realization spaces will be discussed in more depth in Chapter 7, where we'll see that there are several closely related but

5.2 Weak Maps

distinct notions of "realization space." For now it's most convenient to work with *vector realization spaces*, but all of our observations here generalize easily to other types of realization space.

Definition 5.20 Let $\mathcal{M}$ be a rank r oriented matroid on elements $[n]$. The *vector realization space* $\mathrm{Vreal}(\mathcal{M})$ is the set of all $r \times n$ matrices of rank r realizing $\mathcal{M}$. $\mathrm{Vreal}(\mathcal{M})$ is topologized as a subspace of $\mathbb{R}^{r \times n}$.

Proposition 5.21 *Let $\mathcal{M}$ and $\mathcal{N}$ be realizable rank r oriented matroids on elements $[n]$. If $\overline{\mathrm{Vreal}(\mathcal{M})} \cap \mathrm{Vreal}(\mathcal{N}) \neq \emptyset$ then $\mathcal{M} \rightsquigarrow \mathcal{N}$.*

This may be most easily seen by way of chirotopes. Let $\binom{[n]}{r}$ denote the set of r-tuples $(i_1, \ldots, i_r)$ of integers with $1 \leq i_1 < i_2 < \cdots < i_r \leq n$. For a chirotope $\chi: [n]^r \to \{0, +, -\}$, let $\hat{\chi}$ denote its restriction to $\binom{[n]}{r}$. For a function $\rho: \binom{[n]}{r} \to \{0, +, -\}$, let $\mathcal{O}_\rho$ denote the orthant $\{\mathbf{x} \in \mathbb{R}^{\binom{[n]}{r}} : \forall (i_1, \ldots, i_r)\ \mathrm{sign}(x_{i_1, \ldots, i_r}) = \rho(i_1, \ldots, i_r)\}$.

Consider the function

$$d: \mathbb{R}^{r \times n} \to \mathbb{R}^{\binom{[n]}{r}}$$

$$M \to (\det(M_{i_1, \ldots, i_r}) : 1 \leq i_1 < i_2 < \cdots < i_r \leq n).$$

An oriented matroid with chirotope χ has vector realization space $d^{-1}(\mathcal{O}_{\hat{\chi}} \cup \mathcal{O}_{-\hat{\chi}})$. Since d is continuous, $\overline{\mathrm{Vreal}(\mathcal{M})} \subseteq d^{-1}(\overline{\mathcal{O}_{\hat{\chi}_{\mathcal{M}}} \cup \mathcal{O}_{-\hat{\chi}_{\mathcal{M}}}})$. However,

$$\overline{\mathcal{O}_{\hat{\chi}_{\mathcal{M}}} \cup \mathcal{O}_{-\hat{\chi}_{\mathcal{M}}}} = \overline{\mathcal{O}_{\hat{\chi}_{\mathcal{M}}}} \cup \overline{\mathcal{O}_{-\hat{\chi}_{\mathcal{M}}}}$$

$$= \bigcup_{\rho \leq \hat{\chi}_{\mathcal{M}}} \mathcal{O}_\rho \cup \bigcup_{\rho \leq -\hat{\chi}_{\mathcal{M}}} \mathcal{O}_\rho$$

$$= \bigcup_{\rho \leq \hat{\chi}_{\mathcal{M}}} (\mathcal{O}_\rho \cup \mathcal{O}_{-\rho}).$$

Since $\mathcal{N}$ has a vector realization in this space, we see $\mathcal{N}$ has a chirotope $\chi_{\mathcal{N}}$ with $\hat{\chi}_{\mathcal{N}} = \rho$ for some $\rho \leq \hat{\chi}_{\mathcal{M}}$.

Proposition 5.22 *Let $\mathcal{M} \rightsquigarrow \mathcal{N}$ be a weak map of realizable rank r oriented matroids on elements $[n]$.*

1. *$\mathcal{M} \rightsquigarrow \mathcal{N}$ is weakly realizable if and only if $\overline{\mathrm{Vreal}(\mathcal{M})} \cap \mathrm{Vreal}(\mathcal{N}) \neq \emptyset$.*
2. *$\mathcal{M} \rightsquigarrow \mathcal{N}$ is strongly realizable if and only if $\mathrm{Vreal}(\mathcal{N}) \subseteq \overline{\mathrm{Vreal}(\mathcal{M})}$.*

A partition $\{P_i : i \in I\}$ of a topological space is *normal* if $P_i \subseteq \overline{P_j}$ whenever $P_i \cap \overline{P_j} \neq \emptyset$. For instance, the set of open cells in a regular cell decomposition of a space is a normal partition. Our examples in Section 5.2.1 of weak maps that are not strongly realizable demonstrate that the partition of $\mathrm{Mat}(3, n)$ into realization spaces is not normal whenever $n \geq 7$.

5.3 Topological Representations of Maps

Not only does every oriented matroid have a topological representation, but each strong map and each weak map has a topological representation as well. Topological representations of $\mathcal{M}$ in S^{r-1} and of $\mathcal{N}$ in S^{s-1} induce cell decompositions of both spheres. A strong map $\mathcal{M} \to \mathcal{N}$ is represented by an inclusion $S^{s-1} \to S^{r-1}$ taking cells into cells, and a weak map $\mathcal{M} \rightsquigarrow \mathcal{N}$ is represented by a map $S^{r-1} \to S^{s-1}$ taking cells to cells. These results will be much easier to prove than the Topological Representation Theorem, but the discussion will involve the topology of order complexes. We begin by reviewing this topology.

The reader less interested in topology may skip this section – it won't be essential to most of the upcoming chapters.

5.3.1 Some Homotopy Results on Order Complexes

Recall that the order complex ΔP of a poset P is the abstract simplicial complex consisting of all chains in P, and $\|\Delta P\|$ denotes a geometric realization of ΔP.

If a poset P contains an element p that is comparable to every element, then every maximal chain in P contains p. Thus ΔP is a cone with apex p, and $\|\Delta P\|$ is contractible. This is the case, for instance, if P has a unique maximum or a unique minimum.

More generally, if $P = P_1 \dot\cup P_2$ is a poset and every element of P_1 is less than every element of P_2 then ΔP is the join $\Delta P_1 * \Delta P_2$.

The *Contractible Carrier Lemma* (Lemma 5.24) leads to various results on homotopy types of poset maps.

Definition 5.23 Let X be an abstract simplicial complex, T a topological space, and $\mathcal{P}(T)$ the power set of T. A function $C: X \to \mathcal{P}(T)$ is a **contractible carrier** if

1. for each $\sigma \in X$, $C(\sigma)$ is contractible, and
2. $C(\tau) \subseteq C(\sigma)$ whenever $\tau \subseteq \sigma$.

A function $f: \|X\| \to T$ is **carried** by C if $f(\|\sigma\|) \subseteq C(\sigma)$ for every σ.

Lemma 5.24 (Contractible Carrier Lemma) (cf. Walker 1981) *Let $C: X \to T$ be a contractible carrier.*

1. *There is a function $\|X\| \to T$ carried by C.*
2. *Every two functions carried by C are homotopic. If Y is a subcomplex of X and f and g are carried by C and coincide on Y then there is a homotopy from f to g that fixes Y.*

Proof: Every continuous map from the boundary of a closed ball to a contractible space can be extended to the entire ball. Thus a function $\|X\| \to T$ carried by C can be constructed by induction on the skeleta of X. Also, $\|X\| \times I$ has a regular cell decomposition, with maximal cells of the form $\|\sigma\| \times I$ for each maximal element σ of X. So, given functions f and g carried by C, we can construct a homotopy $H: \|X\| \times I \to T$ from f to g by inductively defining H on cells $\|\sigma\| \times I$ of increasing dimension. □

The following result is sometimes called the *Order Homotopy Lemma*.

Proposition 5.25 *Let P and Q be posets, and let f and g be poset maps from P to Q. If $f(x) \leq g(x)$ for every x then $\|\Delta f\| \simeq \|\Delta g\|$. If in addition f and g coincide on $R \subset P$ then there is a homotopy from $\|\Delta f\|$ to $\|\Delta g\|$ that fixes $\|\Delta R\|$.*

Proof: For each $\sigma \in \Delta P$, let $C(\sigma) = \|\Delta Q_{\geq f(\min \sigma)}\|$. Then C is a contractible carrier carrying both $\|\Delta f\|$ and $\|\Delta g\|$. □

The following theorem is one of the most important tools we have to show two posets to be homotopy equivalent. It first appeared in greater generality, as a proof on nerves of categories (Quillen 1973). We give a proof due to Walker (1981).

Theorem 5.26 (Quillen's Theorem A) *If $f: P \to Q$ is a poset map and $\|\Delta f^{-1}(Q_{\leq q})\|$ is contractible for each $q \in Q$ then $\|\Delta f\|$ is a homotopy equivalence.*

Proof: For each $\sigma \in \Delta Q$, let $C(\sigma) = \|\Delta f^{-1}(Q_{\leq \max \sigma})\|$. Then C is a contractible carrier, so there is a $g: \|\Delta Q\| \to \|\Delta P\|$ carried by C.

Let $C_P: \Delta P \to \mathcal{P}(\|\Delta P\|)$ be the map $C_P(\sigma) = \|\Delta(P_{\leq \max \sigma})\|$. Then C_P is a contractible carrier for both $\|\Delta g \circ \Delta f\|$ and the identity map, so these maps are homotopic. Likewise let $C_Q: \Delta Q \to \mathcal{P}(\|\Delta Q\|)$ be the map $C_Q(\sigma) = \|\Delta(Q_{\leq \max \sigma})\|$. Then C_Q is a contractible carrier for both $\|\Delta f \circ \Delta g\|$ and the identity map, so these maps are homotopic. □

Remark 5.27 1. We could as well have taken the hypothesis to be that $\|\Delta f^{-1}(Q_{\geq q})\|$ is contractible for each $q \in Q$.
2. A charming proof of Barmak (2011) shows that $\|\Delta f\|$ is a simple homotopy equivalence.

5.3.2 Topological Representation of Strong Maps

Notation 5.28 If $\mathcal{S} = (S, S^+, S^-)$ is a signed pseudosphere in S^{n-1} and $f: T \to S^{n-1}$ is a function then $f^{-1}(\mathcal{S})$ denotes the triple $(f^{-1}S, f^{-1}S^+, f^{-1}S^-)$.

Proposition 5.29 *Let $\mathcal{M}$ be an oriented matroid of rank m and $\mathcal{N}^r$ an oriented matroid of rank n. The following are equivalent:*

1. $\mathcal{M} \to \mathcal{N}$.
2. *For every topological representation $(\mathcal{S}_e : e \in E)$ of $\mathcal{M}$ in S^{m-1} there exists an embedding $\eta \colon S^{n-1} \to S^{m-1}$ such that $(\eta^{-1}(\mathcal{S}_e) : e \in E)$ is a topological representation of $\mathcal{N}$.*
3. *There exists a topological representation $(\mathcal{S}_e : e \in E)$ of $\mathcal{M}$ in S^{m-1} and an embedding $\eta \colon S^{n-1} \to S^{m-1}$ such that $(\eta^{-1}(\mathcal{S}_e) : e \in E)$ is a topological representation of $\mathcal{N}$.*

Proof: The only implication that requires much thought is (1)⇒(2).

Given a topological representation $(\mathcal{S}_e : e \in E)$ of $\mathcal{M}$ in S^{m-1}, recall $\mathcal{V}^*(\mathcal{M})\setminus\{0\}$ is isomorphic to the poset of cells in the cell decomposition of S^{m-1} induced by $(\mathcal{S}_e : e \in E)$. Proposition 4.18 tells us that $\Delta(\mathcal{V}^*(\mathcal{M})\setminus\{0\})$ is the abstract simplicial complex of a barycentric subdivision of this cell complex, and so there is a homeomorphism $h_{\mathcal{M}} \colon S^{m-1} \to \|\Delta(\mathcal{V}^*(\mathcal{M})\setminus\{0\})\|$ taking each closed cell to the order complex of the appropriate subposet. Similarly, every topological representation of $\mathcal{N}$ induces a homeomorphism $h_{\mathcal{N}} \colon S^{n-1} \to \|\Delta(\mathcal{V}^*(\mathcal{N})\setminus\{0\})\|$. The composition

$$S^{n-1} \stackrel{h_{\mathcal{N}}}{\to} \|\Delta(\mathcal{V}^*(\mathcal{N})\setminus\{0\})\| \hookrightarrow \|\Delta(\mathcal{V}^*(\mathcal{M})\setminus\{0\})\| \stackrel{h_{\mathcal{M}}^{-1}}{\to} S^{m-1}$$

is the desired embedding. □

5.3.3 Topological Representations of Weak Maps

Proposition 5.30 (Anderson 2001) *Let $\mathcal{M}_1$ and $\mathcal{M}_2$ be oriented matroids with the same elements. The following are equivalent:*

1. $\mathcal{M}_1 \rightsquigarrow \mathcal{M}_2$.
2. *There is a poset map $g \colon \mathcal{V}(\mathcal{M}_1)\setminus\{0\} \to \mathcal{V}(\mathcal{M}_2)\setminus\{0\}$ such that $g(X) \leq X$ for every X.*
3. *The collection of all g as above has a unique maximal element.*
4. *There is a surjective poset map $g^* \colon \mathcal{V}^*(\mathcal{M}_1) \to \mathcal{V}^*(\mathcal{M}_2)$ such that $g^*(Y) \leq Y$ for every Y.*
5. *The collection of all g^* as above has a unique maximal element.*

If $\text{rank}(\mathcal{M}_1) = \text{rank}(\mathcal{M}_2)$, then the unique maximal g is surjective and the unique maximal g^* restricts to a surjective map $: \mathcal{V}^*(\mathcal{M}_1)\setminus\{0\} \to \mathcal{V}^*(\mathcal{M}_2)\setminus\{0\}$.

This extends the Topological Representation Theorem to a representation theorem for rank-preserving weak maps. A poset map $\mathcal{V}^*(\mathcal{M}_1)\setminus\{0\} \to \mathcal{V}^*(\mathcal{M}_2)\setminus\{0\}$ induces a simplicial map $\Delta\mathcal{V}^*(\mathcal{M}_1)\setminus\{0\} \to \Delta\mathcal{V}^*(\mathcal{M}_2)\setminus\{0\}$.

5.3 Topological Representations of Maps 151

Also, $\Delta \mathcal{V}^*(\mathcal{M}_i) \setminus \{0\}$ is the abstract simplicial complex of a barycentric subdivision of the cell decomposition of the sphere in a topological representation of $\mathcal{M}_i$. So a rank-preserving weak map of oriented matroids induces a topological map $\|\Delta g^*\|$ of their topological representations. Further, since there is a maximal g^* for each weak map, by the Order Homotopy Lemma (Proposition 5.25) this topological map is unique up to homotopy equivalence.

Proof: Obviously (3) $\Rightarrow$ (2), and the $\mathcal{V}$ characterization of weak maps in Proposition 5.17 shows that (2) $\Rightarrow$ (1). If $\mathcal{M}_1 \rightsquigarrow \mathcal{M}_2$, we get (3) as follows. By the $\mathcal{V}$ characterization of weak maps in Proposition 5.17, we know that for every $X_1 \in \mathcal{V}(\mathcal{M}_1)$ the set $G_{X_1} := \{X \in \mathcal{V}(\mathcal{M}_2) : X \leq X_1\}$ is nonempty. Further, for every two elements X and X' of G_{X_1}, we have $X \circ X' = X' \circ X \in G_{X_1}$. Thus the composition of all elements of G_{X_1}, in whatever order, is the unique maximal element of G_{X_1}. Define $g \colon \mathcal{V}(\mathcal{M}_1) \to \mathcal{V}(\mathcal{M}_2)$ by $g(X) = \max(G_X)$ for every X. This is clearly the maximal map $\mathcal{V}(\mathcal{M}_1) \to \mathcal{V}(\mathcal{M}_2)$ with the property that $g(X) \leq X$ for every X. It's also a poset map: If $X < X'$ are two elements of $\mathcal{V}(\mathcal{M}_1)$ then $G_X \subseteq G_{X'}$, and so $\max(G_X) \leq \max(G_{X'})$. Thus we have (3)$\Rightarrow$(2)$\Rightarrow$(1)$\Rightarrow$(3).

Also, if (4) holds then $\mathcal{M}_1$ and $\mathcal{M}_2$ satisfy the $\mathcal{V}^*$ characterization of weak maps. Conversely, assume $\mathcal{M}_1 \rightsquigarrow \mathcal{M}_2$. For each $Y_1 \in \mathcal{V}^*(\mathcal{M}_1)$ define $G^*_{Y_1} = \{Y \in \mathcal{V}^*(\mathcal{M}_2) : Y \leq Y_1\}$. Then $G^*_{Y_1} \neq \emptyset$ since $0 \in G^*_{Y_1}$, and as in our previous discussion, $G^*_{Y_1}$ has a unique maximal element. Define $g^* \colon \mathcal{V}(\mathcal{M}_1) \to \mathcal{V}(\mathcal{M}_2)$ by $g^*(X) = \max(G^*_X)$. By the same argument as before, we see this is a poset map.

To see that g^* is surjective, induct on $\mathrm{rank}(\mathcal{M}_2)$. First note that in every rank, every $Y_2 \in \mathcal{V}^*(\mathcal{M}_2)$ is in some $G^*_{Y_1}$. If Y_2 is a tope, then it's $g^*(Y_1)$. In particular, g^* is surjective when $\mathrm{rank}(\mathcal{M}_2) = 1$. For larger rank, consider a nontope $Y_2 \in \mathcal{V}^*(\mathcal{M}_2)$. Then $Y_2(e) = 0$ for some nonloop e. So apply the induction hypothesis to $\mathcal{M}_1/e \rightsquigarrow \mathcal{M}_2/e$ to see Y_2 is the maximal element of $g^*(Y_1)$ for some $Y_1 \in \mathcal{V}^*(\mathcal{M}_1)$ with $Y_1(e) = 0$. Thus $g^*(Y_1) = Y_2$.

If $\mathrm{rank}(\mathcal{M}_1) = \mathrm{rank}(\mathcal{M}_2)$, then $\mathcal{M}_1 \rightsquigarrow \mathcal{M}_2$ if and only if $\mathcal{M}_1^* \rightsquigarrow \mathcal{M}_2^*$, and the maximal map g for the relation $\mathcal{M}_1 \rightsquigarrow \mathcal{M}_2$ is the maximal map g^* for the relation $\mathcal{M}_1^* \rightsquigarrow \mathcal{M}_2^*$. Thus we have the final statement. $\square$

Example 5.31 Figure 5.7 shows a rank-preserving weak map $\mathcal{M} \rightsquigarrow \mathcal{N}$ and some preimages of covectors under the map $\mathcal{V}^*(\mathcal{M})\setminus\{0\} \to \mathcal{V}^*(\mathcal{N})\setminus\{0\}$ described in the last part of Proposition 5.30.

Proposition 5.32 (Anderson 2001) *Let $\mathcal{M} \rightsquigarrow \mathcal{N}$ be a rank-preserving weak map and $g^* \colon \mathcal{V}^*(\mathcal{M})\setminus\{0\} \to \mathcal{V}^*(\mathcal{N})\setminus\{0\}$ as in Proposition 5.30. Then*

$$\|\Delta g^*\| \colon \|\Delta(\mathcal{V}^*(\mathcal{M})\setminus\{0\})\| \to \|\Delta(\mathcal{V}^*(\mathcal{N})\setminus\{0\})\|$$

is a homotopy equivalence.

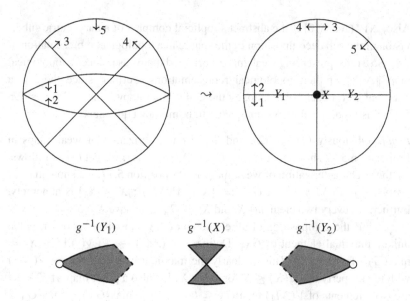

Figure 5.7 Some preimages under $\mathcal{V}^*(\mathcal{M})\backslash\{0\} \to \mathcal{V}^*(\mathcal{N})\backslash\{0\}$.

Lemma 5.33 *Let $\mathcal{M}$ be an oriented matroid on elements E and $X \in \{0, +, -\}^E$ such that $\emptyset \neq \mathcal{V}^*(\mathcal{M})_{\geq X} \not\supseteq \mathcal{V}^*(\mathcal{M})\backslash\{0\}$. Then $\|\Delta\mathcal{V}^*(\mathcal{M})_{\geq X}\|$ is contractible.*

Proof: $\mathcal{V}^*(\mathcal{M})_{\geq X}$ is a nonempty upper order ideal in $\mathcal{V}^*(\mathcal{M})$, so it contains at least one tope. We induct on the number of topes in $\mathcal{V}^*(\mathcal{M})_{\geq X}$. When $\mathcal{V}^*(\mathcal{M})_{\geq X}$ contains only one tope then it has a unique maximum, thus is contractible.

Beyond this case, let T_1 and T_2 be topes in $\mathcal{V}^*(\mathcal{M})_{\geq X}$, and let $e \in S(T_1, T_2)$. Then $X(e) = 0$. Consider the map

$$f : \mathcal{V}^*(\mathcal{M})_{\geq X} \to \mathcal{V}^*(\mathcal{M}\backslash e)_{\geq X\backslash e}$$
$$Y \to Y\backslash e.$$

If $Z \in \mathcal{V}^*(\mathcal{M}\backslash e)_{\geq X\backslash e}$ then $f^{-1}(\mathcal{V}^*(\mathcal{M}\backslash e)_{\geq Z})$ has a unique minimal element, which is one of Ze^0, Ze^+, Ze^-. Thus by Quillen's Theorem A (Theorem 5.26), $\|\Delta f\|$ is a homotopy equivalence □

Proof of Proposition 5.32: Let $X \in \mathcal{V}^*(\mathcal{N})\backslash\{0\}$. Then $\|\Delta(g^*)^{-1}(\mathcal{V}^*(\mathcal{N})_{\geq X})\|$ = $\|\Delta\mathcal{V}^*(\mathcal{M})_{\geq X}\|$, which is nonempty because $\mathcal{M} \rightsquigarrow \mathcal{N}$, hence is contractible by Lemma 5.33. Thus the result follows from Quillen's Theorem A (Theorem 5.26). □

5.4 Some Additional Results on Strong Maps

The maps g^* are not functorial! Let **OM** be the category of oriented matroids and weak maps. Let **Sphere** be the category of simplicial spheres and simplicial homotopy equivalences. The Topological Representation Theorem gives a map from objects of **OM** to objects of **Sphere**, and Proposition 5.30 extends this to a map from morphisms of **OM** to morphisms of **Sphere**. However, this does *not* amount to a functor – see the following problem.

Problem 5.34 Show that in every sequence $\mathcal{M}_1 \rightsquigarrow \mathcal{M}_2 \rightsquigarrow \mathcal{M}_3$ of three distinct rank 2 oriented matroids on three elements the maximal map g^* induced by the relation $\mathcal{M}_1 \rightsquigarrow \mathcal{M}_3$ is not the composition of the maximal maps g^* induced by $\mathcal{M}_1 \rightsquigarrow \mathcal{M}_2$ and $\mathcal{M}_2 \rightsquigarrow \mathcal{M}_3$.

In fact, there is no "nice" functor from **OM** to **Sphere** – that is, no functor that sends each oriented matroid to its topological representation and is well behaved with respect to deletion and contraction. See Anderson (2001) for details.

5.4 Some Additional Results on Strong Maps

5.4.1 Framing a Strong Map Image

Definition 5.35 Let $W \subset V$ be vector spaces over $\mathbb{R}$. Let k be the codimension of W in V. A **framing** of W in V is a set $\{H_1, \ldots, H_k\}$ of hyperplanes in V such that $\bigcap_{i=1}^{k} H_i = W$.

Whenever we describe a codimension k subspace W of $\mathbb{R}^m$ as the solution set to a system of k linear equations, we are implicitly giving a framing of W.

To see the oriented matroid analog to a framing, consider $W \subset V$ vector spaces over $\mathbb{R}$, and let $(\mathcal{H}_e : e \in E)$ be a signed hyperplane arrangement in V with oriented matroid $\mathcal{M}$. Let $\mathcal{N}$ be the oriented matroid of the arrangement $(\mathcal{H}_e \cap W : e \in E)$ in W. Assume $E \cap [k] = \emptyset$. If $\{H_1, \ldots, H_k\}$ is a framing of W, then by choosing a positive side for each H_e we get an extension $(\mathcal{H}_e : e \in E \cup [k])$ of our arrangement and thus an extension $\mathcal{M} \cup [k]$ of $\mathcal{M}$. For each $\mathbf{x} \in W$ with corresponding covector $X = \text{sign}(\mathbf{x}) \in \mathcal{V}^*(\mathcal{N}) \subset \mathcal{V}^*(\mathcal{M})$, since $\mathbf{x} \in \bigcap_{i=1}^{k} H_i$, we have $X[k]^0 \in \mathcal{V}^*(\mathcal{M} \cup [k])$. Further, the elements of $\mathcal{V}^*(\mathcal{M} \cup [k])$ of the form $Y[k]^0$ are exactly the covectors arising from elements of W. Thus we have

$$X \in \mathcal{V}^*(\mathcal{N}) \Leftrightarrow X[k]^0 \in \mathcal{V}^*(\mathcal{M} \cup [k])$$
$$\Leftrightarrow X \in \mathcal{V}^*((\mathcal{M} \cup [k])/[k]).$$

Definition 5.36 Let $\mathcal{M} \to \mathcal{N}$ be a strong map, and let $k = \text{rank}(\mathcal{M}) - \text{rank}(\mathcal{N})$. A **framing** of $\mathcal{M} \to \mathcal{N}$ is an extension $\mathcal{M} \cup \{e_1, \ldots, e_k\}$ of $\mathcal{M}$ such that $\mathcal{N} = (\mathcal{M} \cup \{e_1, \ldots, e_k\})/\{e_1, \ldots, e_k\}$.

Framing might seem like a promising tool for reducing questions about strong map images to questions about extensions and contractions. Sadly, not every strong map has a framing, even if the oriented matroids involved are realizable.

Proposition 5.37 (Wu 2021) *There is a strong map from a uniform realizable oriented matroid $\mathcal{M}$ of rank 4 on eight elements to a uniform oriented matroid $\mathcal{N}$ of rank 2 that does not admit a factorization into strong maps $\mathcal{M} \to \widetilde{\mathcal{M}} \to \mathcal{N}$ with $\widetilde{\mathcal{M}}^3$ of rank 3.*

If this strong map had a framing $\mathcal{M} \cup \{e_1, e_2\}$ then we could factor $\mathcal{M} \to \mathcal{N}$ as $\mathcal{M} \to (\mathcal{M} \cup e_1)/e_1 \to \mathcal{N}$.

Another example, on 12 elements, is given in Richter-Gebert (1993).

By contrast, every strong map of unoriented matroids does have a framing (Higgs 1968).

Every strong map $\mathcal{M} \to \mathcal{N}$ with $\text{rank}(\mathcal{M}) = \text{rank}(\mathcal{N}) + 1$ does have a framing (see Exercise 6.3): In a topological representation of $\mathcal{M} \to \mathcal{N}$, the image of the map $S^{\text{rank}(\mathcal{N})-1} \hookrightarrow S^{\text{rank}(\mathcal{M})-1}$ is a pseudosphere that can be signed to give a topological representation of the extension $\mathcal{M} \cup e_1$.

Remark 5.38 One could hope that the poset P_E of oriented matroids on a fixed set E of elements, ordered by strong maps, has interesting connections to spaces of flags in a vector space. But at least one result tells us that the connections will not be too nice.

Let $\mathcal{G}_n$ be the poset of all linear subspaces of $\mathbb{R}^n$, ordered by inclusion. (This poset is called a *Grassmannian poset* in Živaljević (2016).) $\mathcal{G}_n$ is certainly pure: A maximal chain is $\{0\} \subset V_1 \subset \cdots \subset V_{n-1} \subset \mathbb{R}^n$, where each V_i has dimension i. By contrast, the poset of oriented matroids on elements $[n]$, ordered by strong maps, need not be pure. For example, each maximal chain in $\mathcal{G}_8$ gives a chain in $P_{[8]}$ with nine elements. But a maximal chain in $P_{[8]}$ containing the oriented matroids $\mathcal{M}$ and $\mathcal{N}$ of Proposition 5.37 can have at most eight elements.

5.4.2 The Generalized Levi Enlargement Lemma

Given a set S of k points in an n-dimensional vector space V, with $k \leq n$, we can always find a k-dimensional subspace of V that contains S. This, together with the subspace criterion for strong maps in Proposition 5.8, proves the following.

Proposition 5.39 *If $\mathcal{M}^n$ is realizable, $k \leq n$, and $S = \{P_1, \ldots, P_k\} \subset \mathcal{V}^*(\mathcal{M})$, then there is a rank k oriented matroid $\mathcal{N}^k$ such that $\mathcal{M}^n \to \mathcal{N}^k$ and $S \subseteq \mathcal{V}^*(\mathcal{N})$.*

The case $k = 2$ generalizes to arbitrary oriented matroids. The case $k = 2$ and $n = 3$ of this generalization was essentially proved in Levi (1926), in the context of arrangements of pseudolines. Our proof is adapted from Mandel's thesis (Mandel 1982) and uses the techniques of Chapter 4.

Theorem 5.40 (Mandel 1982) *Let $\mathcal{M}$ be an oriented matroid of rank at least two. Let $P, Q \in \mathcal{V}^*(\mathcal{M})$. There is a rank 2 strong map image $\mathcal{N}$ of $\mathcal{M}$ such that $P, Q \in \mathcal{V}^*(\mathcal{N})$.*

Proof: Let $A = P^0 \cap Q^0$. If we can find a rank 2 strong map image $\mathcal{N}'$ of $\mathcal{M}/A$ such that $P \backslash A, Q \backslash A \in \mathcal{V}^*(\mathcal{N}')$, then the oriented matroid $\mathcal{N}$ obtained from $\mathcal{N}'$ by extending by a loop e_a for each $a \in A$ is the oriented matroid we want. So it's enough to consider the case $P^0 \cap Q^0 = \emptyset$.

In this case $P \circ Q$, $P \circ -Q$, $Q \circ P$, and $Q \circ -P$ are all topes. Also, the set of elements of $\mathcal{M}$ is the disjoint union $S(P, Q) \dot\cup S(P, -Q) \dot\cup P^0 \dot\cup Q^0$.

Let $P \circ Q = X_1 \prec_{P \circ Q} X_2 \prec_{P \circ Q} \cdots \prec_{P \circ Q} X_k = Q \circ P$ be a saturated chain in the tope poset $\mathcal{T}(\mathcal{M}, P \circ Q)$. For each i let $Y_i = X_i \wedge X_{i+1}$ and $E_i = S(X_i, X_{i+1})$. (These are discussed in Proposition 4.46.) Thus $Y_i^0 = E_i$ and $\bigcup_i E_i = S(P, Q)$.

Let $P \circ -Q = X'_1 \prec_{P \circ -Q} X'_2 \prec_{P \circ -Q} \cdots \prec_{P \circ -Q} X'_l = -Q \circ P$ be a saturated chain in the poset $\mathcal{T}(\mathcal{M}, P \circ -Q)$. For each i let $Y'_i = X'_i \wedge X'_{i+1}$ and $E'_i = S(X_i, X_{i+1})$. Thus $(Y'_i)^0 = E'_i$ and $\bigcup_i E'_i = S(P, -Q)$.

Then there is a rank 2 oriented matroid $\mathcal{N}$ with

$$\mathcal{V}^*(\mathcal{N}) = \{0\} \cup \{\pm P, \pm Q\} \cup \bigcup_i \{\pm X_i\} \cup \bigcup_i \{\pm Y_i\} \cup \bigcup_i \{\pm X'_i\} \cup \bigcup_i \{\pm Y'_i\}.$$

Indeed, a realization of $\mathcal{N}$ (omitting the signs) is shown in Figure 5.8. The line labeled by P and $-P$ represents all elements of P^0, the line labeled by Y_i represents all elements of E_i, and so forth. □

Theorem 5.40 does not extend to $k = 3$. This is shown by the existence of rank 4 *non-Euclidean* oriented matroids, a highlight of Chapter 6.

156 Strong Maps and Weak Maps

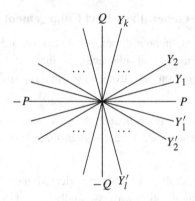

Figure 5.8 Figure for the proof of Theorem. 5.40

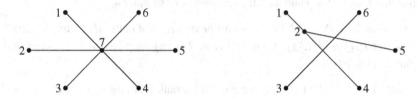

Figure 5.9 Another weak map that is not weakly realizable.

Exercises

5.1 Show by example that the following statement is *false*:

$\mathcal{M} \rightsquigarrow \mathcal{N}$ if and only if for every $Y_2 \in C^*(\mathcal{N})$ there is some $Y_1 \in C^*(\mathcal{M})$ with $Y_2 \leq Y_1$.

5.2 Define a relation ρ on realizable oriented matroids by $M_1 \rho M_2$ if and only if there is a weakly realizable weak map from M_1 to M_2. Prove that ρ is not transitive.

5.3 1. Give an example of realizable rank 3 oriented matroids $\mathcal{M}$ and $\mathcal{N}$ such that $\mathcal{M} \rightsquigarrow \mathcal{N}$, $|\mathrm{supp}(\chi_{\mathcal{M}})| > |\mathrm{supp}(\chi_{\mathcal{N}})| + 1$, and there does not exist an oriented matroid $\mathcal{O}$ such that $\mathcal{M} \rightsquigarrow \mathcal{O} \rightsquigarrow \mathcal{N}$ and $\mathcal{M} \neq \mathcal{O} \neq \mathcal{N}$.
2. Use this example to show that for all sufficiently large n and all $r \geq 3$ the poset MacP(r, n) is not pure.

5.4 (Sturmfels 1989) Let $\mathcal{M}$ and $\mathcal{N}$ be the oriented matroids of affine point arrangements pictured in Figure 5.9. (In the right arrangement, element 7 is a loop.) Prove that $\mathcal{M} \rightsquigarrow \mathcal{N}$ but that this weak map is not weakly realizable.

Exercises

5.5 This exercise will show that products and coproducts are not always defined either in the category of oriented matroids and strong maps or the category of oriented matroids and weak maps.

Recall that for two elements A, B of a category $\mathcal{K}$, a **product** is an object, denoted $A \times B$, together with morphisms $\pi_A: A \times B \to A$ and $\pi_B: A \times B \to B$, satisfying the following universal property: For every object X in $\mathcal{K}$ and every pair of morphisms $f_A: X \to A$ and $f_B: X \to B$, there is a unique morphism $f: X \to A \times B$ such that the following diagram commutes.

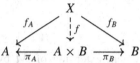

A **coproduct** is an object, denoted $A \coprod B$, together with morphisms $\iota_A: A \to A \coprod B$ and $\iota_B: B \to A \coprod B$, satisfying the following universal property: For every object Y in $\mathcal{K}$ and every pair of morphisms $g_A: A \to Y$ and $g_B: B \to Y$, there is a unique morphism $g: A \coprod B \to Y$ such that the following diagram commutes.

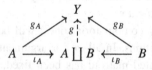

1. Find two rank 2 oriented matroids $\mathcal{M}_1$ and $\mathcal{M}_2$ and distinct rank 3 oriented matroids $\mathcal{N}_1$ and $\mathcal{N}_2$ such that $\mathcal{N}_i \to \mathcal{M}_j$ for each i and j. Conclude that $\mathcal{M}_1$ and $\mathcal{M}_2$ have no product and that $\mathcal{N}_1$ and $\mathcal{N}_2$ have no coproduct in the category of oriented matroids and strong maps.
2. In a similar way, use rank 2 and rank 3 oriented matroids to show that products and coproducts need not exist in the category of oriented matroids and weak maps.

6

Single-Element Extensions

Recall from Section 3.5 that a single-element extension of $\mathcal{M}$ by f is an oriented matroid $\mathcal{M} \cup f$ on elements $E \cup \{f\}$ such that $(\mathcal{M} \cup f)\backslash f = \mathcal{M}$. In this chapter we'll be concerned with single-element extensions $\mathcal{M} \cup f$ in which f is not a coloop. The Topological Representation Theorem will serve us well in examining such single-element extensions in more depth: Given a topological representation of $\mathcal{M}$, a single-element extension by a non-coloop just adds one more pseudosphere to the picture.

Many of our arguments, even purely combinatorial ones, will be guided by intuition from topological representations. In all of our arguments, when a topological representation of an oriented matroid has been fixed, we will feel free to identify a covector with the corresponding cell in the topological representation.

Several of the main proofs in this chapter are new proofs of older results. The new proofs were motivated by a desire for a self-contained presentation that builds on the Topological Representation Theorem. The older proofs built on connections between oriented matroids and other subjects. Notably, Las Vergnas's work on localizations (Section 6.2) was originally built on Crapo's similar work on matroids, and Fukuda's proof of duality of the Euclidean property (Theorem 6.73) was originally proved using the oriented matroid abstraction of linear programming. These connections to other topics are interesting and important, and the reader interested in exploring any of these topics in more depth is encouraged to study the original proofs.

6.1 Preliminaries

Let $\mathcal{M}$ have rank r. In the cell decomposition of S^{r-1} given by a topological representation of $\mathcal{M}$ the 0-cells are signed cocircuits, which are elements X

of $\mathcal{V}^*(\mathcal{M})$ with $\text{rank}(X^0) = r - 1$, and the 1-cells are $X \in \mathcal{V}^*(\mathcal{M})$ with $\text{rank}(X^0) = r - 2$. These two types of covectors form a graph $G(\mathcal{M})$ with vertex set $\mathcal{C}^*(\mathcal{M})$, which we call the **cocircuit graph** of $\mathcal{M}$.

Our first abuse of graph terminology is to define $G(\mathcal{M})$ as a single subset of $\mathcal{V}^*(\mathcal{M})$ rather than a set of vertices together with a set of edges. Our second abuse of terminology is to define a walk in $G(\mathcal{M})$ as a subset of $G(\mathcal{M})$ (containing both vertices and edges) rather than a sequence of edges.

Definition 6.1 An **edge** in $G(\mathcal{M})$ is a nonminimal element of $G(\mathcal{M})$. The **endpoints** of an edge X are the two elements of $\mathcal{C}^*(\mathcal{M})$ that are less than X. A **path** in $G(\mathcal{M})$ between two signed cocircuits X and Y is a subset of $G(\mathcal{M})$ of the form $\{X = X_0, X_1, \ldots, X_{2k} = Y\}$, where all X_i are distinct and each X_i with i odd is an edge with endpoints X_{i-1} and X_{i+1}.

For a path γ we let $\gamma^0 = \gamma \cap \mathcal{C}^*(\mathcal{M})$.

We'll avoid calling elements of γ^0 "vertices," because we'll be using that word in another sense shortly.

Recall (Definition 4.32) that a *closed convex set* in $\mathcal{V}^*(\mathcal{M})$ is $C(Y) := \{X \in \mathcal{V}^*(M) : \forall a \in A\ X(a) \leq Y(a)\}$, for some A and some $Y \in \{0, +, -\}^A$. When we're working with more than one oriented matroid we may denote a convex set in $\mathcal{M}$ by $C_{\mathcal{M}}(Y)$. By convention we say that if $A = \emptyset$ then $Z = \emptyset$ and $C(\emptyset) = \mathcal{V}^*(\mathcal{M})$.

Definition 6.2 The **convex hull** of $S \subseteq \mathcal{V}^*(\mathcal{M})$ is the smallest closed convex set containing S.

Lemma 6.3 *Let* $S \subseteq \mathcal{V}^*(\mathcal{M})$.

$$\text{conv}(S) = \left\{ Y \in \mathcal{V}^*(\mathcal{M}) : Y \leq Y' \text{ for some } Y' \in \boxplus_{X \in S} X \right\}.$$

Proof: Let $A = E \setminus \bigcup_{X,X' \in S} S(X, X')$. Let $Z \in \{0, +, -\}^A$ be defined by $Z(e) = \max_{X \in S} X(e)$. Then $\text{conv}(S) = C(Z)$. But also $\{Y \in \mathcal{V}^*(\mathcal{M}) : Y \leq Y'$ for some $Y' \in \boxplus_{X \in S} X\} = C(Z)$. □

We can interpret Signed Circuit Elimination in $\mathcal{C}^*(\mathcal{M})$ in terms of a topological representation $(\mathcal{S}_e : e \in E)$: If $X(f) = -Y(f) \neq 0$ then an elimination of f between X and Y is a signed cocircuit in $\text{conv}(\{X, Y\}) \cap S_f$. If $X \neq -Y$ then $\text{conv}(\{X, Y\}) \cap G(\mathcal{M})$ is connected, and X and Y lie in different connected components of $\text{conv}(\{X, Y\}) \setminus S_f$. Thus a path from X to Y in $G(\mathcal{M}) \cap \text{conv}(\{X, Y\})$ must contain a vertex Z with $Z(g) = 0$.

6.2 Localizations

If $\mathcal{M} \cup f$ is an extension of $\mathcal{M}$, then for every signed cocircuit X of $\mathcal{M}$ there is a unique signed cocircuit of $\mathcal{M} \cup f$ of the form $Xf^{\sigma(X)}$. We can see this either by pondering the definition of deletion (Definition 2.26) or by considering a topological representation $(\mathcal{S}_e : e \in E \cup \{f\})$ of $\mathcal{M} \cup f$. Deleting the pseudosphere $\mathcal{S}_f$ yields a topological representation of $\mathcal{M}$, and the 0-cell corresponding to X in the cell decomposition of S^{n-1} induced by this smaller arrangement is in $S_f^{\sigma(X)}$ for exactly one $\sigma(X) \in \{0, +, -\}$. The function $\sigma : \mathcal{C}^*(\mathcal{M}) \to \{0, +, -\}$ is called the **localization** of the extension.

Figure 6.1 illustrates a rank 3 oriented matroid $\mathcal{M}$, an extension by an element f, and the value of σ on some signed cocircuits. To simplify the figure we have omitted the arrows indicating the signs on elements of $\mathcal{M}$.

Proposition 6.4 (Las Vergnas 1978) *If $\mathcal{M}'$ is an extension by a non-coloop then the localization σ determines $\mathcal{M}'$, via:*

$$\mathcal{C}^*(\mathcal{M}') = \{Xf^{\sigma(X)} : X \in \mathcal{C}^*(\mathcal{M})\} \cup$$
$$\{(Y_1 \circ Y_2)f^0 : Y_1, Y_2 \in \mathcal{C}^*(\mathcal{M}), \operatorname{rank}(Y_1^0 \cap Y_2^0) = \operatorname{rank}(\mathcal{M}) - 2,$$
$$S(Y_1, Y_2) = \emptyset, \sigma(Y_1) = -\sigma(Y_2) \neq 0\}.$$
(6.1)

The second set on the right-hand side is worth looking at closely. The condition $\operatorname{rank}(Y_1^0 \cap Y_2^0) = \operatorname{rank}(\mathcal{M}) - 2$ says that in a topological representation of $\mathcal{M}$, Y_1 and Y_2 are distinct points on a single pseudocircle. (Such

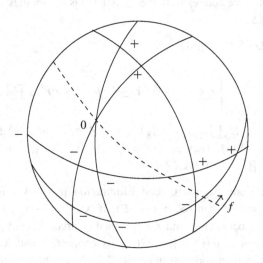

Figure 6.1 An extension and its localization.

a pair is *comodular*. Comodular pairs appeared briefly in Section 2.6.) The condition $S(Y_1, Y_2) = \emptyset$ says that these points are neighboring 0-cells on this pseudocircle, and the condition $\sigma(Y_1) = -\sigma(Y_2) \neq 0$ says that in a topological representation of $\mathcal{M}'$ obtained by extending our topological representation of $\mathcal{M}$ by a signed pseudosphere $\mathcal{S}_f$, these two points are on opposite sides of $\mathcal{S}_f$. We can see several such pairs in the lower right-hand side of Figure 6.1.

Proof: Consider a topological representation of $\mathcal{M}'$ by an arrangement ($\mathcal{S}_e$: $e \in E \cup \{f\}$) of pseudospheres. Thus ($\mathcal{S}_e$: $e \in E$) is a topological representation of $\mathcal{M}$.

For every $X \in \mathcal{C}^*(\mathcal{M}')$, one of the following is true:

1. X lies in a 0-sphere of the form S_A with $f \notin A$. In this case $X \backslash f \in \mathcal{C}^*(\mathcal{M})$ and $X = (X \backslash f) f^{\sigma(X')}$.
2. (1) fails, and X lies in a 0-sphere of the form $S_{A \cup \{f\}}$ with S_A a pseudocircle. The definition of oriented pseudosphere arrangements tells us that one side of this pseudocircle is in S_f^+ and the other side is in S_f^-. Thus X's signed cocircuit neighbors on the pseudocircle S_A have the form $Y_1 f^-$ and $Y_2 f^+$, and $X = (Y_1 \circ Y_2) f^0$. Since Y_1 and Y_2 are distinct 0-cells on a common pseudocircle, we have that $\text{rank}(Y_1^0 \cap Y_2^0) = \text{rank}(\mathcal{M}) - 2$, and since $\mathcal{S}_f$ is the only pseudosphere separating $Y_1 f^-$ and $Y_2 f^+$, we have that $S(Y_1, Y_2) = \emptyset$.

Thus we have inclusion of the left-hand side of Equation (6.1) into the right-hand side. The first set in the right-hand side is a subset of $\mathcal{C}^*(\mathcal{M}')$ by definition. For every element $(Y_1 \circ Y_2) f^0$ in the second set on the right-hand side, we have that $Y_1 f^{\sigma(Y_1)}$ and $Y_2 f^{\sigma(Y_2)}$ are elements of $\mathcal{C}^*(\mathcal{M}')$, and $(Y_1 \circ Y_2) f^0$ is the elimination of f between these two elements. Thus we also have inclusion of the right-hand side into the left-hand side. □

Remark 6.5 If $\mathcal{M} \cup f$ is an extension of $\mathcal{M}$ by a coloop then $\mathcal{C}^*(\mathcal{M}') = \{Xf^0 : X \in \mathcal{C}^*(\mathcal{M})\} \cup \{f^+, f^-\}$. Thus the localization σ is identically 0, the same as the localization of the extension by a loop, and the conclusion of Proposition 6.4 fails.

6.2.1 The Las Vergnas Condition for Localizations

If $\mathcal{N} = (\mathcal{M}/A) \backslash B$ is a minor of $\mathcal{M}$, then every signed cocircuit of $\mathcal{N}$ has the form $X \backslash (A \cup B)$ for exactly one signed cocircuit X of $\mathcal{M}$. (One can see from the definitions, but also this is clear when we consider a topological representation ($\mathcal{S}_e$: $e \in E$) of $\mathcal{M}$. Once we delete all signed pseudospheres

$\mathcal{S}_b$ with $b \in B$, the signed cocircuits of $\mathcal{N}$ correspond to 0-cells in $\bigcap_{a \in A} \mathcal{S}_a$, and each such 0-cell corresponds to exactly one 0-cell in the topological representation of $\mathcal{M}$.) Thus a function $\sigma \colon \mathcal{C}^*(\mathcal{M}) \to \{0, +, -\}$ induces a function $\sigma_{\mathcal{N}}$ on $\mathcal{C}^*(\mathcal{N})$, taking $X \backslash (A \cup B)$ to $\sigma(X)$.

Proposition 6.6 *Let* $\mathcal{M} \cup f$ *be an extension of* $\mathcal{M}$ *with localization* σ, *and let* $\mathcal{N} = (\mathcal{M}/A) \backslash B$. *Consider the induced function* $\sigma_{\mathcal{N}} \colon \mathcal{C}^*(\mathcal{N}) \to \{0, +, -\}$.

1. *If* f *is a coloop in* $((\mathcal{M} \cup f)/A) \backslash B$ *then* $\sigma_{\mathcal{N}}$ *is identically 0.*
2. *Otherwise* $\sigma_{\mathcal{N}}$ *is the localization of the extension* $((\mathcal{M} \cup f)/A) \backslash B$ *of* $\mathcal{N}$.

The proof is just a matter of checking definitions. In either case of the proposition, $\sigma_{\mathcal{N}}$ is a localization: If f is a coloop in $((\mathcal{M} \cup f)/A) \backslash B$ then $\sigma_{\mathcal{N}}$ is the localization of the extension of $\mathcal{N}$ by a loop.

Thus we see a necessary condition for a function $\sigma \colon \mathcal{C}^*(\mathcal{M}) \to \{0, +, -\}$ to be a localization: For each minor of $\mathcal{M}$, the induced function on $\mathcal{C}^*(\mathcal{N})$ must also be a localization.

In Section 6.3 we'll prove the following necessary and sufficient condition for σ to be a localization, and in later sections we'll apply it often.

Theorem 6.7 (The Las Vergnas Condition) (Las Vergnas 1978) *Let* $\mathcal{M}$ *be an oriented matroid and* $\sigma \colon \mathcal{C}^*(\mathcal{M}) \to \{0, +, -\}$ *be a function such that* $\sigma(-X) = -\sigma(X)$ *for all* X. *Then* σ *is a localization if and only if, for each rank 2 minor* $\mathcal{N}$ *of* $\mathcal{M}$ *with three elements, the induced function on* $\mathcal{C}^*(\mathcal{N})$ *is a localization.*

A quick examination of the topological representations of rank 2 oriented matroids $\mathcal{N}$ on three elements reveals that there are only three types of function $\sigma_{\mathcal{N}} \colon \mathcal{C}^*(\mathcal{N}) \to \{0, +, -\}$ that satisfy $\sigma_{\mathcal{N}}(-X) = -\sigma_{\mathcal{N}}(X)$ for all X but are not localizations on $\mathcal{N}$. These three "forbidden subconfigurations" are shown in Figure 6.2.

Corollary 6.8 *Let* $\sigma \colon \mathcal{C}^*(\mathcal{M}) \to \{0, +, -\}$ *such that* $\sigma(-X) = -\sigma(X)$ *for every* X. *Then* σ *is a localization if and only if, for every rank 2 minor* $\mathcal{N}$ *of* $\mathcal{M}$

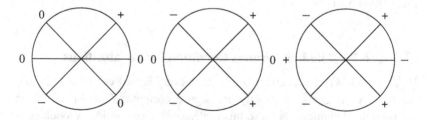

Figure 6.2 The three forbidden configurations.

on three elements, the induced function on $C^*(\mathcal{N})$ has none of the three forms shown in Figure 6.2.

Lemma 6.9 *Let* $\sigma : C^*(\mathcal{M}) \to \{0, +, -\}$ *be a function such that for every rank 2 minor $\mathcal{N}$ with three elements, the induced function $\sigma_{\mathcal{N}}$ is a localization. Then for every rank 2 minor $\mathcal{N}'$, the induced function $\sigma_{\mathcal{N}'}$ is a localization.*

Proof: Let $\mathcal{N}'$ be a rank 2 minor of $\mathcal{M}$. $\mathcal{N}'$ has a centrally symmetric topological representation in S^1. The induced function $\sigma_{\mathcal{N}'}$ is a localization if and only if $\sigma_{\mathcal{N}'}^{-1}(+)$ lies in one open half-circle in the topological representation and $\sigma_{\mathcal{N}'}^{-1}(-)$ lies in the negative of this open half-circle. Clearly, if this fails to be true, we can find one of the three forbidden subconfigurations. □

6.3 Proof of the Las Vergnas Characterization

Proposition 6.6 already proved one direction of Theorem 6.7. We briefly sketch the proof of the hard direction. If $\sigma : C^*(\mathcal{M}) \to \{0, +, -\}$ is a function, then Equation (6.1) gives a candidate $\mathcal{D}$ for the signed cocircuit set of an extension with localization σ. If σ induces a localization on every rank 2 minor with three elements, then the only difficult part in showing that $\mathcal{D}$ is indeed a signed cocircuit set is proving that it satisfies Signed Circuit Elimination. We'll do this using the topological approach to Signed Circuit Elimination described in Section 6.1.

Our function σ puts signs on the vertices of the cocircuit graph $G(\mathcal{M})$. Lemma 6.23 will show that proving Signed Circuit Elimination for $\mathcal{D}$ amounts to proving the existence of certain paths in sets $G(\mathcal{M}) \cap \text{conv}(\{X, Y\})$ that are "good" with respect to σ. If σ induces a localization on each rank 2 contraction of $\mathcal{M}$, then we'll show the existence of good paths by an argument that says

- $\text{conv}(\{X, Y\})$ behaves like a convex polytope, and
- the restriction of σ to the "vertices" of $\text{conv}(\{X, Y\})$ behaves like the restriction of a linear function to the vertices of a convex polytope.

This idea of convex hulls in $\mathcal{V}^*(\mathcal{M})$ behaving like convex polytopes will arise again in the proof of Theorem 6.73.

6.3.1 Polytopal Sets in $\mathcal{V}^*(\mathcal{M})$

Definition 6.10 The **boundary** $\partial C(Z)$ of a closed convex set $C(Z)$ is the set of all $X \in C(Z)$ such that, for some $e \in E$ and $\epsilon \in \{+, -\}$, we have $X(e) = 0$, $Y(e) = \epsilon$ for some $Y \in C(Z)$, and $Y'(e) \in \{0, \epsilon\}$ for all $Y' \in C(Z)$.

The **interior** of a closed convex set $C(Z)$ is the complement of the boundary of $C(Z)$.

The Topological Representation Theorem tells us that, in a topological representation of $\mathcal{M}$, the union of the cells in a closed convex set $C(Z)$ is a topological ball or topological sphere. A nonzero element X of $C(Z)$ is in the boundary of $C(Z)$ in our sense if and only if this union is a ball and the cell X is in the boundary of this ball.

Different sign vectors Z can give the same $C(Z)$. For instance, if $Z \in \{0, +, -\}^A$, $e \in E \backslash A$, and $C(Z) \subseteq C(e^+)$ then $C(Z) = C(Ze^+)$. However, the following proposition shows a relationship between $C(Z)$ and A that is independent of the choice of Z.

Proposition 6.11 *If $A, A' \subseteq E$, $Z \in \{0, +, -\}^A$, $Z' \subseteq \{0, +, -\}^{A'}$, and $C(Z) = C(Z')$ then*

1. $\operatorname{rank}(A) = \operatorname{rank}(A')$, and
2. $\operatorname{rank}(A \cap X^0) = \operatorname{rank}(A' \cap X^0)$ *for every $X \in C(Z)$.*

Lemma 6.12 *Let $A \subseteq E$, $Z \in \{0, +, -\}^A$, and $X \in C(Z)$. X is in the interior of $C(Z)$ if and only if, for each $Y \in C(Z)$, both $X \circ Y$ and $X \circ -Y$ are in $C(Z)$.*

Proof: Certainly $C(Z)$ is closed under composition, so $X \circ Y \in C(Z)$. If $X, Y \in C(Z)$ and $X \circ -Y \notin C(Z)$, then there is some $e \in A$ such that $X \circ -Y(e) = -Z(e) \neq 0$. Since $X(e) \leq Z(e)$, this implies $X(e) = 0$ and $Y(e) \neq 0$, hence X is in the boundary of $C(Z)$.

Conversely, if X is in the boundary of $C(Z)$ then for the Y of Definition 6.10 we have $X \circ -Y \notin C(Z)$. □

Lemma 6.13 *If $C(Z) \not\subseteq C(b^0)$ and there is a covector X in the interior of $C(Z)$ such that $X(b) = 0$, then $\emptyset \neq C(Zb^\epsilon) \subsetneq C(Z)$ for each $\epsilon \in \{0, +, -\}$.*

Proof: Let $Y \in C(Z)$ such that $Y(b) \neq 0$. Then $X \circ Y$ and $X \circ -Y$ are in $C(Z)$. □

In a topological representation of $\mathcal{M}$ in S^{r-1}, the set of cells in S_b is the set of cells indexed by $C(b^0)$. Consider the ball or sphere B in S^{r-1} corresponding to $C(Z)$. Lemma 6.13 says that if $B \not\subseteq S_b$ and S_b intersects the interior of B then both S_b^+ and S_b^- intersect B.

Proof of Proposition 6.11: For such an A, A', Z, and Z', define an element Z'' of $\{0, +, -\}^{A \cup A'}$ by

$$Z''(e) = \begin{cases} Z(e) \wedge Z'(e) & \text{if } e \in A \cap A', \\ Z(e) & \text{if } e \in A \backslash A', \\ Z'(e) & \text{if } e \in A' \backslash A. \end{cases}$$

6.3 Proof of the Las Vergnas Characterization

Then $C(Z) = C(Z'')$. Thus it's enough to consider the case when $A \subset A'$.

1. We'll show that for every $b \in E \setminus A$, if $\text{rank}(A \cup \{b\}) \neq \text{rank}(A)$ then $C(Zb^\epsilon) \neq C(Z)$ for each $\epsilon \in \{0, +, -\}$.

This hypothesis on rank implies that $C(A^0) \not\subseteq C(b^0)$, and so by Lemma 6.13 $C(A^0 b^+)$ and $C(A^0 b^-)$ are both nonempty proper subsets of $C(A^0)$. But $C(A^0) \subseteq C(Z)$, and so

$$C(Z) \cap C(A^0) = C(A^0)$$
$$\supsetneq C(A^0 b^\epsilon)$$
$$= C(Zb^\epsilon) \cap C(A^0).$$

Thus $C(Z) \cap C(A^0) \neq C(Zb^\epsilon) \cap C(A^0)$, and so $C(Z) \neq C(Zb^\epsilon)$.

2. This proof is similar: We show that for every $b \in X^0 \setminus A$, if $\text{rank}(A \cap X^0) \neq \text{rank}((A \cup \{b\}) \cap X^0)$ then $C(Zb^\epsilon) \neq C(Z)$ for each $\epsilon \in \{0, +, -\}$.

Define $\hat{Z} \in \{0, +, -\}^A$ by

$$\hat{Z}(e) = \begin{cases} 0 & \text{if } e \in X^0, \\ Z(e) & \text{otherwise.} \end{cases}$$

The hypothesis on rank implies that $C((A \cap X^0)^0) \not\subseteq C(b^0)$. Let $Y \in C((A \cap X^0)^0)$ with $Y(b) \neq 0$. Then $X \circ Y \in C(\hat{Z})$ and $X \circ Y(b) \neq 0$, so $C(\hat{Z}) \not\subseteq C(b^0)$. So by Lemma 6.13 $C(\hat{Z}b^+)$ and $C(\hat{Z}b^-)$ are both nonempty proper subsets of $C(\hat{Z})$. Thus

$$C(Zb^+) \cap C((A \cap X^0)^0) = C(\hat{Z}b^+)$$
$$\neq C(\hat{Z})$$
$$= C(Z) \cap C((A \cap X^0)^0),$$

and so $C(Zb^+) \neq C(Z)$. □

Definition 6.14 Let $C(Z)$ be a convex set in $\mathcal{V}^*(\mathcal{M})$ with $Z \in \{0, +, -\}^A$. A **vertex** of $C(Z)$ is a $V \in C(Z) \setminus \{0\}$ such that $\text{rank}(V^0 \cap A) = \text{rank}(\mathcal{M}) - 1$.

For a vertex V, since $V \neq \mathbf{0}$ the rank of V^0 is not $\text{rank}(\mathcal{M})$, so it must be $\text{rank}(\mathcal{M}) - 1$. Thus by Corollary 3.41 $V \in \mathcal{C}^*(\mathcal{M})$, and so in a topological representation V corresponds to a 0-cell.

Example 6.15 Figure 6.3 shows convex sets in two oriented matroids, as well as the vertices of each.

Problem 6.16 Show that if $C(Z)$ has elements that are neither $\mathbf{0}$ nor signed cocircuits then every vertex of $C(Z)$ is in $\partial C(Z)$.

Single-Element Extensions

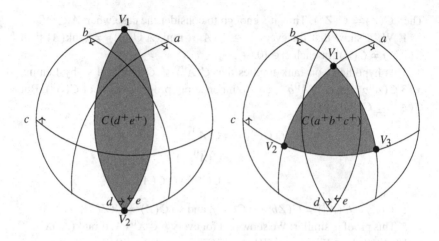

Figure 6.3 Vertices of convex sets.

Definition 6.17 A convex set $C(Z)$ is **polytopal** if $C(Z)$ is the convex hull of its vertex set.

A foundational result for convex polytopes (cf. chapter 1 in Ziegler 1995) says that a convex polytope can be described as either a bounded intersection of closed half-spaces or as the convex hull of its vertex set. A set $C(Z)$ is analogous to an intersection of closed half-spaces. $C(Z)$ is polytopal if it has a description analogous to the vertex description of convex polytopes.

Example 6.18 In the oriented matroid pictured in Figure 6.3 the convex set $C(d^+e^+)$ is not polytopal, because the convex hull of its vertex set is $\{0, V_1, V_2\}$. The convex set $C(a^+b^+c^+)$ is polytopal.

Lemma 6.19 Let $C(Z)$ be a convex set in $\mathcal{V}^*(\mathcal{M})$ with $Z \in \{0, +, -\}^A$. Then $C(Z)$ is polytopal if and only if one of the following holds:

1. $C(Z) = \{0\}$.
2. $C(Z) = \{X, 0\}$ for some $X \in \mathcal{C}^*(\mathcal{M})$.
3. $C(Z) = \{X, -X, 0\}$ for some $X \in \mathcal{C}^*(\mathcal{M})$.
4. $\text{rank}(A) = \text{rank}(\mathcal{M})$.

Proof: ($\Leftarrow$) If $C(Z) = \{0\}$ then $C(Z)$ has no vertices and $C(Z)$ is the convex hull of the empty set.

If $C(Z) = \{X, 0\}$ or $C(Z) = \{X, -X, 0\}$ for some $X \in \mathcal{C}^*(\mathcal{M})$, then let $A' = X^0 \cup A$, and define $Z' \in \{0, +, -\}^{A'}$ by

6.3 Proof of the Las Vergnas Characterization

$$Z'(e) = \begin{cases} 0 & \text{if } e \in X^0, \\ Z(e) & \text{if } e \notin X^0. \end{cases}$$

Then $C(Z) = C(Z')$, so by Proposition 6.11,

$$\begin{aligned} \text{rank}(A \cap X^0) &= \text{rank}(A' \cap X^0) \\ &= \text{rank}(X^0) \\ &= \text{rank}(\mathcal{M}) - 1, \end{aligned}$$

and so X is a vertex of $C(Z)$.

If $\text{rank}(A) = \text{rank}(\mathcal{M})$ then we induct on the number of maximal elements of $C(Z)$.

If $C(Z)$ has only one maximal element T then we check that the vertices of $C(Z)$ are the signed cocircuits conformal with T. If V is a vertex of $C(Z)$ then V is a signed cocircuit and $V \circ T$ is a tope and is in $C(Z)$. Thus $V \circ T = T$, and so V is indeed a signed cocircuit conformal with T. Conversely, let $X \in \mathcal{V}^*(\mathcal{M})$ be conformal with T. We show by induction on $\text{rank}(X^0)$ that $\text{rank}(X^0 \cap A) = \text{rank}(X^0)$. If $\text{rank}(X^0) = 1$ then since $X \circ -T \notin C(Z)$ we have that $X^0 \cap A$ contains a nonloop, so has rank 1. For X with X^0 of higher rank, let X_1, X_2 be atoms of the interval $[X, T]$ in $\mathcal{V}^*(\mathcal{M})$. Then $X^0 = X_1^0 \cup X_2^0$, so $X^0 \cap A = (X_1^0 \cap A) \cup (X_2^0 \cap A)$. By the induction hypothesis $\text{rank}(X_1^0 \cap A) = \text{rank}(X_1) = \text{rank}(X^0) - 1$, and $X_2 \cap A$ contains an element a that is not in the flat X_1^0. Thus $\text{rank}(X^0 \cap A) > \text{rank}(X_1^0 \cap A) = \text{rank}(X^0) - 1$ and so $\text{rank}(X^0 \cap A) = \text{rank}(X^0)$. Since T is a composition of the signed cocircuits conformal with T, we conclude that T is in the convex hull of the vertices of $C(Z)$. Thus $C(Z) = \mathcal{V}^*(\mathcal{M})_{\leq T}$ is the convex hull of the vertices of $C(Z)$.

If $C(Z)$ has more than one maximal element, then since all maximal elements of $C(Z)$ have the same support there is an $e \in E$ and maximal elements X_+ and X_- such that $X_+(e) = +$ and $X_-(e) = -$. Thus $C(Ze^+)$ and $C(Ze^-)$ each have fewer maximal elements than $C(Z)$, and so by the induction hypothesis these sets are polytopal. In particular, since $X_+(e) \in C(Ze^+)$ and $X_+(e) = +$, there is a vertex V_+ of $C(Ze^+)$ such that $V_+(e) = +$. Since $r - 1 = \text{rank}(V_+^0 \cap (A \cup \{e\}))$ and $e \notin V_+^0$ we have that V_+ is a vertex of $C(Z)$. Similarly V_- is a vertex of $C(Z)$.

Now let P_e be the set of all elements parallel or antiparallel to e, and consider $C_{\mathcal{M} \setminus P_e}(Z)$. It has fewer maximal elements than $C_{\mathcal{M}}(Z)$, and so by the induction hypothesis $C_{\mathcal{M} \setminus P_e}(Z)$ is the convex hull of its vertices. Because $\text{rank}(A) = \text{rank}(\mathcal{M})$, we know that $\text{rank}(\mathcal{M} \setminus P_e) = \text{rank}(\mathcal{M})$, so each vertex of $C_{\mathcal{M} \setminus P_e}(Z)$ has the form $V \setminus P_e$ for some signed cocircuit V of $\mathcal{M}$. Each such V is a vertex of $C_{\mathcal{M}}(Z)$.

Let $X \in C_{\mathcal{M}}(Z)$. Then $X \backslash P_e \in C_{\mathcal{M} \backslash P_e}(Z)$, so $X \backslash P_e$ is in the convex hull of the vertices $V \backslash P_e$ of $C_{\mathcal{M} \backslash e}(Z)$. Since V_+ and V_- are vertices of $C_{\mathcal{M}}(Z)$, by our earlier observations on vertices we conclude that X is in the convex hull of the vertices of $C_{\mathcal{M}}(Z)$.

Thus by Lemma 6.3 $C_{\mathcal{M}}(Z)$ is contained in the convex hull C' of the vertices of $C_{\mathcal{M}}(Z)$, and by the definition of convex hull $C' \subseteq C_{\mathcal{M}}(Z)$.

($\Rightarrow$) If $C(Z)$ is polytopal and the first three alternatives don't hold, then $C(Z)$ has vertices X_1 and X_2 with different support. Since X_1 and X_2 are vertices, $\text{rank}(X_i^0) = \text{rank}(\mathcal{M}) - 1 = \text{rank}(X_i^0 \cap A)$, and so X_i^0 is the closure of $X_i^0 \cap A$. Since these closures are different, there is an $e \in X_2^0 \cap A$ not in the closure of $X_1^0 \cap A$, and so $\text{rank}((X_1^0 \cup X_2^0) \cap A) = \text{rank}(\mathcal{M})$. □

Our interest in polytopal sets arises from the following lemma to the proof of Theorem 6.7.

Lemma 6.20 *Let C be a polytopal set in $\mathcal{V}^*(\mathcal{M})$ and Y a signed cocircuit in the interior of C. Let σ be the localization of an extension of $\mathcal{M}$.*

1. *If $\sigma(Y) = \epsilon \neq 0$ then there is a vertex Z of C so that $\sigma(Z) = \epsilon$.*
2. *If $\sigma(Y) = 0$ then either $\sigma(X) = 0$ for every signed cocircuit $X \in C$ or there are vertices X_+ and X_- of C so that $\sigma(X_+) = +$ and $\sigma(X_-) = -$.*

The first part of Lemma 6.20 is analogous to the observation that a linear function on a convex polytope achieves its extreme values at vertices.

Proof: Let $\mathcal{M} \cup f$ be the extension with localization σ. Fix a Z so that $C = C_{\mathcal{M}}(Z)$. The vertices of $C_{\mathcal{M} \cup f}(Z)$ are exactly $Xf^{\sigma(X)}$ such that X is a vertex of $C_{\mathcal{M}}(Z)$. Also $C_{\mathcal{M} \cup f}(Z)$ is polytopal, by Lemma 6.19. Thus:

- We see the contrapositive of (1): If $\sigma(X) \in \{0, -\epsilon\}$ for every vertex X of C then $C_{\mathcal{M} \cup f}(Z) = \text{conv}(Xf^{\sigma(X)} : X \text{ a vertex of } C_{\mathcal{M}}(Z)) \subseteq C(Zf^{-\epsilon})$. Since $Yf^{\sigma(Y)} \in C_{\mathcal{M} \cup f}(Z)$, we conclude $\sigma(Y) \neq \epsilon$.
- We see the contrapositive of (2): If $\sigma(X) \in \{0, \epsilon\}$ for each vertex X of $C_{\mathcal{M}}(Z)$ and $\sigma(X_0) = \epsilon$ for some vertex X_0, then $C_{\mathcal{M} \cup f}(Z) = \text{conv}(Xf^{\sigma(X)} : X \text{ a vertex of } C_{\mathcal{M}}(Z)) = C(Zf^{-\epsilon})$, but by Lemma 6.12, $Yf^{\sigma(Y)} \circ -(X_0 f^\epsilon) \in C_{\mathcal{M} \cup f}(Z)$. Thus $\sigma(Y) = -\epsilon$.

□

6.3.2 Good Paths

Our proof of Theorem 6.7 will be in terms of paths in $G(\mathcal{M}) \cap C$ for various convex sets C.

Definition 6.21 Let $\sigma : \mathcal{C}^*(\mathcal{M}) \to \{0, +, -\}$, and let $X, Y \in \mathcal{C}^*(\mathcal{M})$ such that $X \neq -Y$.

6.3 Proof of the Las Vergnas Characterization

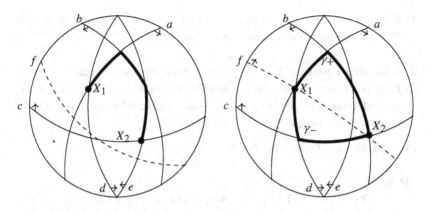

Figure 6.4 A good path and a good path pair.

A path γ in $G(\mathcal{M})$ from X to Y is **good** (with respect to σ) if $\gamma \subseteq \text{conv}(\{X,Y\})$ and for each $Z \in \gamma^0$, $\sigma(Z) \in (\sigma(X) \boxplus \sigma(Y)) \cup \{0\}$.

A pair γ_+, γ_- of paths in $G(\mathcal{M})$ from X to Y is a **good path pair** (with respect to σ) if $\gamma_+ \cup \gamma_- \subseteq \text{conv}(\{X,Y\})$, $\sigma(Z) \in \{0,+\}$ for each $Z \in \gamma_+^0$, and $\sigma(Z) \in \{0,-\}$ for each $Z \in \gamma_-^0$.

Example 6.22 The left-hand side of Figure 6.4 shows a good path from X_1 to X_2 in a case with $\sigma(X_1) = \sigma(X_2) \neq 0$. The right-hand side of Figure 6.4 shows a good path pair from X_1 to X_2 in a case with $\sigma(X_1) = \sigma(X_2) = 0$.

Lemma 6.23 (Good Path Lemma) *Let* $\sigma : \mathcal{C}^*(\mathcal{M}) \to \{0, +, -\}$ *satisfy all of the following:*

1. $\sigma(-X) = -\sigma(X)$ *for all* X.
2. *For each* $X_1, X_2 \in \mathcal{C}^*(\mathcal{M})$ *with* $X_1 \neq -X_2$ *and* $\{\sigma(X_1), \sigma(X_2)\} \neq \{0\}$, *there is a good path from* X_1 *to* X_2.
3. *For each* $X_1, X_2 \in \mathcal{C}^*(\mathcal{M})$ *with* $X_1 \neq -X_2$ *and* $\sigma(X_1) = \sigma(X_2) = 0$, *there is either a good path from* X_1 *to* X_2 *or a good path pair from* X_1 *to* X_2.

Then σ *is a localization.*

The proof of Lemma 6.23 has a dispiriting number of cases, but a simple idea: Given the function σ, Proposition 6.4 gives us a candidate $\mathcal{D}$ for the signed cocircuit set of an extension with localization σ, and this $\mathcal{D}$ clearly satisfies all of the signed circuit axioms except Elimination. The Elimination axiom for $\mathcal{D}$ will follow in various cases from the same observations about a good path γ from X_1 to X_2:

170   *Single-Element Extensions*

1. if $\sigma(X_1) = -\sigma(X_2) \neq 0$ then at some point on the path the value of σ changes, leading to an elimination of f between $X_1 f^{\sigma(X_1)}$ and $X_2 f^{\sigma(X_2)}$, and
2. if for some e we have $X_1(e) = -X_2(e) \neq 0$ then at some point on the path there is a vertex Z with $Z(e) = 0$. (This is true because if Z_1 and Z_2 are the endpoints of an edge in γ then Z_1 and Z_2 are conformal.) Then $Zf^{\sigma(Z)}$ eliminates e between $X_1 f^{\sigma(X_1)}$ and $X_2 f^{\sigma(X_2)}$.

Proof: Following Proposition 6.4, let

$$\mathcal{D} = \{Xf^{\sigma(X)} : X \in \mathcal{C}^*(\mathcal{M})\} \cup$$
$$\{(Y_1 \circ Y_2)f^0 : Y_1, Y_2 \in \mathcal{C}^*(\mathcal{M}), \text{rank}(Y_1^0 \cap Y_2^0) = \text{rank}(\mathcal{M}) - 2,$$
$$S(Y_1, Y_2) = \emptyset, \sigma(Y_1) = -\sigma(Y_2) \neq 0\}.$$

$\mathcal{D}$ satisfies the Symmetry and Incomparability Axioms. We shall see that $\mathcal{D}$ satisfies Signed Circuit Elimination, so $\mathcal{D} = \mathcal{C}^*(\mathcal{M}')$ for some oriented matroid $\mathcal{M}'$ on elements $E \cup \{f\}$. Since $\mathcal{C}^*(\mathcal{M}'\backslash f) = \min(\mathcal{D}\backslash f) = \mathcal{C}^*(\mathcal{M})$, we see that $\mathcal{M}'$ is an extension of $\mathcal{M}$ with localization σ.

Let $X_1, X_2 \in \mathcal{D}$ and $e \in E \cup \{f\}$ such that $X_1 \neq -X_2$ and $X_1(e) = -X_2(e) \neq 0$.

Case 1: $e = f$. Then $X_1 = (X_1\backslash f)f^{\sigma(X_1\backslash f)}$ and $X_2 = (X_2\backslash f)f^{\sigma(X_2\backslash f)}$. In this case, σ is not constant on a good path from $X_1\backslash f$ to $X_2\backslash f$, so there are two adjacent vertices of the path on which σ has different values. If $\sigma(Z) = 0$ for one of these vertices, then Zf^0 is our desired elimination. Otherwise, the two vertices Y_1, Y_2 satisfy $S(Y_1, Y_2) = \emptyset$, $\sigma(Y_1) = -\sigma(Y_2) \neq 0$, and, since the vertices are connected by an edge, $\text{rank}(Y_1^0 \cap Y_2^0) = r - 2$. Thus $(Y_1 \circ Y_2)f^0$ is our desired elimination.

Case 2: $e \neq f$, $X_1\backslash f, X_2\backslash f \in \mathcal{C}^*(\mathcal{M})$, and there is a good path γ from $X_1\backslash f$ to $X_2\backslash f$. Since $X_1(e) = -X_2(e) \neq 0$, there is a $Z \in \gamma^0$ such that $Z(e) = 0$. Then $Zf^{\sigma(Z)}$ is our desired elimination.

Case 3: $e \neq f$, $X_1\backslash f, X_2\backslash f \in \mathcal{C}^*(\mathcal{M})$, and there is a good path pair γ_+, γ_- from $X_1\backslash f$ to $X_2\backslash f$. Thus $\sigma(X_1\backslash f) = \sigma(X_2\backslash f) = 0$. (This case is illustrated in Figure 6.5.) Then as in our other cases we can find $Z_+ \in \gamma_+^0$ and $Z_- \in \gamma_-^0$ with $Z_+(e) = Z_-(e) = 0$. If $\sigma(Z_+) = 0$ then $Z_+ f^0$ is our desired elimination, and if $\sigma(Z_-) = 0$ then $Z_- f^0$ is our desired elimination. Otherwise, let γ be a good path from Z_+ to Z_-. Then since $\sigma(Z_+) = -\sigma(Z_-) \neq 0$, either there is a $Z_0 \in \gamma^0$ with $\sigma(Z_0) = 0$, in which case $Z_0 f^0$ is our desired elimination, or there is an edge in γ with endpoints Y_1, Y_2 such that $\sigma(Y_1) = -\sigma(Y_2) \neq 0$, in which case $(Y_1 \circ Y_2)f^0$ is our desired elimination.

Case 4: $e \neq f$ and exactly one of X_1 and X_2 has the form $(Y_1 \circ Y_2)f^0$ with $Y_1, Y_2 \in \mathcal{C}^*(\mathcal{M})$. Without loss of generality we will assume it's X_2 that has

6.3 Proof of the Las Vergnas Characterization

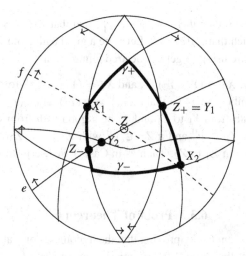

Figure 6.5 Case 3 in the proof of Lemma 6.23.

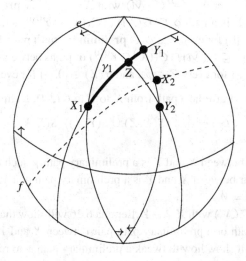

Figure 6.6 Case 4.1 in the proof of Lemma 6.23.

this form. Then there are good paths γ_1 from $X_1 \backslash f$ to Y_1 and γ_2 from $X_1 \backslash f$ to Y_2. We now have two subcases.

1. If $\sigma(X_1 \backslash f) \neq 0$ then without loss of generality assume $\sigma(X_1 \backslash f) = \sigma(Y_1)$. (This case is illustrated in Figure 6.6.) If $Y_1(e) = 0$ then $Y_1 f^{\sigma(Y_1)}$ is our desired elimination. Otherwise $Y_1(e) = X_2(e) = -(X_1 \backslash f)(e)$, so there is a $Z \in \gamma_1^0$ with $Z(e) = 0$. Then $Zf^{\sigma(Z)}$ is our desired elimination.

2. If $\sigma(X_1 \setminus f) = 0$ then there is a $Z_1 \in \gamma_1^0$ such that $Z_1(e) = 0$, and there is a $Z_2 \in \gamma_2^0$ such that $Z_2(e) = 0$. Let γ be a good path from Z_1 to Z_2. Then as in Case 3 we use γ to get our desired elimination.

Case 5: $e \neq f$, $X_1 = (Y_1 \circ Y_2) f^0$, and $X_2 = (Y_3 \circ Y_4) f^0$ for $Y_1, Y_2, Y_3, Y_4 \in C^*(\mathcal{M})$. We number the Y_i so that $\sigma(Y_1) = \sigma(Y_3)$ and $\sigma(Y_2) = \sigma(Y_4)$. Let γ_1 be a good path from Y_1 to Y_3 and γ_2 be a good path from Y_2 to Y_4. Once again, there is a $Z_1 \in \gamma_1^0$ and a $Z_2 \in \gamma_2^0$ with $Z_1(e) = Z_2(e) = 0$. Let γ_3 be a good path from Z_1 to Z_2. Then as in Case 3 we use γ_3 to get our desired elimination. □

6.3.3 Proof of Theorem 6.7

We'll prove Theorem 6.7 by proving that the hypotheses of that theorem imply the hypotheses of the Good Path Lemma (Lemma 6.23).

Definition 6.24 Let $X, Y \in C^*(\mathcal{M})$ with $X \neq -Y$. A **preliminary path** between X and Y is a path $\phi \subseteq G(\mathcal{M}) \cap C((X^0 \cap Y^0)^0)$ such that $\sigma(Z) \in (\sigma(X) \boxplus \sigma(Y)) \cup \{0\}$ for each $Z \in \phi$. A **preliminary path pair** between X and Y is a pair $\phi_+, \phi_- \subseteq G(\mathcal{M}) \cap C((X^0 \cap Y^0)^0)$ of paths between X and Y such that $\sigma(Z) \in \{0, +\}$ for every $Z \in \phi_+$ and $\sigma(Z) \in \{0, -\}$ for every $Z \in \phi_-$.

Definition 6.25 Let γ be a path from X to Y in $G(\mathcal{M})$. Define

$$S(\gamma) = \bigcup_{Z \in \gamma^0} (S(X, Z) \cup S(Y, Z)) \setminus S(X, Y).$$

A good path between X and Y is a preliminary path γ such that $S(\gamma) = \emptyset$. A good path pair between X and Y is a preliminary path pair γ_+, γ_- such that $S(\gamma_+) = S(\gamma_-) = \emptyset$.

Let $X, Y \in C^*(\mathcal{M})$ with $X \neq -Y$. Lemma 6.26 will show that there is either a preliminary path or a preliminary path pair between X and Y. Our proof of Theorem 6.7 will show how to tweak a preliminary path ϕ to reduce $S(\phi)$ and how to tweak a preliminary path pair ϕ_+, ϕ_- to reduce $S(\phi_+) \cup S(\phi_-)$.

Lemma 6.26 *Let $\sigma \colon C^*(\mathcal{M}) \to \{0, +, -\}$ be a function such that every restriction to a rank 2 minor is a localization. Let $X, Y \in C^*(\mathcal{M})$ with $X \neq -Y$.*

1. *If $\{\sigma(X), \sigma(Y)\} \neq \{0\}$ then there is a preliminary path between X and Y.*
2. *If $\{\sigma(X), \sigma(Y)\} = \{0\}$ then there is either a preliminary path between X and Y or a preliminary path pair between X and Y.*

Proof: We induct on the rank of $\mathcal{M}$, with the rank 2 case following from our hypothesis.

6.3 Proof of the Las Vergnas Characterization

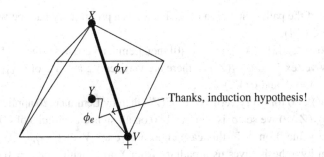

Figure 6.7 Constructing a preliminary path from X to Y.

If at least one of $\sigma(X)$ and $\sigma(Y)$ is zero, we will assume $\sigma(X) = 0$.

If $X^0 \cap Y^0$ contains nonloops then the induction hypothesis gives us a preliminary path or preliminary path pair between $X \backslash (X^0 \cap Y^0)$ and $Y \backslash (X^0 \cap Y^0)$ in $\mathcal{M}/(X^0 \cap Y^0)$, from which we get a preliminary path or preliminary path pair between X and Y in $\mathcal{M}$. So we will assume that $X^0 \cap Y^0$ contains no nonloops.

Let $A = E \backslash S(X, Y)$, so that $\mathrm{conv}(\{X, Y\}) = C(Z)$ for some $Z \in \{0, +, -\}^A$. Let $e \in Y^0 \backslash X^0$, and let $B = X^0 \cup \{e\}$: note $B \subseteq A$. Let Z' be the restriction of Z to B. Then since $\mathrm{rank}(B) = \mathrm{rank}(\mathcal{M})$, Lemma 6.19 tells us that $C_{\mathcal{M}}(Z')$ and $C_{\mathcal{M}/e}(Z' \backslash e)$ are polytopal. There is an obvious isomorphism between $C_{\mathcal{M}/e}(Z' \backslash e)$ and $C_{\mathcal{M}}((Z' \backslash e)e^0)$. Further, since $X^0 \cap Y^0$ contains no nonloops, by Lemma 6.12 Y is in the interior of $C((Z' \backslash e)e^0)$.

In a topological representation of $\mathcal{M}$, if we ignore pseudospheres not indexed by B then $C_{\mathcal{M}}(Z')$ looks like a cone with apex X and base $C_{\mathcal{M}}((Z' \backslash e)e^0)$ in S_e, and Y is in the interior of this base. The idea of the remainder of the proof is given in Figure 6.7, which illustrates $C_{\mathcal{M}}(Z')$.

To prove (1): The vertices of $C_{\mathcal{M}}((Z' \backslash e)e^0)$ are exactly the signed cocircuits V of $\mathcal{M}$ such that $V \backslash e$ is a vertex of $C_{\mathcal{M}/e}(Z' \backslash e)$. Since $\mathcal{C}^*(\mathcal{M}/e) \cong \{X \in \mathcal{C}^*(\mathcal{M}) : X(e) = 0\}$, the map σ induces a function on $\mathcal{C}^*(\mathcal{M}/e)$, which we denote σ_e. We apply Lemma 6.20 to σ_e and $C_{\mathcal{M}/e}(Z' \backslash e)$. Since $\sigma_e(Y \backslash e) \neq 0$, by Lemma 6.20 there is a vertex V of $C_{\mathcal{M}}((Z' \backslash e)e^0)$ such that $\sigma_e(V \backslash e) = \sigma_e(Y \backslash e)$, and hence $\sigma(V) = \sigma(Y)$. The induction hypothesis tells us there is a path in $G(\mathcal{M}/e)$ between $Y \backslash e$ and $V \backslash e$. Thus there is a path ϕ_e in $G(\mathcal{M})$ between Y and V in which $Z(e) = 0$ for all elements of the path. Also, $V^0 \backslash e$ contains a rank $r - 2$ subset B' of B, and so $V \backslash B'$ and $X \backslash B'$ are signed cocircuits of the rank 2 oriented matroid $\mathcal{M}/B'$. The set of nonzero elements of $C_{\mathcal{M}/B'}(Z \backslash B')$ is a good path from $V \backslash B'$ to $X \backslash B'$ in $G(\mathcal{M}/B')$. Thus there is a good path ϕ_V between V and X in $G(\mathcal{M})$ in which $Z(B') = 0$ for all elements of the path. The union of ϕ_e and ϕ_V is a preliminary path between X and Y in $G(\mathcal{M})$.

To prove (2): If $\{\sigma(X), \sigma(Y)\} = \{0\}$ then Lemma 6.20 says either $\sigma(V) = 0$ for every vertex V of $C_\mathcal{M}(Z')$ or there are vertices V_+ and V_- of $C_\mathcal{M}(Z')$ so that $\sigma(V_+) = +$ and $\sigma(V_-) = -$.

If $\sigma(V) = 0$ for every vertex V of $C(Z')$ then by Lemma 6.20 applied to σ_e and $C_{\mathcal{M}/e}(Z' \backslash e)$ we see σ is identically 0 on all signed cocircuits of $C_\mathcal{M}(Z')$ that have value 0 on e. In this case choose a vertex V of $C_{\mathcal{M}/e}(Z' \backslash e)$. The induction hypothesis gives us a path ϕ_e from Y to V with $\sigma(Z) = 0$ for all $Z \in \phi_e^0$. Also, since $V \neq -X$ but $\sigma(V) = \sigma(X) = 0$, we have a good path ϕ_V from V to X with $\sigma(Z) = 0$ for all $Z \in \phi_V^0$, as in Case 1. Thus $\sigma(Z) = 0$ for all $Z \in \phi^0$.

If we have vertices V_+ and V_-, then as in Case 1 we get a path ϕ_+ between X and Y, through Y_+, so that $\sigma(Z) = +$ for each $Z \in \phi_+^0 \backslash \{X, Y\}$, and a path ϕ_- from X to Y, through Y_-, so that $\sigma(Z) = -$ for each $Z \in \phi_-^0 \backslash \{X, Y\}$. □

Proof of Theorem 6.7: Necessity of the conditions on σ was proved in Proposition 6.6. To see sufficiency, we use the Good Path Lemma (Lemma 6.23). We'll induct on rank to see that for every $X, Y \in C^*(\mathcal{M})$ with $X \neq -Y$ there is a good path from X to Y.

By Lemma 6.26, either there is a preliminary path ϕ between X and Y, or $\{\sigma(X), \sigma(Y)\} = \{0\}$ and there is a preliminary path pair ϕ_+, ϕ_- between X and Y.

If $S(\phi) = \emptyset$ in the former case, then ϕ is our desired good path, and if $S(\phi_+) = S(\phi_-) = \emptyset$ in the latter case, then ϕ_+, ϕ_- is our desired good path pair. So the remainder of the proof shows how to reduce $S(\phi)$ resp. $S(\phi_+) \cup S(\phi_-)$ while maintaining the property of being a preliminary path or path pair.

Case 1: $\{\sigma(X), \sigma(Y)\} \neq \{0\}$. Let $s \in S(\phi)$. In a topological representation of $\mathcal{M}$, the path ϕ crosses S_s in at least two points. If Z_1 and Z_2 are the first and last crossings, then we use the induction hypothesis in $\mathcal{M}/s$ to get a good path ν from Z_1 to Z_2. Let ϕ' be the path obtained from ϕ by replacing the part of ϕ between Z_1 and Z_2 with ν. Then $S(\phi') \subseteq S(\phi) \backslash s$, because ν is a good path with $Z(s) = 0$ for all $Z \in \nu^0$, and ν has replaced all the elements of ϕ that have the wrong sign on s. Also, elements of $(\phi')^0$ satisfy the conditions for ϕ' to be a preliminary path because elements of both ϕ^0 and ν^0 satisfy these conditions.

Case 2: $\{\sigma(X), \sigma(Y)\} = \{0\}$ and there is a preliminary path ϕ between X and Y. Let $s \in S(\phi)$. Once again we let Z_1 and Z_2 be the first and last points where ϕ crosses S_s. The induction hypothesis tells us that either there is a good path ν from Z_1 to Z_2 or there is a good path pair ν_+, ν_- from Z_1 to Z_2.

In the former case, once again we let ϕ' be the path obtained from ϕ by replacing the part of ϕ between Z_1 and Z_2 with ν. Then $S(\phi') \subseteq S(\phi)\setminus s$, and elements of $(\phi')^0$ satisfy the conditions for ϕ' to be a preliminary path because elements of both ϕ^0 and ν^0 satisfy these conditions.

In the latter case we let ϕ_+ be the path obtained from ϕ by replacing the part of ϕ between Z_1 and Z_2 with ν_+, and we let ϕ_- be the path obtained from ϕ by replacing the part of ϕ between Z_1 and Z_2 with ν_-. Then $S(\phi_-) \cup S(\phi_+) \subseteq S(\phi)\setminus s$, and elements of ϕ_+^0 and ϕ_-^0 satisfy the conditions for ϕ_+, ϕ_- to be a preliminary path pair because elements of ϕ^0, ν_+^0, and ν_-^0 satisfy these conditions.

Case 3: $\{\sigma(X), \sigma(Y)\} = \{0\}$ and there is a preliminary path pair ϕ_+, ϕ_- between X and Y. Let $s \in S(\phi_+) \cup S(\phi_-)$.

If $s \in S(\phi_+)$, the same argument as in Case 1 lets us modify ϕ_+ to get a path ϕ'_+ between X and Y such that $S(\phi'_+) \subseteq S(\phi_+)\setminus s$ and ϕ'_+, ϕ_- is a preliminary path pair. Similarly if $s \in S(\phi_-)$, we can modify ϕ_- to get a path ϕ'_1 between X and Y such that $S(\phi'_-) \subseteq S(\phi_-)\setminus s$ and ϕ_+, ϕ'_- is a preliminary path pair. □

6.4 Lexicographic Extensions

A localization on $\mathcal{M}$ is a sign vector (an element of $\{0, +, -\}^{\mathcal{C}^*(\mathcal{M})}$), so the composition of two localizations on $\mathcal{M}$ makes sense.

Proposition 6.27 *If σ_1 and σ_2 are localizations on $\mathcal{M}$ then $\sigma_1 \circ \sigma_2$ is a localization on $\mathcal{M}$.*

Proof: By Theorem 6.7, it's enough to check that $\sigma_1 \circ \sigma_2$ induces a localization on each rank 2 contraction $\mathcal{N}$ of $\mathcal{M}$ with three elements.

Let $\sigma_1^{\mathcal{N}}$ and $\sigma_2^{\mathcal{N}}$ denote the localizations on $\mathcal{N}$ induced by σ_1 and σ_2. If $\sigma_1^{\mathcal{N}}$ is nonzero on every signed cocircuit then the function on $\mathcal{N}$ induced by $\sigma_1 \circ \sigma_2$ is just $\sigma_1^{\mathcal{N}}$, hence is a localization. If $\sigma_1^{\mathcal{N}}$ is identically zero then the function on $\mathcal{N}$ induced by $\sigma_1 \circ \sigma_2$ is just $\sigma_2^{\mathcal{N}}$, hence is a localization.

In the remaining case, in a topological representation of $\mathcal{N}$ the values of $\sigma_1^{\mathcal{N}}$ are as shown in Figure 6.8. The values of the function induced by $\sigma_1 \circ \sigma_2$ are the same except with the values at X and $-X$ given by σ_2. Since $\sigma_2(-Z) = -\sigma_2(Z)$ for all Z, this function is a localization. □

There is one easy way to produce many single-element extensions of a given $\mathcal{M}$, which we will describe in terms of composition.

For each element e of $\mathcal{M}$, an extension of $\mathcal{M}$ by an element parallel to e has localization that we'll denote $\sigma[e^+]$, given by $\sigma[e^+](X) = X(e)$ for all

Single-Element Extensions

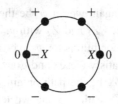

Figure 6.8 Figure for the proof of Proposition 6.27.

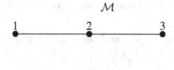

Figure 6.9 The oriented matroid $\mathcal{M}$ for Example 6.29.

X. An extension of $\mathcal{M}$ by an element antiparallel to e has localization $\sigma[e^-]$ given by $\sigma[e^-](X) = -X(e)$ for all X.

Definition 6.28 (Las Vergnas 1978) Let $\mathcal{M}$ be an oriented matroid on elements E, let $e_1, \ldots, e_k$ in E, and let $s_1, \ldots, s_k$ in $\{+, -\}$. A **lexicographic extension** of $\mathcal{M}$ given by $[e_1^{s_1}, \ldots, e_k^{s_k}]$ is an extension, denoted $\mathcal{M}[e_1^{s_1}, \ldots, e_k^{s_k}]$, whose localization is $\sigma[e_1^{s_1}] \circ \sigma[e_2^{s_2}] \circ \cdots \circ \sigma[e_k^{s_k}]$.

If $\mathcal{M}$ is realized by an arrangement ($\mathbf{v}_e : e \in E$) then we can see a realization ($\mathbf{v}_e : e \in E \cup \{f\}$) of $\mathcal{M}[e_1^{s_1}, \ldots, e_k^{s_k}]$ recursively: $\mathcal{M}[e_1^{s_1}]$ is realized with $\mathbf{v}_f$ either $\mathbf{v}_{e_1}$ or $-\mathbf{v}_{e_1}$, and given a realization ($\mathbf{v}_e : e \in E \cup \{g\}$) of $\mathcal{M}[e_1^{s_1}, \ldots, e_{k-1}^{s_{k-1}}]$, we get $\mathbf{v}_f$ to realize $\mathcal{M}[e_1^{s_1}, \ldots, e_k^{s_k}]$ as $\mathbf{v}_f = \mathbf{v}_g \pm \epsilon \mathbf{v}_{e_k}$, where $\epsilon > 0$ is small and the sign $\pm$ depends on s_k.

Example 6.29 The affine point arrangement in Figure 6.9 has corresponding oriented matroid $\mathcal{M}$ on five elements. The point arrangements in Figures 6.10 and 6.11 give two lexicographic extensions of $\mathcal{M}$, each of which can be described as a lexicographic extension in multiple ways.

Example 6.30 Figure 6.12 shows an extension of the oriented matroid from Example 6.29 that is not lexicographic.

6.4 Lexicographic Extensions

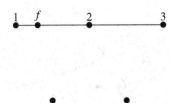

Figure 6.10 $\mathcal{M}[1^+2^+] = \mathcal{M}[1^+3^+] = \mathcal{M}[2^+1^+] = \mathcal{M}[2^+3^-]$.

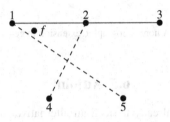

Figure 6.11 $\mathcal{M}[1^+2^+4^+] = \mathcal{M}[1^+3^+4^+] = \mathcal{M}[1^+2^+5^+] = \mathcal{M}[1^+3^+5^+] = \mathcal{M}[1^+5^+2^+] = \mathcal{M}[1^+5^+3^+] = \mathcal{M}[2^+3^-4^+]\ldots$

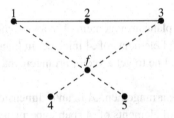

Figure 6.12 A nonlexicographic extension.

Definition 6.31 An element e of $\mathcal{M}$ is in **general position** if e is not contained in any unsigned circuit of size less than rank$(\mathcal{M}) + 1$. An extension $\mathcal{M} \cup p$ of $\mathcal{M}$ is in **general position** if p is in general position in $\mathcal{M} \cup p$.

Example 6.32 Figure 6.13 shows a configuration of five points giving a rank 3 oriented matroid $\mathcal{M}$ together with a sixth point f defining an extension of $\mathcal{M}$ in general position that is not lexicographic.

If $\mathcal{M} \cup f = \mathcal{M}[e_1^{s_1}, \ldots, e_k^{s_k}]$ and S is a maximal independent subset of $\{e_1, \ldots, e_k\}$ then $S \cup \{f\}$ is an unsigned circuit. Thus the extension $\mathcal{M}[e_i^{s_1}, \ldots, e_k^{s_k}]$ is in general position if and only if $\{e_1, \ldots, e_k\}$ contains a basis.

178   Single-Element Extensions

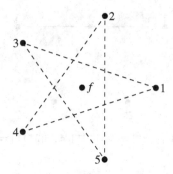

Figure 6.13 A nonlexicographic extension in general position.

6.5 Adjoints

The reader may have already noticed an alternative route to create a vector arrangement from a hyperplane arrangement (besides just taking orthogonal complements) and a route to create a hyperplane arrangement from a vector arrangement:

- Given a rank r hyperplane arrangement $\mathcal{A}$ in an r-dimensional vector space, some $(r-1)$-tuples of elements of $\mathcal{A}$ intersect in lines: Choose a nonzero element of each such line to get a vector arrangement, which we'll call an **adjoint** to $\mathcal{A}$.
- Given a rank r vector arrangement $\mathcal{A}$ in an r-dimensional vector space, some $(r-1)$-tuples of elements of $\mathcal{A}$ span hyperplanes: Take the arrangement of all such hyperplanes. Choose signs on the hyperplanes arbitrarily to get a signed hyperplane arrangement, which we'll call an **adjoint** to $\mathcal{A}$.

Adjointness is not a duality: The adjoint of an adjoint of $\mathcal{A}$ typically has more elements than $\mathcal{A}$. The notions of adjoint for vector arrangements and signed hyperplane arrangements are essentially the same thing: If $\mathcal{A} = (\mathbf{v}_e : e \in E)$ has adjoint $(\mathbf{w}_f^\perp : f \in F)$ then $(\mathbf{v}_e^\perp : e \in E)$ has adjoint $(\mathbf{w}_f : f \in F)$, and vice versa.

The relationship between arrangements and their adjoints is reflected in the oriented matroids.

Definition 6.33 (Bachem and Kern 1986) Let $\mathcal{M}$ be an oriented matroid on elements E with no loops. An **adjoint** of $\mathcal{M}$ is an oriented matroid $\mathcal{M}_{\text{adj}}$ of the same rank as $\mathcal{M}$ satisfying both of the following:

6.5 Adjoints

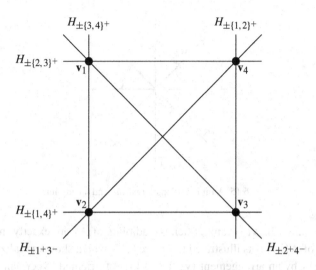

Figure 6.14 A vector realization of $\mathcal{M}$ and a hyperplane realization of $\mathcal{M}_{\text{adj}}$.

1. The set E_{adj} of elements of $\mathcal{M}_{\text{adj}}$ is a subset of $\mathcal{C}^*(\mathcal{M})$ containing exactly one element of each pair $\{X, -X\}$ of signed cocircuits of $\mathcal{M}$.
2. For each $e \in E$, the sign vector $Y_e \in \{0, +, -\}^{E_{\text{adj}}}$ defined by $Y_e(X) = X(e)$ is an element of $\mathcal{C}^*(\mathcal{M}_{\text{adj}})$.

Example 6.34 Figure 6.14 shows an arrangement $(v_e : e \in [4])$ of points in the affine plane realizing a rank 3 oriented matroid $\mathcal{M}$ and an arrangement of hyperplanes that, together with a choice of signs, realizes an adjoint of $\mathcal{M}$.

Every reorientation of an adjoint of $\mathcal{M}$ is also an adjoint of $\mathcal{M}$. Even beyond reorientation, adjoints of $\mathcal{M}$ are typically not unique. For instance, consider two realizations of $\mathcal{M}_{\text{hex}}$, the rank 3 oriented matroid on elements [6] realized by the vertices of a convex hexagon in the affine plane, numbered consecutively. If we choose a realization $(v_e : e \in [6])$ as the vertices of a regular hexagon, then the hyperplanes spanned by $\{v_1, v_4\}$, $\{v_2, v_5\}$, and $\{v_3, v_6\}$ intersect at a point, and the resulting adjoint for $\mathcal{M}_{\text{hex}}$ will have a signed cocircuit Z that is 0 on each of the three elements corresponding to these hyperplanes. But this realization is special: A generic realization $\mathcal{M}_{\text{hex}}$ by a vector arrangement will have no three of the resulting hyperplanes intersecting at a point, and so the resulting adjoint will be uniform.

Problem 6.35 Find a similar argument for the nonuniqueness of adjoints based on realizations of $\mathcal{M}_{\text{hex}}$ by signed hyperplane arrangements.

Single-Element Extensions

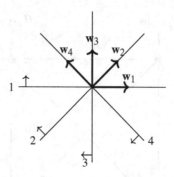

Figure 6.15 A rank 2 oriented matroid and an adjoint.

Example 6.36 If $\mathcal{M}$ is rank 2 then the adjoints of $\mathcal{M}$ are exactly the reorientations of $\mathcal{M}$. This is illustrated in Figure 6.15, which shows a realization of a rank 2 $\mathcal{M}$ by an arrangement ($\mathbf{v}_e^\perp : \mathbf{v} \in [4]$) of oriented hyperplanes and a realization of an adjoint of $\mathcal{M}$ by an arrangement ($\mathbf{w}_e : e \in [4]$) of vectors. The arrangement ($\mathbf{w}_e : e \in [4]$) is just a $\pi/2$ rotation of the arrangement ($\mathbf{v}_e : e \in [4]$).

There is a nice way to describe adjoints by way of matrices of signs.

Definition 6.37 For a matrix $M = (m_{ij}) \in \mathbb{R}^{C \times E}$, the **sign** of $\mathcal{M}$ is $\mathrm{sign}(M) := (\mathrm{sign}(m_{ij})) \in \{0, +, -\}^{C \times E}$.

If M is a matrix of signs and B is the set of indices on some columns of M, we will say that M is in **reduced row-echelon form** with respect to B if there is a bijection β from B to the set of rows of M such that each $\beta(b)$ has nonzero entry in the column indexed by b and 0 entries in all columns indexed by $B \backslash \{b\}$.

Notation 6.38 If M is a matrix and F is the set of indices on some rows of M then $M(F)$ denotes the submatrix comprising the rows indexed by F.

Recall from Proposition 1.14 that if M is a matrix of real numbers in reduced row-echelon form with oriented matroid $\mathcal{M}$ then each nonzero row of $\mathrm{sign}(M)$ is an element of $\mathcal{C}^*(\mathcal{M})$, and that for every $X \in \mathcal{C}^*(\mathcal{M})$ at least one of $X, -X$ arises in this way from a matrix that is row-equivalent to M and in reduced row-echelon form. So we've already seen matrices of signs with columns indexed by the elements of $\mathcal{M}$ and rows indexed by some of the signed cocircuits of $\mathcal{M}$. Now let $\mathcal{M}$ be an oriented matroid on elements E, choose a set E_{adj} comprised of one element from each pair $X, -X$ of signed cocircuits of $\mathcal{M}$, and let $[\mathcal{M}] \in \{0, +, -\}^{E_{\mathrm{adj}} \times E}$ be a matrix whose rows are the elements of E_{adj}. Then an adjoint for $\mathcal{M}$ with elements E_{adj} is an oriented

6.5 Adjoints

matroid with the same rank as $\mathcal{M}$ whose set of signed cocircuits includes all of the columns of $[\mathcal{M}]$.

To see the condition $\text{rank}(\mathcal{M}_{\text{adj}}) = \text{rank}(\mathcal{M})$ in terms of $[\mathcal{M}]$, let $\mathcal{N}$ be an oriented matroid on E_{adj} that has all of the columns of $[\mathcal{M}]$ as signed cocircuits, and let's look at what $[\mathcal{M}]$ tells us about independence in $\mathcal{M}$ and in $\mathcal{N}$. A set I of elements of an oriented matroid is independent if and only if there is a set $\{X_i : i \in I\}$ of signed cocircuits such that $\text{supp}(X_i) \cap I = \{i\}$ for each i. Thus $I \subseteq E$ is independent in $\mathcal{M}$ if and only if there is a submatrix $[\mathcal{M}](F)$ of $[\mathcal{M}]$ that's in reduced row-echelon form with respect to I, and the bases of $\mathcal{M}$ are the maximal subsets of E with respect to which $[\mathcal{M}]$ has a submatrix in reduced row-echelon form. Since the columns of $[\mathcal{M}]$ probably aren't all of the signed cocircuits of $\mathcal{N}$, $[\mathcal{M}]$ doesn't tell us as much about independence in $\mathcal{N}$: All we can say is that if $[\mathcal{M}](F)$ is in reduced row-echelon form then F is independent. Thus $\text{rank}(\mathcal{N}) \geq \text{rank}(\mathcal{M})$, and $\text{rank}(\mathcal{N}) = \text{rank}(\mathcal{M})$ if and only if every F with $[\mathcal{M}](F)$ a maximal reduced row-echelon form is a basis of $\mathcal{N}$.

Example 6.39 Let $\mathcal{M}$ be the rank 3 oriented matroid on five elements realized by the affine point arrangement in Figure 6.16. If we choose E_{adj} to be the signed cocircuits of $\mathcal{M}$ with first coordinate being positive, then (up to reordering rows)

$$[\mathcal{M}] = \begin{pmatrix} 0 & 0 & + & + & + \\ + & 0 & 0 & + & + \\ + & + & 0 & 0 & + \\ 0 & + & + & + & 0 \\ 0 & + & 0 & - & 0 \\ + & 0 & - & 0 & 0 \end{pmatrix}.$$

One basis for $\mathcal{M}$ is $\{1,2,3\}$. The submatrix of $[\mathcal{M}]$ comprising the first, second, and fifth rows is in reduced echelon form with respect to this basis. Thus an oriented matroid $\mathcal{N}$ with elements E_{adj} is an adjoint for $\mathcal{M}$ if and only if every column of this matrix is a signed cocircuit of $\mathcal{N}$ and the set of elements of E_{adj} indexing the first, second, and fifth rows is a basis.

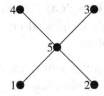

Figure 6.16 Illustration for Example 6.39.

Every rank 3 oriented matroid has an adjoint (see Exercise 6.7). Section 6.6.6 will give an example of a rank 4 oriented matroid with no adjoint. As we have seen, realizable oriented matroids have adjoints. Beyond this, we have no clear characterization of oriented matroids with adjoints.

When $\mathcal{M}$ does have an adjoint, such an adjoint gives us insight on extensions of $\mathcal{M}$.

Definition 6.40 Let $\mathcal{M}$ be an oriented matroid on elements E, and let $f \notin E$. The **extension poset** $\mathcal{E}(\mathcal{M})$ of $\mathcal{M}$ is the poset of all proper single-element extensions $\mathcal{M} \cup f$ of $\mathcal{M}$, ordered by $\mathcal{N}_1 \geq \mathcal{N}_2$ if $\mathcal{N}_1 \rightsquigarrow \mathcal{N}_2$.

Let $\mathcal{A} = (\mathbf{v}_e : e \in E)$ be a vector arrangement in $\mathbb{R}^r$ and $f \notin E$. The **extension poset** $\mathcal{E}(\mathcal{A})$ of $\mathcal{A}$ is the set of all oriented matroids of extensions $(\mathbf{v}_e : e \in E \cup \{f\})$ of $\mathcal{A}$ by a nonloop $\mathbf{v}_f$ in $\mathbb{R}^r$, also ordered by weak maps.

Notation 6.41 Let $\mathcal{A} = (\mathbf{v}_e : e \in E)$ be a vector arrangement in $\mathbb{R}^r$ and $f \notin E$. For $\mathbf{x} \in \mathbb{R}^r$ we let $\mathcal{A} \cup \mathbf{x}$ denote the arrangement $(\mathbf{v}_e : e \in E \cup \{f\})$ with $\mathbf{v}_f = \mathbf{x}$.

Let $\mathcal{A}$ be a vector arrangement with oriented matroid $\mathcal{M}$, let $\mathcal{A}_{\text{adj}}$ be an adjoint for $\mathcal{A}$ and $\mathcal{M}_{\text{adj}}$ its oriented matroid, and let χ be a chirotope for $\mathcal{M}$. The relationship between $\mathcal{E}(\mathcal{A})$ and $\mathcal{M}_{\text{adj}}$ is clear if we look at a picture such as Figure 6.14, showing an arrangement of points in $\mathbb{A} \subset \mathbb{R}^r$ and an adjoint to that arrangement. The elements of $\mathcal{V}^*(\mathcal{M}_{\text{adj}}) \backslash \{0\}$ index the non-$\{0\}$ cones in $\mathbb{R}^r$ cut out by the adjoint arrangement. But also each of these cones corresponds to an element of $\mathcal{E}(\mathcal{A})$. To see this, let $\mathbf{x}, \mathbf{y} \in \mathbb{R}^r$, and let χ_x and χ_y be the extensions of χ to chirotopes of the oriented matroids of $\mathcal{A} \cup \mathbf{x}$ and $\mathcal{A} \cup \mathbf{y}$. Then

$\quad$ $\mathbf{x}$ and $\mathbf{y}$ are in the same cone.

$\Leftrightarrow$ For every independent $I = \{\mathbf{v}_{e_2}, \ldots, \mathbf{v}_{e_r}\}$ from $\mathcal{A}$, $\mathbf{x}$ and $\mathbf{y}$ are on the same side of the hyperplane spanned by I.

$\Leftrightarrow$ For every independent $\{e_2, \ldots, e_r\} \subseteq E$,
$$\chi_x(f, e_2, \ldots, e_r) = \chi_y(f, e_2, \ldots, e_r).$$
$\Leftrightarrow \chi_x = \chi_y$.

Thus we have $\mathcal{V}^*(\mathcal{M}_{\text{adj}}) \backslash \{0\} \cong \mathcal{E}(\mathcal{A}) \subseteq \mathcal{E}(\mathcal{M})$.

The same conclusion holds for general oriented matroids with adjoints, though the proof is more involved.

Proposition 6.42 (Bachem and Kern 1986, Cordovil 1987) *For every adjoint $\mathcal{M}_{\text{adj}}$ of $\mathcal{M}$ we have an injective poset map*

$$\mathcal{V}^*(\mathcal{M}_{\text{adj}}) \backslash \{0\} \rightarrow \mathcal{E}(\mathcal{M}),$$

taking each Y to an extension with localization

$$\sigma_Y(X) = \begin{cases} Y(X) & \text{if } X \in E_{\text{adj}}, \\ -Y(-X) & \text{if } -X \in E_{\text{adj}}. \end{cases}$$

The map of Proposition 6.42 need not be a bijection. For instance, as we have seen, different realizations $\mathcal{A}$ of $\mathcal{M}_{\text{hex}}$ lead to different adjoints. But one might conjecture this map to induce a homotopy equivalence of order complexes. This conjecture is a special case of the *Extension Space Conjecture*. This special case was only disproved in 2016: See Section 10.2 for more discussion.

Lemma 6.43 *Let A be a flat of $\mathcal{M}$, and let $\mathcal{M}_{\text{adj}}$ be an adjoint for $\mathcal{M}$ with elements E_{adj}. Let*

$$E_1 = \{X \in E_{\text{adj}} : X(A) = 0\},$$
$$\mathcal{M}_1 = \mathcal{M}_{\text{adj}}(E_1),$$
$$E_2 = \{X \backslash A : X \in E_1\},$$

and let $\mathcal{M}_2$ be the oriented matroid isomorphic to $\mathcal{M}_1$ by the isomorphism

$$E_1 \to E_2$$
$$X \to X \backslash A.$$

Then $\mathcal{M}_2$ is an adjoint for $\mathcal{M}/A$.

The proof is a matter of bookkeeping, which we can carry out by way of the matrix representation of the adjoint conditions.

Proof: First, notice that $[\mathcal{M}/A]$ is defined with respect to a choice of elements of $\mathcal{C}^*(\mathcal{M}/A)$ to be the elements of our adjoint. Each pair $\pm X$ in $\mathcal{C}^*(\mathcal{M}/A)$ is $\pm Y \backslash A$ for some $\pm Y$ in $\mathcal{C}^*(\mathcal{M})$: If Y is the element of this pair that we chose to be in E_{adj}, then $Y \backslash A$ will be our choice to be in an adjoint for $\mathcal{M}/A$.

The columns of $[\mathcal{M}](E_1)$ indexed by elements of A are all $\mathbf{0}$: Let $[\mathcal{M}]_1$ be the submatrix of $[\mathcal{M}]$ with rows indexed by E_1 and columns indexed by $E \backslash A$. Then $\mathcal{M}_1$ has every column of $[\mathcal{M}]_1$ as a signed cocircuit. $[\mathcal{M}/A]$ is the matrix obtained from $[\mathcal{M}]_1$ by renaming the indices on rows via the bijection $E_1 \to E_2$. From the definition of $\mathcal{M}_2$ we see that the elements of $\mathcal{M}_2$ are indeed the indices on rows of $[\mathcal{M}/A]$ and that each column of $[\mathcal{M}/A]$ is indeed a signed cocircuit of $\mathcal{M}_2$.

If B is a basis for $\mathcal{M}/A$ and $B' \supset B$ is a basis for $\mathcal{M}$, then $[\mathcal{M}]$ has a submatrix $[\mathcal{M}](F)$ that is maximal with respect to being in reduced row-echelon form with respect to B', and $[\mathcal{M}](F \cap E_1)$ is maximal in $[\mathcal{M}]_1$ with respect to being in reduced row-echelon form with respect to B'. In particular, $|F \cap E_1| = |B| = \text{rank}(\mathcal{M}/A)$. F is a basis of $\mathcal{M}_{\text{adj}}$, and so $F \cap E_1$ is a basis in $\mathcal{M}(E_1)$. We conclude $\text{rank}(\mathcal{M}_2) = \text{rank}(\mathcal{M}(E_1)) = |F \cap E_1| = \text{rank}(\mathcal{M}/A)$. □

Proof of Proposition 6.42: The only issue is showing that each σ_Y is indeed a localization. So let $\mathcal{M}/A$ be a rank 2 contraction, with A a flat. By Lemma 6.43, $\mathcal{M}_{\text{adj}}(\{X \in \mathcal{C}^*(\mathcal{M}) : X(A) = 0\})$ is isomorphic to an adjoint for $\mathcal{M}/A$. As we saw in Example 6.36, this adjoint is just a reorientation of $\mathcal{M}/A$. Fix a vector arrangement $\mathcal{A}$ that we view as a realization of both a reorientation of $\mathcal{M}_{\text{adj}}(\{X \in \mathcal{C}^*(\mathcal{M}) : X(A) = 0\})$ and $\mathcal{M}/A$, by the canonical identification $X \leftrightarrow X \backslash A$ of elements. The restriction of Y to $\{X \in \mathcal{C}^*(\mathcal{M}) : X(A) = 0\}$ is a covector of $\mathcal{M}_{\text{adj}}(\{X \in \mathcal{C}^*(\mathcal{M}) : X(A) = 0\})$, and so it arises by a signed hyperplane partitioning the arrangement $\mathcal{A}$. The same hyperplane describes the restriction of σ_Y to $\mathcal{C}^*(\mathcal{M}/A)$. Thus σ_Y cannot give rise to a forbidden subconfiguration. □

In Sections 2.7.1 and 2.7.2 we had a somewhat fuzzy discussion of "point arrangements in a distorted affine plane." Now we'll make that discussion more precise and relate it to adjoints.

Definition 6.44 (Björner et al. 1999) A rank r **pseudoconfiguration of points** is a pair $(\mathcal{A}, P)$, where $\mathcal{A} = (\mathcal{S}_e : e \in E)$ is an essential rank r arrangement of signed pseudospheres and P is a subset of the set of 0-cells in the resulting cell complex satisfying all of the following:

1. $|P| \geq r$.
2. Every subset of P of size $r - 1$ is contained in S_e for some $e \in E$.
3. For each $e \neq f \in E$ $P \cap S_e \not\subseteq P \cap S_f$.

Definition 6.44 was the idea lurking behind the examples in Sections 2.7.1 and 2.7.2. There we chose the elements of P to all lie in one open hemisphere in S^2, we projected that hemisphere radially onto an affine plane and we only drew some of the pseudospheres in $\mathcal{A}$.

Proposition 6.45 (Björner et al. 1999) *Let $\mathcal{A} = (\mathcal{S}_e : e \in E)$ be an essential rank r arrangement of signed pseudospheres, and let $(\mathcal{A}, P)$ be a pseudoconfiguration of points. For each $e \in E$ define $X_e \in \{0, +, -\}^P$ by defining $X_e(p) = \epsilon$, where $p \in S_e^\epsilon$.*

1. $\bigcup_{e \in E} \{X_e, -X_e\}$ is the set of signed cocircuits of a rank r oriented matroid $\mathcal{M}$.
2. If $\mathcal{N}$ is the oriented matroid of $\mathcal{A}$ then $\mathcal{N}$ is an adjoint of $\mathcal{M}$.

(We omit the proof.)

Definition 6.46 (Björner et al. 1999) If $\mathcal{M}$ is as in Proposition 6.45 then $(\mathcal{A}, P)$ is called a **representation of $\mathcal{M}$ by a pseudoconfiguration of points**.

Corollary 6.47 *An oriented matroid has a representation as a pseudoconfiguration of points if and only if it has an adjoint.*

Proof: One direction is given by Proposition 6.45. Conversely, if $\mathcal{M}_{adj}$ is an adjoint of $\mathcal{M}$, $\mathcal{A}$ is a topological representation of $\mathcal{M}_{adj}$, and P is the set of 0-cells in the resulting cell complex corresponding to $\{Y_e : e \in E\} \subset \mathcal{C}^*(\mathcal{M}_{adj})$ then $(\mathcal{A}, P)$ is a representation of $\mathcal{M}$ by a pseudoconfiguration of points. □

Remark 6.48 Adjoints were studied at the matroid level before the dawn of oriented matroids (cf. Oxley 1992).

Remark 6.49 Consider the broad question of where to draw the line between weird and nonweird oriented matroids. Here an oriented matroid is "weird" if it is not only nonrealizable, it also fails to have some obviously desirable combinatorial property that all realizable oriented matroids have. The existence of a representation by a pseudoconfiguration of points is one such property, and in Section 6.6 we'll see a flavor of additional properties that hold for all oriented matroids with adjoints. Thus, for many purposes, oriented matroids with adjoints aren't very weird.

6.6 Intersection Properties

The aim of this section is to examine generalizations (or lack thereof) of easy results about hyperplane arrangements of the following form.

Given a hyperplane arrangement $\mathcal{A}$ in a vector space V and [a specified set S of intersections of elements of $\mathcal{A}$], there exists a hyperplane H in V that contains all of the elements of S.

The oriented matroid version of such a result would claim existence of an extension $\mathcal{M} \cup f$ of a given $\mathcal{M}$ such that $\text{rank}(A \cup \{f\}) = \text{rank}(A)$ for certain sets A. These are conjectures that make sense at the matroid level, not just the oriented matroid level: They were laid out at the matroid level in Bachem and Kern (1986). At the oriented matroid level they take on a topological flavor.

In fact, none of the properties of this type that we will examine hold for arbitrary oriented matroids: Each is an *intersection property* that an oriented matroid may or may not have. All of the intersection properties we examine obviously hold for realizable oriented matroids, and all of them can be seen to hold for matroids with adjoints by way of Proposition 6.42.

6.6.1 Levi's Intersection Property

Definition 6.50 (Bachem and Kern 1986) A rank r oriented matroid $\mathcal{M}$ satisfies **Levi's intersection property** if for each $X_1, \ldots, X_{r-1} \in \mathcal{C}^*(\mathcal{M})$ there is an extension $\mathcal{M} \cup f$ by a nonloop such that $X_i f^0 \in \mathcal{C}^*(\mathcal{M})$ for each i.

If $\mathcal{M}$ is realized by a hyperplane arrangement $\mathcal{A}$, then each of the X_i corresponds to a one-dimensional intersection of elements of $\mathcal{A}$. Every collection of $r - 1$ lines in an r-dimensional vector space is contained in a hyperplane in that vector space, and so Levi's intersection property holds for realizable oriented matroids.

The Generalized Levi Enlargement Lemma (Theorem 5.40) tells us that all rank 3 oriented matroids satisfy Levi's intersection property.

More generally, if $\mathcal{M}$ has an adjoint $\mathcal{M}_{\text{adj}}$, then $S = \{X_1, \ldots, X_{r-1}\}$ is a set of elements of $\mathcal{M}_{\text{adj}}$ of rank at most $r - 1$, so there is a $Y \in \mathcal{C}^*(\mathcal{M}_{\text{adj}})$ such that $Y(S) = 0$. The embedding $\mathcal{V}^*(\mathcal{M}_{\text{adj}}) \to \mathcal{E}(\mathcal{M})$ given by Proposition 6.42 sends Y to an extension $\mathcal{M} \cup f$ satisfying the conclusion of Levi's intersection property. Thus

$$\{\mathcal{M} : \mathcal{M} \text{ is realizable}\} \subset \{\mathcal{M} : \mathcal{M} \text{ has an adjoint}\}$$
$$\subseteq \{\mathcal{M} : \mathcal{M} \text{ satisfies Levi's intersection property}\}.$$

Section 6.6.6 will give an example of a rank 4 oriented matroid that does not satisfy Levi's intersection property.

6.6.2 The Generalized Euclidean Property

Two subspaces V, V' of a real vector space will have nonzero elements in their intersection if and only if $\text{rank}(V) + \text{rank}(V') > \text{rank}(V \cup V')$. So, given a vector arrangement $(\mathbf{v}_e : e \in E)$ and $G, H \subseteq E$ such that $\text{rank}(\mathbf{v}_g : g \in G) + \text{rank}(\mathbf{v}_h : h \in H) > \text{rank}(\mathbf{v}_e : e \in G \cup H)$, by choosing a nonzero element of $\langle \mathbf{v}_g : g \in G \rangle \cap \langle \mathbf{v}_h : h \in H \rangle$ and naming it $\mathbf{v}_f$, we get a vector arrangement $(\mathbf{v}_e : e \in E \cup \{f\})$ such that $\text{rank}(\mathbf{v}_g : g \in G) = \text{rank}(\mathbf{v}_g : g \in G \cup \{f\})$ and $\text{rank}(\mathbf{v}_h : h \in H) = \text{rank}(\mathbf{v}_h : h \in H \cup \{f\})$.

To say this in terms of the hyperplane arrangement $(\mathbf{v}_e^\perp : e \in E)$, let $V_G = \bigcap_{e \in G} \mathbf{v}_e^\perp$ and $V_H = \bigcap_{e \in H} \mathbf{v}_e^\perp$. Let r be the dimension of the vector space in which our arrangement lies. Then V_G has dimension $r - \text{rank}\{\mathbf{v}_g : g \in G\}$ and V_H has dimension $r - \text{rank}\{\mathbf{v}_h : h \in H\}$. So our statement is that if $\dim(V_G) + \dim(V_H) < r$ then there is a hyperplane containing $V_G \cup V_H$.

Definition 6.51 (Bachem and Kern 1986) An oriented matroid $\mathcal{M}$ satisfies the *generalized Euclidean property* if, for each pair G, H of subsets of E such

6.6 Intersection Properties

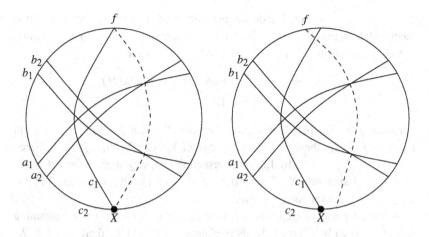

Figure 6.17 Illustration for Remark 6.52.

that $\operatorname{rank}(G) + \operatorname{rank}(H) > \operatorname{rank}(G \cup H)$, there is an extension $\mathcal{M} \cup \{f\}$ in which f is not a loop, $\operatorname{rank}(G \cup \{f\}) = \operatorname{rank}(G)$, and $\operatorname{rank}(H \cup \{f\}) = \operatorname{rank}(H)$.

We'll see the justification for the name in Section 6.6.3.

We will often think of intersection properties in terms of pseudosphere arrangements. Let $(\mathcal{S}_e : e \in E)$ be a topological representation of $\mathcal{M}$ in $S^{\operatorname{rank}(\mathcal{M})-1}$. A pair G, H of subsets of E satisfies $\operatorname{rank}(G) + \operatorname{rank}(H) > \operatorname{rank}(G \cup H)$ if $\dim(S_G) + \dim(S_H) < \operatorname{rank}(\mathcal{M}) - 1$. An extension as in Definition 6.51 is given by a nondegenerate signed pseudosphere $\mathcal{S}_f$ such that $S_G \cup S_H \subseteq S_f$.

Remark 6.52 The extension $\mathcal{M} \cup \{f\}$ in Definition 6.51 is far from being unique, even in the realizable case and even when $\operatorname{rank}(G) + \operatorname{rank}(H) = \operatorname{rank}(\mathcal{M}) + 1$. Figure 6.17 shows a rank 3 oriented matroid $\mathcal{M}$ with subsets $A = \{a_1, a_2\}$ and $B = \{b_1, b_2\}$ along with two extensions $\mathcal{M} \cup \{f\}$ satisfying the conclusion of Definition 6.51 but with different localizations at the signed cocircuit X.

Proposition 6.53 *If $\mathcal{M}$ satisfies Levi's Intersection Property then $\mathcal{M}$ satisfies the generalized Euclidean property.*

Proof: Let G and H be subsets of E such that $\operatorname{rank}(G) + \operatorname{rank}(H) > \operatorname{rank}(G \cup H)$. Let I' be a maximal independent set in $\mathcal{M}/(G \cup H)$. By Exercise 6.4 it's enough to find an extension $(\mathcal{M}/I') \cup f$ of $\mathcal{M}/I'$ in which $\operatorname{rank}(G \cup \{f\}) = \operatorname{rank}(G)$ and $\operatorname{rank}(H \cup \{f\}) = \operatorname{rank}(H)$.

Let I_G be a maximal independent subset of G and I_H be a maximal independent subset of H. Let $B_G = I_G \dot\cup J_G$ and $B_H = I_H \dot\cup J_H$ be maximal independent sets in $G \cup H$. Let $r = \text{rank}(G \cup H)$. Then

$$|J_G| + |J_H| = r - \text{rank}(G) + r - \text{rank}(H)$$
$$\leq r - 1.$$

For each $j \in J_G$ there is a signed cocircuit X_j with $X_j(B_G \setminus j) = 0$, and for each $j \in J_H$ there is a signed cocircuit Y_j with $Y_j(B_H \setminus j) = 0$. Since $|J_G| + |J_H| \leq r - 1$, by Levi's Intersection Property there is an extension $(\mathcal{M}/I') \cup f$ such that $X_j f^0 \in \mathcal{C}^*((\mathcal{M}/I') \cup f)$ for each $j \in F_G$ and $Y_j f^0 \in \mathcal{C}^*((\mathcal{M}/I') \cup f)$ for each $j \in F_H$.

Assume by way of contradiction there is a basis of $(\mathcal{M}/I') \cup f$ containing $I_G \cup \{f\}$. Then by (Signed) Basis Exchange with $I_G \cup J_G$ there is a $j \in J_G$ such that $(I_G \cup J_G \cup \{f\}) \setminus \{j\}$ is a basis. But the signed cocircuit $X_j f^0$ is 0 on $(I_G \cup J_G \cup \{f\}) \setminus \{j\}$, and so $\text{rank}((I_G \cup J_G \cup \{f\}) \setminus \{j\}) \leq r - 1$, a contradiction. Thus $\text{rank}(I_G \cup \{f\}) = \text{rank}(I_G)$, and so $\text{rank}(G \cup \{f\}) = \text{rank}(G)$. By the same argument $\text{rank}(H \cup \{f\}) = \text{rank}(H)$. □

One can restrict the generalized Euclidean property to a property about subsets G, H of E with fixed ranks.

Definition 6.54 Let k, l be positive integers. An oriented matroid $\mathcal{M}$ satisfies the (k,l)-**intersection property** if, for every pair A, B of subsets of E with $\text{rank}(A) = k$, $\text{rank}(B) = l$, and $\text{rank}(A \cup B) < k + l$, there is an extension $\mathcal{M} \cup \{f\}$ in which f is not a loop, $\text{rank}(A \cup \{f\}) = \text{rank}(A)$, and $\text{rank}(B \cup f) = \text{rank}(B)$.

All oriented matroids satisfy every $(1,l)$-intersection property, and if $\mathcal{M}$ satisfies the (k,l)-intersection property for each k,l such that $k + l \leq \text{rank}(\mathcal{M}) + 1$ then $\mathcal{M}$ satisfies all (k,l)-intersection properties. Because all rank 3 oriented matroids have adjoints (Exercise 6.7), they satisfy all intersection properties. Thus the interesting questions are about $(k, r - k + 1)$-intersection properties for rank r oriented matroids, with $k \geq 2$ and $r \geq 4$. The remainder of this chapter explores the case $k = 2$, which is the only case that has been much studied. As we shall see, this case leads us to unexpected connections to Euclid's Axioms and to linear programming.

6.6.3 Euclidean Oriented Matroids

Definition 6.55 A rank r oriented matroid $\mathcal{M}$ on elements E is **Euclidean** if it satisfies the $(2, r-1)$-intersection property.

6.6 Intersection Properties

In other words, $\mathcal{M}$ is Euclidean if, for each pair A, B of subsets of E such that $\text{rank}(A) = 2$ and $\text{rank}(B) = r - 1$, there is an extension $\mathcal{M} \cup f$ by a nonloop in which $\text{rank}(A \cup \{f\}) = 2$ and $\text{rank}(B \cup \{f\}) = r - 1$.

Our favorite interpretations of the Euclidean property will come from choosing the rank 2 set A to be an independent set $\{e_\infty, e_0\}$ and the rank $(r-1)$-set B to be a flat. Thus $B = X^0$ for some signed cocircuit X. If $X(e_0) = 0$ or $X(e_\infty) = 0$ then an extension $\mathcal{M} \cup f$ of $\mathcal{M}$ with f parallel to e_0 resp. e_∞ will satisfy our conclusion $\text{rank}(\{e_\infty, e_0, f\}) = 2$ and $\text{rank}(B \cup \{f\}) = r - 1$. So we will assume $X(e_0) \neq 0$ and $X(e_\infty) \neq 0$.

We can justify the name "Euclidean" by pondering the realizable case. If $\mathcal{M}$ has a realization by an arrangement of signed hyperplanes ($\mathcal{H}_e : e \in E$), then a signed hyperplane $\mathcal{H}_f$ defines an extension $\mathcal{M} \cup \{f\}$ with $\text{rank}\{e_\infty, e_0, f\} = 2$ and $\text{rank}(B \cup \{f\}) = r-1$ if and only if $H^0_{e_0} \cap H^0_f \subseteq H^0_{e_\infty}$ and $\bigcap_{b \in B} H^0_b \subseteq H^0_f$. Let $\mathbb{A}$ be an affine hyperplane parallel to (but not equal to) $H^0_{e_\infty}$. Then our conditions on H^0_f can be expressed as: $H^0_f \cap \mathbb{A}$ is parallel to $H^0_e \cap \mathbb{A}$ and H^0_f contains the point $\bigcap_{b \in B} H^0_b \cap \mathbb{A}$. Thus $H^0_f \cap \mathbb{A}$ demonstrates Euclid's Parallel Postulate: For every affine hyperplane H and every point p, there is a hyperplane that is parallel to H and contains p.

Notice why we named one element e_∞: Its corresponding hyperplane plays the role of the "hyperplane at infinity." Also notice that our discussion of a hyperplane arrangement ($\mathcal{H}_e : e \in E$) and affine objects in $\mathbb{A}$ could as well have been phrased as being about a pseudosphere arrangement ($\mathcal{S}_e : e \in e$) and objects in the hemisphere $S^+_{e_\infty}$. This is the starting point for expanding our discussion to general oriented matroids.

Definition 6.56 An **affine oriented matroid** is a pair $(\mathcal{M}, e_\infty)$ where $\mathcal{M}$ is an oriented matroid and e_∞ is a nonloop in $\mathcal{M}$. The **affine space** of $(\mathcal{M}, e_\infty)$ is $\{X \in \mathcal{V}^*(\mathcal{M}) : X(e_\infty) = +\}$. The **hyperplane at infinity** of $(\mathcal{M}, e_\infty)$ is $\{X \in \mathcal{V}^*(\mathcal{M}) : X(e_\infty) = 0\}$.

When we refer to rank, elements, and so on of an affine oriented matroid $(\mathcal{M}, e_\infty)$, we mean the rank, elements, and so on of $\mathcal{M}$.

If $(\mathcal{M}, e_\infty)$ is an affine rank r oriented matroid, then there is a topological representation $(\mathcal{S}_e : e \in E)$ of $\mathcal{M}$ in the unit sphere of $\mathbb{R}^r$ in which S_{e_∞} is the equator $\{\mathbf{x} \in S^{r-1} : x_r = 0\}$. The affine space of $(\mathcal{M}, e_\infty)$ indexes the cells in the open hemisphere $\{\mathbf{x} \in S^{r-1} : x_r > 0\}$. The homeomorphism

$$\rho : \{\mathbf{x} \in S^{r-1} : x_r > 0\} \to \{\mathbf{x} : x_r = 1\},$$
$$\rho(\mathbf{x}) = x_r^{-1} \mathbf{x}$$

lets us identify the affine space of $(\mathcal{M}, e_\infty)$ with the parts in a partition of the affine space $\{\mathbf{x} : x_r = 1\}$. For each $e \in E\backslash\{e_\infty\}$, let $\rho(\mathcal{S}_e) = (\rho(S_e \cap \{\mathbf{x} : x_r > 0\}), \rho(S_e^+ \cap \{\mathbf{x} : x_r > 0\}), \rho(S_e^- \cap \{\mathbf{x} : x_r > 0\}))$. We call $(\rho(\mathcal{S}_e) : e \in E\backslash\{e_\infty\})$ a **topological representation of** $(\mathcal{M}, e_\infty)$.

We call a nonempty set $\rho(S_A)$ a **pseudosubspace** of the representation. If $\rho(S_A)$ is zero-dimensional then we call it a **vertex**, if $\rho(S_A)$ is one-dimensional then we call it a **pseudoline**, and if $\rho(S_A)$ is $(r-2)$-dimensional then we call it a **pseudohyperplane**. Two pseudohyperplanes are **parallel** if they are disjoint.

Definition 6.57 Let $(\mathcal{M}, e_\infty)$ be an affine oriented matroid. Two elements f, g of $E\backslash\{e_\infty\}$ are **parallel** resp. **antiparallel** in $(\mathcal{M}, e_\infty)$ if they are parallel resp. antiparallel in $\mathcal{M}/e_\infty$.

Thus f and g are either parallel or antiparallel in $(\mathcal{M}, e_\infty)$ if and only if $S_f \cap S_g \subseteq S_{e_\infty}$, and this is true if and only if the corresponding pseudohyperplanes in a topological representation of $(\mathcal{M}, e_\infty)$ are parallel.

We can rephrase the Euclidean property for oriented matroids analogously to the Parallel Postulate.

Proposition 6.58 *An oriented matroid $\mathcal{M}$ is Euclidean if and only if, for each affine $(\mathcal{M}, e_\infty)$, and for each pseudohyperplane H_g^0 and vertex v in a topological representation $(\mathcal{H}_e : e \in E)$ of $(\mathcal{M}, e_\infty)$, the representation can be extended to $(\mathcal{H}_e : e \in E \cup \{f\})$ such that H_f^0 is parallel to H_g^0 and $v \in H_f^0$.*

6.6.4 Oriented Matroid Programs and Directions

To briefly review some affine geometry: Consider an affine space $\mathbb{A} = \{\mathbf{x} \in \mathbb{R}^n : \mathbf{x} \cdot \mathbf{v}_\infty = 1\}$. Its hyperplane at infinity is $\{\mathbf{x} \in \mathbb{R}^n : \mathbf{x} \cdot \mathbf{v}_\infty = 0\}$. The **sphere at infinity** is the set of all rays $\mathbf{x}\mathbb{R}_{>0} \neq \{\mathbf{0}\}$ in the hyperplane at infinity. There is an obvious bijection between the sphere at infinity and the set of parallelism classes of rays in $\mathbb{A}$. So when we work with an affine space $\mathbb{A}$ without considering the ambient space $\mathbb{R}^n$, we often identify the sphere at infinity with the set of directions in $\mathbb{A}$.

Let $\mathbf{w} \in \mathbb{R}^n$, and let $\phi : \mathbb{A} \to \mathbb{R}$ be the function $\phi(\mathbf{x}) = \mathbf{w} \cdot \mathbf{x}$. The directions in which ϕ is increasing are the directions $\mathbf{d}\mathbb{R}_{>0}$ with $\mathbf{d} \in (\mathbf{w}^\perp)^+$.

All of these observations on directions generalize to affine oriented matroids, although the analogs to linear functions will be somewhat restricted: We will only have one "linear function" for each element of $E\backslash\{e_\infty\}$. If the oriented matroid is realized by a vector arrangement $(\mathbf{v}_e : e \in E)$ in $\mathbb{R}^n$, then for each $e_0 \in E\backslash\{e_\infty\}$ the "linear function" of e_0 in the oriented matroid corresponds to the function $\phi(\mathbf{x}) = \mathbf{x} \cdot \mathbf{v}_{e_0}$.

6.6 Intersection Properties

Definition 6.59 A **direction** of an affine oriented matroid $(\mathcal{M}, e_\infty)$ is a nonzero element of the hyperplane at infinity.

If $e \in E \setminus \{e_\infty\}$ and X is a direction for $(\mathcal{M}, e_\infty)$ then we say e is **increasing in the direction** X if $X(e) = +$, and e is **decreasing in the direction** X if $X(e) = -$.

Recall the cocircuit graph $G(\mathcal{M}) \subseteq \mathcal{V}^*(\mathcal{M})$ and the various abuses of graph theory terminology introduced in Section 6.3.

Definition 6.60 The **graph** $G(\mathcal{M}, e_\infty)$ of the affine oriented matroid $(\mathcal{M}, e_\infty)$ is the induced subgraph of $G(\mathcal{M})$ whose vertices are the vertices X of $G(\mathcal{M})$ with $X(e_\infty) = +$.

In other words, $G(\mathcal{M})$ is a graph whose vertex set is $\{X \in \mathcal{C}^*(\mathcal{M}) : X(e_\infty) = +\}$, with an edge $Z_1 \circ Z_2$ between vertices Z_1 and Z_2 if Z_1 and Z_2 are conformal and $Z_1 \circ Z_2$ has rank 2 in $\mathcal{V}^*(\mathcal{M})$.

Let $e_0 \in E \setminus \{e_\infty\}$. Then e_0 determines a partial direction on $G(\mathcal{M}, e_\infty)$ (i.e., directions on a subset of the edges of $G(\mathcal{M}, e_\infty)$) as follows. Let Y be an edge with endpoints Z_1 and Z_2. We define the direction on Y in three equivalent ways.

1. Let X be the elimination of e_∞ between Z_1 and $-Z_2$. (This is unique because $\mathcal{M}/(Z_1^0 \cap Z_2^0)$ is rank 2.) Then we direct Y from Z_1 to Z_2 if e_0 is decreasing in the direction X, and we direct Y from Z_2 to Z_1 if e_0 is increasing in the direction X.
2. Notice that $-X$ is the elimination of e_∞ between Z_2 and $-Z_1$. We direct Y from Z_1 to Z_2 if e_0 is increasing in the direction $-X$, and we direct Y from Z_2 to Z_1 if e_0 is decreasing in the direction $-X$.
3. Let $A = Y^0 = Z_1^0 \cap Z_2^0$. Then in every centrally symmetric topological representation $(\mathcal{S}_e : e \in E)$ of $\mathcal{M}$, S_A is a topological circle, and $S_A \cap (S_{e_\infty} \cup S_{e_\infty}^+)$ is a curve whose endpoints are antipodal points in S_{e_∞}. If one endpoint X is in $S_{e_0}^-$ then the other endpoint $-X$ is in $S_{e_0}^+$. In this case we direct the curve from X to $-X$, and we direct Y consistently with this direction. (Thus we direct the edge from X_1 to X_2 if the cell corresponding to X_1 comes before the cell corresponding to X_2 in this direction.)

Example 6.61 Figure 6.18 shows an oriented matroid program $(\mathcal{M}, e_\infty)$, its graph (indicated by bold edges), and the partial direction on the graph induced by an element e_0.

Definition 6.62 An **oriented matroid program** is a triple $(\mathcal{M}, e_\infty, e_0)$, where $\mathcal{M}$ is an oriented matroid, e_∞ is a nonloop of $\mathcal{M}$, e_0 is a non-coloop of $\mathcal{M}$, and $e_\infty \neq e_0$.

192   Single-Element Extensions

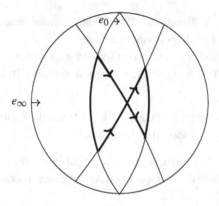

Figure 6.18 Directions in $G(\mathcal{M}, e_\infty)$.

The program is **nontrivial** if $\{e_\infty, e_0\}$ is independent in $\mathcal{M}$.

A nontrivial program $(\mathcal{M}, e_\infty, e_0)$ is **Euclidean** if, for each signed cocircuit X of the affine space of $(\mathcal{M}, e_\infty)$, there is an extension $\mathcal{M} \cup f$ by a nonloop f such that f is parallel to e_0 in $(\mathcal{M} \cup f, e_\infty)$ and $Xf^0 \in \mathcal{V}^*(\mathcal{M} \cup f)$.

Thus an oriented matroid is Euclidean if and only if each of its nontrivial programs is Euclidean.

Let $G(\mathcal{M}, e_\infty, e_0)$ denote the graph $G(\mathcal{M}, e_\infty)$ with the partial direction given by e_0.

Definition 6.63   A **walk** in $G(\mathcal{M}, e_\infty, e_0)$ is a sequence $X_1, X_1 \circ X_2, X_2, \ldots, X_{k-1}$, $X_{k-1} \circ X_k, X_k$ of vertices and edges in $G(\mathcal{M}, e_\infty)$. The walk is **closed** if $X_1 = X_k$. This closed walk is a **cycle** if $X_1 = X_k$ is the only repeated vertex.

This walk is **partially directed** if

- each edge $X_i \circ X_{i+1}$ is either undirected or is directed from X_i to X_{i+1}, and
- at least one edge is directed.

This walk is **wholly undirected** if each edge is undirected.

Our intuition from realizable oriented matroids is that $G(\mathcal{M}, e_\infty, e_0)$ should have no partially directed closed walks, because in the realizable case the value of the linear function $\phi(\mathbf{x}) = \mathbf{v}_{e_0} \cdot \mathbf{x}$ should increase along a partially directed walk. An extremely cool theorem says that this intuition is correct for Euclidean programs, and no others.

Theorem 6.64   (Mandel 1982) *Let* $(\mathcal{M}, e_\infty, e_0)$ *be a nontrivial oriented matroid program, and let X be a vertex of $G(\mathcal{M}, e_\infty, e_0)$. The following are equivalent:*

6.6 Intersection Properties

1. There is an extension $\mathcal{M} \cup f$ of $\mathcal{M}$ such that f is parallel to e_0 in $(\mathcal{M}, e_\infty)$ and $Xf^0 \in \mathcal{V}^*(\mathcal{M} \cup f)$.
2. There is no partially directed closed walk in $G(\mathcal{M}, e_\infty, e_0)$ with X as a vertex.

If $G(\mathcal{M}, e_\infty, e_0)$ has a partially directed closed walk then it also has a partially directed cycle.

Corollary 6.65 $(\mathcal{M}, e_\infty, e_0)$ is Euclidean if and only if $G(\mathcal{M}, e_\infty, e_0)$ has no partially directed cycles.

Definition 6.66 A partially directed cycle in $(\mathcal{M}, e_\infty, e_0)$ is called a **non-Euclidean cycle**.

Lemma 6.67 Let $(\mathcal{M}, e_\infty, e_0)$ be a nontrivial oriented matroid program, and let $\mathcal{M} \cup f$ be an extension of $\mathcal{M}$ with localization σ such that e_0 and f are parallel in $(\mathcal{M} \cup f, e_\infty)$. Let Y be an edge of $G(\mathcal{M}, e_\infty, e_0)$ with endpoints X_1 and X_2.

1. If $\sigma(X_1) = 0$ then:

    a. Y is undirected if and only if $\sigma(X_2) = 0$.
    b. Y is directed from X_1 to X_2 if and only if $\sigma(X_2) = +$.
    c. Y is directed from X_2 to X_1 if and only if $\sigma(X_2) = -$.

2. If $\sigma(X_1) = +$ and Y is directed from X_1 to X_2 then $\sigma(X_2) = +$.

Proof: 1. The elimination of e_∞ between $-X_1 f^0$ and $X_2 f^{\sigma(X_2)}$ in $\mathcal{V}^*(\mathcal{M} \cup f)$ is $Z f^{\sigma(X_2)}$, where Z is the elimination of e_∞ between $-X_1$ and X_2 in $\mathcal{V}^*(\mathcal{M})$. From this and parallelism of e_0 and f we see that $\sigma(X_2) = \sigma(Z) = Z(e_0)$. Thus

$$Y \text{ is undirected} \Leftrightarrow Z(e_0) = 0$$
$$\Leftrightarrow \sigma(X_2) = 0,$$

and the other two statements follow similarly.

2. Once again let Z be the elimination of e_∞ between $-X_1$ and X_2 in $\mathcal{V}^*(\mathcal{M})$. Then $Z(e_0) = +$ and $\sigma(Z) = + = \sigma(X_1)$. Thus $\sigma(X_2) = +$ since otherwise $\{\pm Z, \pm X_1, \pm X_2\}$ would give us a "forbidden subconfiguration" (Figure 6.2). □

Proof of Theorem 6.64: (1)$\Rightarrow$ (2): Consider a partially directed walk $X_1, (X_1 \circ X_2), X_2, \ldots, X_k$ in $G(\mathcal{M}, e_\infty, e_0)$ with X as a vertex. Lemma 6.67 tells us that

- if $\sigma(X_i) = +$ for some i then $\sigma(X_j) = +$ for every $j > i$,
- if $\sigma(X_i) = -$ for some i then $\sigma(X_j) = -$ for every $j < i$,

- if $\sigma(X_i) = \sigma(X_l) = 0$ for some $i < l$ then all edges of the walk between X_i and X_l are undirected.

Since $\sigma(X) = 0$, we conclude that $\sigma(X_1) \in \{0,-\}$, $\sigma(X_k) \in \{0,+\}$, and since at least one edge is directed at least one of $\sigma(X_1)$ and $\sigma(X_2)$ is nonzero. Thus $X_1 \neq X_k$.

(2)$\Rightarrow$ (1): Define an equivalence relation on the vertices of $G(\mathcal{M}, e_\infty, e_0)$ in which $Y \sim Y'$ if either there is a wholly undirected walk from Y to Y' or there is a partially directed cycle with both Y and Y' as vertices. For each Y let $[Y]$ denote its equivalence class.

Notice that if there is a wholly undirected walk γ from Y to Y' and Y is in a partially directed closed walk, then by concatenating the partially directed closed walk with γ and its reverse we get a partially directed closed walk with both Y and Y' as vertices. In particular, if $Z \in \mathcal{C}^*(\mathcal{M})$ and $Z \sim X$ then Z is not in a partially directed closed walk.

Also notice that for equivalence classes $C_1 \neq C_2$, the following are equivalent:

1. For *some* $Y_1 \in C_1$ and $Y_2 \in C_2$, there is a wholly undirected or partially directed walk from Y_1 to Y_2.
2. For *all* $Y_1 \in C_1$ and $Y_2 \in C_2$, there is a wholly undirected or partially directed walk from Y_1 to Y_2.

Define a partial order on equivalence classes by $C_1 \leq C_2$ if these equivalent conditions hold.

Let $\preceq$ be an extension of this partial order to a linear order. Define $\sigma: \mathcal{C}^*(\mathcal{M}) \to \{0, +, -\}$ as follows:

1. If $Z(e_\infty) = 0$ then $\sigma(Z) = Z(e_0)$.
2. If $Z(e_\infty) = +$ then

$$\sigma(Z) = \begin{cases} 0 & \text{if } Z \sim X, \\ - & \text{if } [Z] \prec [X], \\ + & \text{if } [Z] \succ [X]. \end{cases}$$

3. If $Z(e_\infty) = -$ then $\sigma(Z) = -\sigma(-Z)$.

Figure 6.19 illustrates this. The reader should verify that the signs show values of σ that are forced by the partial order on equivalence classes, and that the value of σ at Y could be 0, $+$, or $-$, depending on the choice of $\preceq$.

We will show that σ is a localization using the Las Vergnas condition (Theorem 6.7). Once we have shown this, it's clear that the extension $\mathcal{M} \cup f$ satisfies our conditions.

6.6 Intersection Properties

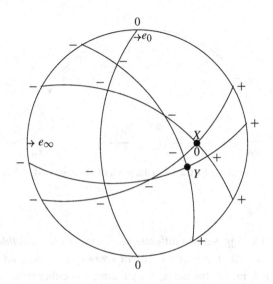

Figure 6.19 The localization in the proof of Theorem 6.64.

So, consider a rank 2 contraction $\mathcal{M}/A$ of $\mathcal{M}$, where A is a flat.

If $e_\infty \in A$ then the induced function σ_A on $C^*(\mathcal{M}/A)$ satisfies $\sigma_A(Z) = Z(e_0)$ for all Z. Thus σ_A is the localization of an extension by an element parallel to e_0.

Otherwise there are exactly two elements $\pm Z_0$ of $C^*(\mathcal{M}/A)$ such that $Z_0(e_\infty) = 0$. We then have three cases:

1. If $Z_0(e_0) = 0$ then for each maximal element Y of $\mathcal{V}^*(\mathcal{M}/A)$ with $Y(e_\infty) = +$ the edge YA^0 of $G(\mathcal{M}, e_\infty, e_0)$ is undirected, and so all vertices of $G(\mathcal{M}, e_\infty, e_0)$ of the form ZA^0 are equivalent. Thus σ_A has a constant value s on $\{Z \in C^*(\mathcal{M}/A) : Z(e_\infty) = +\}$ and constant value $-s$ on $\{Z \in C^*(\mathcal{M}/A) : Z(e_\infty) = -\}$, and life is good.
2. If $Z_0(e_0) \neq 0$ and there is an $X' \sim X$ such that $X'\backslash A \in C^*(\mathcal{M}/A)$, then without loss of generality say $Z_0(e_0) = +$. Then the edges in $G(\mathcal{M}, e_\infty, e_0)$ of the form YA^0 are directed to make a partially directed walk from $-Z_0$ to Z_0.

    Since X' is not in a partially directed closed walk, $\pm X'\backslash A$ are the only two signed cocircuits of $\mathcal{M}/A$ on which σ_A has value 0. Our partial order then tells us that the remaining values of σ_A are as shown in Figure 6.20.
3. If $Z(e_0) \neq 0$ and there is no $X' \sim X$ such that $X'\backslash A \notin C^*(\mathcal{M}/A)$, then our argument is as in the previous case, except that σ_A never has value 0.

□

Single-Element Extensions

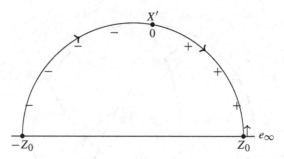

Figure 6.20 Case 2 in the proof of Theorem 6.64.

Theorem 6.64 suggests two different senses in which *a Euclidean oriented matroid is one in which we have a global sense of direction*. At a point **p** in an affine space $\mathbb{A}$ in $\mathbb{R}^n$, one can specify a direction either by a ray $\mathbf{p} + \mathbf{v}\mathbb{R}_{>0}$ emanating from **p** or by the parallel translate of $\mathbf{v}^\perp \cap \mathbb{A}$ through **p**. Either way, this direction extends globally: To every element of $\mathbb{A}$, we can associate a parallel translate of the ray at **p** and a parallel translate of $\mathbf{v}^\perp \cap \mathbb{A}$. In the oriented matroid program $(\mathcal{M}, e_\infty, e_0)$, the orientation on an edge of $G(\mathcal{M}, e_0, e_\infty)$ can be thought of as a local approximate indication of the direction given by e_0; the existence of a non-Euclidean cycle demonstrates that this local sense of direction doesn't extend to a global one. In a topological representation $(\mathcal{S}_e : e \in E)$ of $\mathcal{M}$, the half-pseudosphere $S_{e_0} \cap S^+_{e_\infty}$ is analogous to $\mathbf{v}^\perp \cap \mathbb{A}$, and the nonexistence of an extension $\mathcal{M} \cup f$ as in the first equivalent statement of Theorem 6.64 demonstrates that there is no "parallel translate of e_0 through X."

So if a program $(\mathcal{M}, e_\infty, e_0)$ is non-Euclidean then finding a family of "parallel translates of e_0" through all signed cocircuits in the affine space of $(\mathcal{M}, e_\infty)$ is impossible. Conversely, if $(\mathcal{M}, e_\infty, e_0)$ is Euclidean then the following proposition shows that we can take successive extensions to get a family of parallel translates through all signed cocircuits in the affine space.

Proposition 6.68 *If $(\mathcal{M}, e_\infty, e_0)$ is Euclidean and $\mathcal{M} \cup f$ is an extension such that f is parallel to e_0 in $(\mathcal{M}, e_\infty)$ then $(\mathcal{M} \cup f, e_\infty, e_0)$ is Euclidean.*

Proof: Parallelism of f and e_0 tells us that the partially directed graphs $G(\mathcal{M} \cup f, e_\infty, e_0)$ and $G(\mathcal{M} \cup f, e_\infty, f)$ are equal.

Assume by way of contradiction that $G(\mathcal{M} \cup f, e_\infty, f)$ has a partially directed closed walk. For every vertex X of the closed walk we must have $X(f) \neq 0$: Otherwise we could take an extension of $\mathcal{M} \cup f$ by a parallel copy of f and violate Theorem 6.64. Thus by deleting f from each element of the closed walk we get a closed walk in $G(\mathcal{M}, e_\infty, e_0)$, a contradiction. □

True story: The author once had a nightmare in which she was trapped in a non-Euclidean affine oriented matroid. She tried to run in a consistent direction to escape, but she kept ending up back where she started.

6.6.5 Oriented Matroid Programs and the Simplex Algorithm

Oriented matroid programs first appeared in Fukuda's thesis (Fukuda 1982), which built on the work of Bland (1977) and Rockafellar (1969) to cast linear programming in oriented matroid terms. Consider the linear programming problem: Given a linear function $\phi \colon \mathbb{R}^n \to \mathbb{R}$ (called the *objective function*), an $m \times n$ matrix A, and $\mathbf{b} \in \mathbb{R}^m$, find $\mathbf{x}$ to maximize $\phi(\mathbf{x})$ subject to the constraint $A\mathbf{x} \geq \mathbf{b}$. To say this in linear terms rather than affine: Given $\mathbf{v} \in \mathbb{R}^{n+1}$ and a $m \times (n+1)$ matrix $(A|-\mathbf{b})$, find $\binom{\mathbf{x}}{1}$ to maximize $\mathbf{v}\binom{\mathbf{x}}{1}$ subject to the constraint $(A|-\mathbf{b})\binom{\mathbf{x}}{1} \geq \mathbf{0}$.

Phrased this way, the problem is about oriented hyperplanes in $\mathbb{R}^{n+1}$. Let $\mathbf{r}_1, \ldots, \mathbf{r}_m$ be the rows of $(A|-\mathbf{b})$, and let $\mathcal{A} = (\mathbf{r}_i^\perp : i \in [m])$. The constraint $(A|-\mathbf{b})\binom{\mathbf{x}}{1} \geq \mathbf{0}$ says that $\binom{\mathbf{x}}{1}$ lies in the intersection of the closed positive sides of the elements of $\mathcal{A}$. The solution to our problem (if there is one) lies in the affine hyperplane $\mathbb{A} := \{\mathbf{y} \in \mathbb{R}^{n+1} : y_{n+1} = 1\}$, for which the hyperplane at infinity is $\mathbf{e}_{n+1}^\perp$. The set of points $\binom{\mathbf{x}}{1}$ satisfying our constraints is called the *feasible region* for the problem.

Now let's think about the function $\phi(\mathbf{x}) = \mathbf{v}\binom{\mathbf{x}}{1}$ that we're trying to maximize. The level sets of this function in the affine plane $\mathbb{A}$ form a family of hyperplanes parallel to $(\mathbf{v}^\perp)^0 \cap \mathbb{A}$. Figure 6.21 indicates how we can visualize our LP problem.

The value of $\mathbf{v}\binom{\mathbf{x}}{1}$ increases as $\binom{\mathbf{x}}{1}$ moves to the right. The goal of the problem is to find the rightmost nontrivial intersection of a level set with the feasible region. (Of course, there might be no such point, if the feasible region is empty or unbounded. And there may be multiple such points, as in Figure 6.21.)

An alternative way to visualize our problem is to consider the affine subspace $(\mathbf{v}^\perp)^0 \cap \mathbb{A}$ (which we assume to be nonempty, since otherwise the whole problem is trivial) and all of the lines of the form $\bigcap_{i \in S}(\mathbf{r}_i^\perp)^0 \cap \mathbb{A}$. If such an affine line is parallel to or contained in $(\mathbf{v}^\perp)^0 \cap \mathbb{A}$ (i.e., if $\bigcap_{i \in S}(\mathbf{r}_i^\perp)^0 \cap (\mathbf{e}_{n+1}^\perp)^0 \subseteq (\mathbf{v}^\perp)^0 \cap (\mathbf{e}_{n+1}^\perp)^0$) then the value of $\mathbf{v}\binom{\mathbf{x}}{1}$ is constant on this line,

198 Single-Element Extensions

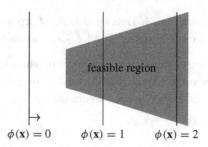

$\phi(\mathbf{x}) = 0$ $\phi(\mathbf{x}) = 1$ $\phi(\mathbf{x}) = 2$

Figure 6.21 Level sets for a linear programming problem.

and we will view this line as undirected. Otherwise, we think of this line as directed in the direction in which $\mathbf{v}\binom{\mathbf{x}}{1}$ is increasing.

It's intuitive (and even true!) that if our problem has a solution, then it has a solution that's in the intersection of the feasible region with one of these directed lines. In fact, the *Simplex Algorithm* for finding a solution to a linear programming problem can be summarized as follows:

1. Find a vertex $\binom{\mathbf{z}}{1}$ of the feasible region (i.e., a zero-dimensional intersection $\bigcap_{i \in A}(\mathbf{r}_i^\perp)^0 \cap \mathbb{A}$ in the boundary of the feasible region).
2. Consider the lines $\bigcap_{i \in A'}(\mathbf{r}_i^\perp)^0 \cap \mathbb{A}$ containing $\binom{\mathbf{z}}{1}$. If one of these lines is directed so that, starting at $\binom{\mathbf{z}}{1}$, we can walk along this line in the increasing direction while staying in the feasible region, then we set out along one such line, in the increasing direction, as far as we can while staying in the feasible region. (There may have been several choices of line to walk along. If so, we can pick one arbitrarily.)

   If there is no such line, then $\mathbf{z}$ is a solution to our original problem.
3. If we did set out on a walk in the previous step, one possibility is that we walk on forever without leaving the feasible region. In this case our feasible region is unbounded and our problem has no solution. The other possibility is that our walk ended at another vertex of the feasible region. In this case we repeat our process of looking for a directed line to walk along.

A theorem which we won't prove says that if the linear programming problem has a solution then the Simplex Algorithm will terminate in one such solution. Even without proving this theorem, it's clear that this algorithm can't get into infinite loops, because the value of $\phi(\mathbf{z})$ increases at each step of the algorithm.

6.6 Intersection Properties

Bland (1977) described how to generalize the Simplex Algorithm to problems on general oriented matroids, leaving open the question of whether the algorithm could get into infinite loops. In Fukuda's thesis (Fukuda 1982) he described the algorithm in the terms we've been using, with $\mathcal{M}$ given by the signed hyperplane $\mathcal{H}_{e_\infty} = \mathbf{e}_{n+1}^\perp$, the signed hyperplane $\mathcal{H}_{e_0} = \mathbf{v}^\perp$, and the signed hyperplanes $\mathbf{r}_i^\perp$. In these terms, the Simplex Algorithm follows a partially directed walk in $G(\mathcal{M}, e_\infty, e_0)$. Using the non-Euclidean oriented matroid EFM[8] described in Section 6.6.6, he showed that the generalization of the Simplex Algorithm to oriented matroids can indeed get into infinite loops.

6.6.6 A Non-Euclidean Oriented Matroid

We've seen that all oriented matroids of rank at most three are Euclidean, and in Section 6.6.7 we'll see that Euclideanness is preserved under duality. Thus all rank 4 oriented matroids with at most seven elements are Euclidean. In this section we'll give two very different descriptions of the same non-Euclidean rank 4 oriented matroid on eight elements. The first description is purely combinatorial and uses matroid-theoretic reasoning to see directly that it violates the Euclidean property. The second description is topological and finds an oriented matroid program with a non-Euclidean cycle.

First Description: An Orientation of the Vámos Matroid

Intersection properties make sense at the matroid level, and so we can talk about an unoriented matroid being Euclidean. Here we'll introduce the *Vámos matroid*, a matroid which is orientable and which is also non-Euclidean as a matroid. (Bland and Las Vergnas 1979 first showed that the Vámos matroid was orientable.)

To construct the Vámos matroid, we begin with the set of vertices of the three-dimensional cube in affine space. We then move two of the vertices a, a' to get the set of points in Figure 6.22. For each $\{x, y\} \subset \{a, b, c, d\}$, the set $\{x, x', y, y'\}$ is coplanar. There are six such four-element sets, corresponding to the four faces of the cube and two planes that cut the cube in half. Let $\mathcal{M}_0$ be the oriented matroid of this arrangement $\mathcal{A}$ of points. Thus $\mathcal{M}_0$ is a realizable rank 4 oriented matroid on eight elements. One of its signed circuits is $X_0 = \{b, d'\}^+ \{b', d\}^-$, and one of its signed cocircuits is $Y_0 = \{a, a'\}^+ \{c, c'\}^-$.

The Vámos matroid is obtained from the underlying matroid of $\mathcal{M}_0$ by declaring $\{b, b', d, d'\}$ to be independent. There are two ways one might try to tweak a chirotope for $\mathcal{M}_0$ to get an orientation for the Vámos matroid, and either one works.

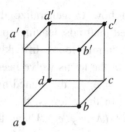

Figure 6.22 The Vámos matroid.

Proposition 6.69 *Let χ_0 be a chirotope for $\mathcal{M}_0$. Let $\chi_{\text{Vam}} : E^4 \to \{0, +, -\}$ be either of the two alternating functions that coincides with χ_0 except on orderings of b, b', d, d' and satisfies $\chi_{\text{Vam}}(b, b', d, d') \neq 0$. Then χ_{Vam} is a chirotope.*

Proof: We check that χ_{Vam} satisfies Signed Basis Exchange. Since χ_0 is a chirotope, it's enough to consider the case when one of our bases is $\{b, b', d, d'\}$. Since $\{a, a', c, c'\}$ is not a basis, our second basis is contained in $E \backslash \{x\}$ for some $x \in \{a, a', c, c'\}$. But for each of these four values of x, the restriction of χ_{Vam} to $(E \backslash \{x\})^4$ is the chirotope of a realizable oriented matroid: A realization is obtained from the arrangement in Figure 6.22 by deleting x and moving b slightly out of the plane spanned by $\{b', d, d'\}$ while staying in the plane spanned by $\{b', a, a'\}$ (if $x \in \{c, c'\}$) or staying in the plane spanned by $\{b', c, c'\}$ (if $x \in \{a, a'\}$). □

Proposition 6.69 tells us that there is at least one oriented matroid whose underlying matroid is the Vámos matroid.

Proposition 6.70 (Björner et al. 1999, proposition 7.5.1) *If $\mathcal{M}$ is an oriented matroid whose underlying matroid is the Vámos matroid then $\mathcal{M}$ is non-Euclidean.*

The reader familiar with ordinary matroids will observe that, by stripping away references to signs in the proof of Proposition 6.70, we get a proof of something a bit stronger: Even within the realm of nonorientable matroids, the Vámos matroid lacks the extensions required by the Euclidean property.

Proposition 6.71 *The Vámos matroid is non-Euclidean.*

Proof of Proposition 6.70: Assume by way of contradiction that there is an extension $\mathcal{M} \cup f$ by a nonloop in which $\{b, b', f\}$ is rank 2 and $\{a, a', d, d', f\}$ is rank 3. Thus also $\{a, a', d, f\}$ is rank 3. We draw the following conclusions:

6.6 Intersection Properties

1. There is a signed circuit with support $\{b, b', f\}$ and a signed circuit with support $\{a, a', b, b'\}$, so we can eliminate b' to get a signed circuit Z_1 with support $\{a, a', b, f\}$. Thus $\{a, a', b, f\}$ has rank 3.
2. Since $\{a, a', b, f\}$ and $\{a, a', d, f\}$ are rank 3 in $\mathcal{M}$, in $\mathcal{M}/\{a, a'\}$ the sets $\{b, f\}$ and $\{d, f\}$ are rank 1. But $\{b, d\}$ has rank 2 in $\mathcal{M}/\{a, a'\}$, so we conclude that f is a loop in $\mathcal{M}/\{a, a'\}$. Thus $\{a, a', f\}$ is dependent in $\mathcal{M}$. Since $\{a, a', c\}$ is a maximal independent subset of $\{a, a', c, c', f\}$, we conclude that $\{a, a', c, c', f\}$ has rank 3.
3. Since $\{a, a', c, c', f\}$ has rank 3 and $\{b, b', c, c', f\}$ has rank 3 in $\mathcal{M}$, in $\mathcal{M}/\{c, c'\}$ the sets $\{a, a', f\}$ and $\{b, b', f\}$ are rank 1. But $\{a, b\}$ has rank 2 in $\mathcal{M}/\{c, c'\}$, so we conclude that f is a loop in $\mathcal{M}/\{c, c'\}$. Thus $\{c, c', f\}$ has rank 2 in $\mathcal{M}$, and $\{c, c', d, d', f\}$ has rank 3.
4. Since $\{a, a', d, d', f\}$ has rank 3 and $\{c, c', d, d', f\}$ has rank 3 in $\mathcal{M}$, by the same reasoning we conclude that $\{d, d', f\}$ has rank 2 in $\mathcal{M}$.
5. Since $\{b, b', f\}$ and $\{d, d', f\}$ are both rank 2 in $\mathcal{M}$, there are signed circuits with support $\{b, b', f\}$ and $\{d, d', f\}$. By eliminating f we see that $\{b, b', d, d'\}$ has rank 3, contradicting the definition of $\mathcal{M}$.

□

Second Description: By Topological Representation

We begin with an arrangement $(\mathcal{H}_1, \ldots, \mathcal{H}_6, \mathcal{H}_{e_0}, \mathcal{H}_{e_\infty})$ of signed hyperplanes in $\mathbb{R}^4$. Figure 6.23 shows a convex polytope in the affine three-space $\mathbb{A} := \{\mathbf{x} \in \mathbb{R}^4 : x_4 = 1\}$. The coordinates of the vertices are as follows:

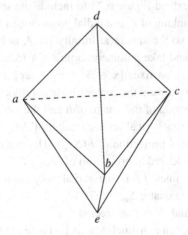

Figure 6.23 Six affine planes in $\mathbb{A}$.

Single-Element Extensions

$$a = (-1, 1, 0, 1),$$
$$b = (0, -1, 0, 1),$$
$$c = (1, 1, 0, 1),$$
$$d = (0, 0, 1, 1),$$
$$e = (0, 0, -1, 1).$$

The six facets of the polytope define six hyperplanes in $\mathbb{R}^4$:

$$H_1^0 = \langle a, b, d \rangle,$$
$$H_2^0 = \langle b, c, d \rangle,$$
$$H_3^0 = \langle a, c, d \rangle,$$
$$H_4^0 = \langle a, b, e \rangle,$$
$$H_5^0 = \langle b, c, e \rangle,$$
$$H_6^0 = \langle a, c, e \rangle.$$

Put signs on each of these hyperplanes arbitrarily to define $\mathcal{H}_1, \ldots, \mathcal{H}_6$. Let $\mathcal{H}_{e_\infty} = (0, 0, 0, 1)^\perp$, and let $\mathcal{H}_{e_0} = (0, 0, 1, 2)^\perp$. Thus $H_{e_0}^0 \cap \mathbb{A}$ would appear in Figure 6.23 as the plane $\{(x_1, x_2, -2, 1) \in \mathbb{R}^4\}$, which lies below the polytope and is parallel to the plane spanned by a, b, and c. Let $E = \{1, 2, 3, 4, 5, 6, e_\infty, e_0\}$, let $\mathcal{M}$ be the oriented matroid of this arrangement, and let $(\mathcal{S}_e : e \in E)$ be the associated arrangement of equators in S^3.

Let L_1 be the line through a and b, L_2 be the line through b and c, and L_3 be the line through a and c.

Now we'd like to extend Figure 6.23 to include the sphere at infinity. One way to do this is by thinking of $\mathbb{A}$ as a radial projection of the open hemisphere $\{\mathbf{x} \in S^3 : x_3 > 0\}$, so the sphere at infinity for $\mathbb{A}$ is $\{\mathbf{x} \in S^3 : x_3 = 0\}$. Let $\epsilon > 0$ be small, and take a homeomorphism h from $\{\mathbf{x} \in S^3 : x_3 \geq 0\}$ to $\{\mathbf{x} \in S^3 : x_3 \geq \epsilon\}$ that fixes $\{\mathbf{x} \in S^3 : x_3 \geq 2\epsilon\}$ and pulls elements of $\{\mathbf{x} \in S^3 : 0 \leq x_3 < 2\epsilon\}$ toward the center of our polytope radially. Figure 6.24 shows the radial projection of the image of h and the most salient parts of the image under h of the topological representation of $\mathcal{M}$. The enclosing sphere is the image under radial projection of $h(\mathcal{S}_{e_\infty})$. The equator indicated on this sphere is the image under radial projection of $h(\mathcal{S}_{e_\infty} \cap \mathcal{S}_{e_0})$.

Since each of the lines L_i is the radial projection of part of a circle intersecting $\mathcal{S}_{e_\infty}$ in the equator $\mathcal{S}_{e_0}$, in $G(\mathcal{M}, e_\infty, e_0)$ the edges corresponding to the 1-cells $\overline{ab}$, $\overline{bc}$, and $\overline{ac}$ are undirected.

Let's tweak $\mathcal{M}$ to get a non-Euclidean $\mathcal{M}'$. There are two equivalent ways to describe the tweak. Note the signed cocircuits $\pm P$ indicated in Figure 6.24.

6.6 Intersection Properties

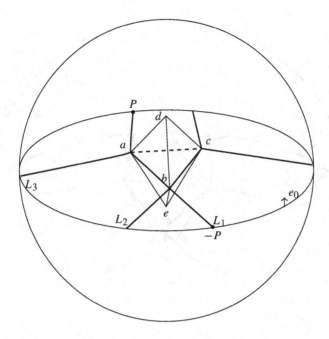

Figure 6.24 Some elements in the affine space of $\mathcal{M}$.

1. $\mathcal{M}'$ is obtained from $\mathcal{M}$ by deleting e_0 and replacing it with a new element, also called e_0, with localization

$$\sigma(X \backslash e_0) = \begin{cases} X(e_0) & \text{if } X \notin \{P, -P\}, \\ - & \text{if } X = P, \\ + & \text{if } X = -P. \end{cases}$$

We can see this is a localization because in our topological representation of $\mathcal{M}$, the only sets $A \subseteq E$ such that S_A is a pseudocircle contained in S_{e_0} are sets containing e_0. Thus the arrangement $(S_e : e \in E \backslash \{e_0\})$ has no pseudocircles that intersect S_{e_0} in more than two points. So we see that in every rank 2 contraction of $\mathcal{M} \backslash e_0$, the only difference between the localization induced by σ and the localization induced by our "old" e_0 is that possibly the localization of the "old" e_0 had value 0 at exactly two points, while σ has values + resp. − at those two points.
2. To see this tweak topologically: We tweak $(S_e : e \in E)$. The tweak keeps everything fixed outside of a small neighborhood of the two points representing P and $-P$ in $L_1 \cap S_{e_\infty}$. Within this neighborhood, we bend

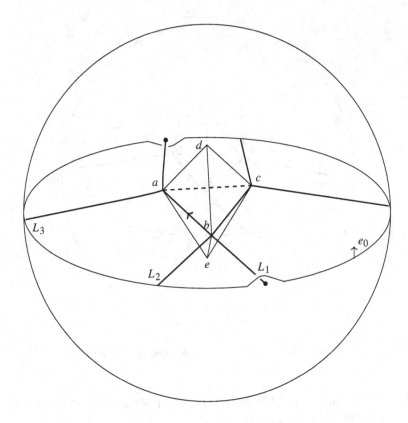

Figure 6.25 Some elements in the affine space of $\mathcal{M}'$.

$\mathcal{S}_{e_0}$ slightly to get a new signed pseudosphere $\mathcal{S}'_{e_0}$, so that P is in $(S'_{e_0})^+$ and $-P$ is in $(S'_{e_0})^-$. See Figure 6.25.

It is easy to check that we still have an arrangement of signed pseudospheres. The set of elements of our original arrangement that intersect the small neighborhood we changed is $\{\mathcal{S}_{e_\infty}, \mathcal{S}_{e_0}, \mathcal{S}_1, \mathcal{S}_4\}$. We check directly that for each subset of this set containing $\mathcal{S}_{e_0}$, the change in their intersections when we replace $\mathcal{S}_{e_0}$ with $\mathcal{S}'_{e_0}$ replaces a topological disk with another topological disk with the same boundary, and hence does not change the topology of the intersection.

Thus our new arrangement of signed pseudospheres defines a new oriented matroid $\mathcal{M}'$. But in $G(\mathcal{M}', e_\infty, e_0)$ the edge $\overline{ab}$ is now directed as indicated in Figure 6.25, and so the triangle with vertices a, b, c is a partially directed cycle.

Of course, there was nothing special about L_1. We could go further and perturb $\mathcal{S}_{e_0}$ at the intersections of each line L_i with $S^-_{e_\infty} \cap S_{e_0}$ to get an oriented matroid program $(\mathcal{M}'', e_\infty, e_0)$ so that each edge of the triangle with vertices a, b, c is directed to make a directed cycle in $G(\mathcal{M}'', e_\infty, e_0)$. The resulting oriented matroid is known as the *Edmonds–Fukuda–Mandel oriented matroid* EFM[8]. The underlying matroid of EFM[8] is uniform, and hence realizable. This demonstrates that Euclideanness of an oriented matroid is not determined by the underlying matroid.

Problem 6.72 Find an isomorphism between $\mathcal{M}'$ and an orientation of the Vamos matroid.

6.6.7 Euclieanness and Duality

Theorem 6.73 $(\mathcal{M}, e_\infty, e_0)$ *is Euclidean if and only if* $(\mathcal{M}^*, e_0, e_\infty)$ *is Euclidean.*

This was originally proven in Fukuda's thesis (Fukuda 1982), by an approach based on Bland's work (Bland 1977) showing that linear programming duality generalizes to oriented matroids, and hence nondegenerate cycling in the program $(\mathcal{M}, e_\infty, e_0)$ implies nondegenerate cycling in $(\mathcal{M}^*, e_0, e_\infty)$. For the reader familiar with (or interested in) the machinery of linear programming, such as tableaux and pivot rules, proofs using this approach can be found in Fukuda (1982), Björner et al. (1999), and Bachem and Kern (1992).

The proof we give here will take another approach. Recall that proper single-element extensions are dual to proper liftings. So, if $(\mathcal{M}, e_\infty, e_0)$ is Euclidean then to see that $(\mathcal{M}^*, e_0, e_\infty)$ is Euclidean as well, we consider $X \in \mathcal{C}(\mathcal{M})$ such that $X(e_0) = +$. We will find a lifting $\widehat{\mathcal{M}}$ of $\mathcal{M}$ whose dual is an extension $\mathcal{M}^* \cup g$ of $\mathcal{M}^*$ such that g is parallel to e_∞ in $(\mathcal{M}^* \cup g, e_0)$ (i.e., there is a signed cocircuit of $\widehat{\mathcal{M}}$ with support either $\{g, e_\infty\}$ or $\{g, e_\infty, e_0\}$) and $Xg^0 \in \mathcal{C}^*(M^* \cup g)$ (i.e., $Xg^0 \in \mathcal{C}(\widehat{\mathcal{M}})$).

If $e_\infty \notin \mathrm{supp}(X)$ then the desired extension $\mathcal{M}^* \cup g$ is just the extension by an element parallel to e_∞. So assume $\{e_0, e_\infty, \} \subseteq \mathrm{supp}(X)$, and so, by Lemma 2.51, there is a unique $\hat{X} \in \mathcal{C}^*(\mathcal{M})$ such that $\mathrm{supp}(X) \cap \mathrm{supp}(\hat{X}) = \{e_0, e_\infty\}$ and $\hat{X}(e_\infty) = +$. Since $(\mathcal{M}, e_\infty, e_0)$ is Euclidean, there is an extension $\mathcal{M} \cup f$ of $\mathcal{M}$ such that $\hat{X} f^0 \in \mathcal{C}^*(\mathcal{M} \cup f)$ and f is parallel to e_0 in $(\mathcal{M} \cup f)/e_\infty$. We will use the localization of this extension to construct the desired lifting $\widehat{\mathcal{M}}$.

To get the idea, we describe the lifting when $\mathcal{M} \cup f$ is realized by a rank 3 arrangement $(\mathcal{H}_e : e \in E \cup \{f\})$ of signed hyperplanes. We treat $\mathcal{H}_{e_\infty}$ as the hyperplane at infinity, so the hyperplanes not parallel to $H_{e_\infty}^0$ can be pictured as an arrangement of lines in an affine plane P in $H_{e_\infty}^+$. $\hat{X}$ corresponds to a

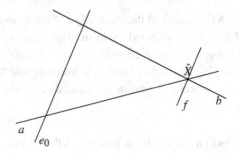

Figure 6.26 Initial affine picture for rank 3 example.

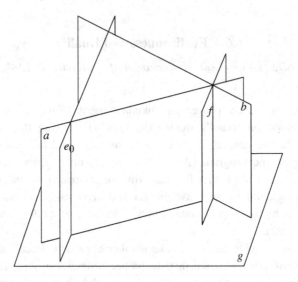

Figure 6.27 The oriented matroid $\mathcal{M}_1$.

point in this plane, and $H_f^0 \cap P$ is a line containing this point and parallel to $H_{e_0}^0 \cap P$. Figure 6.26 shows the picture in P for the simplest case.

We now construct the lifting in four steps:

1. We define $\mathcal{M}_1$ to be an extension of $\mathcal{M}$ by a coloop g. An affine representation of $\mathcal{M}_1$ in an affine three-space $T \supset P$ is obtained from our previous affine line arrangement by replacing each line with the plane through that line and perpendicular to P. P itself coincides with the affine hyperplane corresponding to g. (See Figure 6.27.)

   Each parallelism class of lines in T corresponds to a pair of antipodal points on the sphere at infinity. The parallelism class of lines perpendicular to P intersects the sphere at infinity at the two points corresponding to the

6.6 Intersection Properties

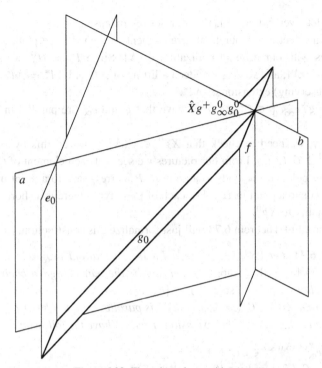

Figure 6.28 The oriented matroid $\mathcal{M}_3$.

covectors g^+ and g^-. In what follows we'll call these two points our *poles*, and we'll refer to them simply as g^+ and g^-. Let $(\mathcal{G}_e : e \in E \cup \{f, g\})$ be our arrangement of signed hyperplanes realizing $\mathcal{M}_1$.

2. Next, we introduce a new signed hyperplane $\mathcal{G}_{g_\infty} = (G^0_{g_\infty}, G^+_{g_\infty}, G^-_{g_\infty})$ so that $G^0_{g_\infty} \cap T$ is parallel to P and in between plane P and pole g^+. The sides of $\mathcal{G}_{g_\infty}$ are chosen so that the pole g^+ is in $G^-_{g_\infty}$. We let $\mathcal{M}_2$ denote the oriented matroid of this arrangement. Thus $\pm g^+ g^-_\infty \in \mathcal{C}^*(\mathcal{M}_2)$. We let P' denote the plane $G^0_{g_\infty} \cap T$.

3. Thirdly, we notice that the lines $G^0_{e_0} \cap P$ and $G^0_{f} \cap P'$ are parallel, so there is an affine plane containing both of them. We introduce a new signed hyperplane $\mathcal{G}_{g_0} = (G^0_{g_0}, G^+_{g_0}, G^-_{g_0})$ so that $G^0_{g_0} \cap T$ is this plane, signed so that $G^+_{g_0} \cap P = G^+_{e_0} \cap P$. The oriented matroid of this arrangement is $\mathcal{M}_3$. Thus for some α we have $\pm g^+ g^-_\infty g^\alpha_0 \in \mathcal{C}^*(\mathcal{M}_3)$. Figure 6.28 shows the picture between the parallel planes P and P'.

4. Finally, $\mathcal{M}_4$ is obtained from $\mathcal{M}_3$ by deleting e_∞ and e_0, renaming g_∞ as e_∞, and renaming g_0 as e_0. Thus $\pm g^+ e^-_\infty e^\alpha_0 \in \mathcal{C}^*(\mathcal{M}_4)$.

Now let's verify that $(\mathcal{M}_4)^*$ is our desired extension of $\mathcal{M}^*$.

The intersection of our final arrangement of signed hyperplanes with P coincides with our original realization of $\mathcal{M}$. Since $P = H_g^0$, we see that $(\mathcal{M}_4)/g = \mathcal{M}$, and so $\mathcal{M}_4$ is indeed a lifting of $\mathcal{M}$, and $\mathcal{M}_4^* = \mathcal{M}^* \cup g$ is a (rank-preserving) extension of $\mathcal{M}^*$.

Since $g^+ e_\infty^- e_0^\alpha \in \mathcal{C}^*(\mathcal{M}_4)$ we have that g and e_∞ are parallel in $(\mathcal{M}^* \cup g, e_0)$.

Finally, we need to check that $Xg^0 \in \mathcal{C}(\mathcal{M}_4)$. We do this by verifying that $Xg^0 \perp \mathcal{C}^*(\mathcal{M}_4)$. From our pictures we see that each element of $\mathcal{C}^*(\mathcal{M}_4)$ either is a pole, corresponds to a point in P, corresponds to a point in P', or corresponds to a point in $H_{e_0}^0$. For each of these types there is a short proof of orthogonality to Xg^0.

The proof of Theorem 6.73 will just formalize this construction.

Lemma 6.74 *Let $(\mathcal{M}, e_\infty, e_0)$ be a Euclidean matroid program. Let $X \in \mathcal{C}(\mathcal{M})$ with $\{e_\infty, e_0\} \subseteq \text{supp}(X)$. Let $\pm \hat{X}$ be the pair of signed cocircuits of $\mathcal{M}$ such that $\text{supp}(X) \cap \text{supp}(\hat{X}) = \{e_\infty, e_0\}$. Let $\mathcal{M} \cup f$ be an extension of $\mathcal{M}$ such that $\hat{X} f^0 \in \mathcal{C}^*(\mathcal{M} \cup f)$ and f is parallel to e_0 in $(\mathcal{M} \cup f)/e_\infty$. Let σ_f be the localization of this extension. Then we have the following:*

- $e_0^+ f^- e_\infty^{X(e_0)X(e_\infty)} \in \mathcal{C}(\mathcal{M} \cup f)$.
- *Let $Y \in \mathcal{C}^*(\mathcal{M})$ and $\epsilon \in \{0, +, -\}$. If $(Y \backslash \{e_\infty, e_0\}) \cdot (X \backslash \{e_\infty, e_0\})) = \{\epsilon\}$ then $\sigma_f(Y) = -\epsilon X(e_0)$.*

Proof: 1. Since e_0 is parallel to f in $(\mathcal{M} \cup f, /e_\infty)$, we know that $\mathcal{C}(\mathcal{M} \cup f)$ has an element of the form $e_0^+ f^- e_\infty^\beta$. Further $\hat{X} f^0 \in \mathcal{C}^*(\mathcal{M} \cup f)$. Thus

$$0 \in e_0^+ f^- e_\infty^\beta \cdot \hat{X} f^0 = \hat{X}(e_0) \boxplus \beta \hat{X}(e_\infty).$$

Thus $\beta = -\hat{X}(e_0)\hat{X}(e_\infty)$, and $e_0^+ f^- e_\infty^{-\hat{X}(e_0)\hat{X}(e_\infty)} \in \mathcal{C}(\mathcal{M} \cup f)$. Since $\hat{X} \perp X$ and $\hat{X} \cdot X = X(e_\infty)\hat{X}(e_\infty) \boxplus \hat{X}(e_0)X(e_0)$ we see $\hat{X}(e_0)\hat{X}(e_\infty) = -X(e_0)X(e_\infty)$, and so $e_0^+ f^- e_\infty^{X(e_0)X(e_\infty)} \in \mathcal{C}(\mathcal{M} \cup f)$.

2. If $\epsilon = 0$ then $\text{supp}(Y) \cap \text{supp}(X) \subseteq \{e_\infty, e_0\}$ and so $Y = \pm \hat{X}$ and $\sigma_f(Y) = 0$. Otherwise, let Y' be the restriction of Y to $\hat{X}^0 \cup \{e_\infty\}$. Then Y is in the interior of the convex set $C_{Y'}$ by Lemma 6.12, and $C_{Y'}$ is polytopal by Lemma 6.19. We'll show for every vertex V of $C_{Y'}$ that $\sigma_f(V) \in \{0, -\epsilon X(e_0)\}$, and that σ_f has value $-\epsilon X(e_0)$ for at least one vertex of $C_{Y'}$. Thus by Lemma 6.20 $\sigma_f(Y) = -\epsilon X(e_0)$.

One vertex of $C_{Y'}$ is $\pm \hat{X}$, and $\sigma_f(\pm \hat{X}) = 0$. For every remaining vertex V we have $V(e_\infty) = 0$. For every vertex V such that $V(e_\infty) = 0$ and every $e \in \text{supp}(X) \backslash e_0$, we have $e \in \hat{X}^0 \cup \{e_\infty\}$, so $V(e) \leq Y'(e) = Y(e)$. Thus $(V \backslash e_0) \cdot (X \backslash e_0)$ is $\{0\}$ or $\{\epsilon\}$. Since $V \perp X$, we have

6.6 Intersection Properties

$$V(e_0)X(e_0) = \begin{cases} 0 & \text{if } (V\backslash e_0) \cdot (X\backslash e_0) = \{0\}, \\ -\epsilon & \text{if } (V\backslash e_0) \cdot (X\backslash e_0) = \{\epsilon\}. \end{cases} \quad (6.2)$$

Further, since $Vf^{\sigma_f(V)} \in \mathcal{C}^*(\mathcal{M} \cup f)$ and $e_0^+ f^- e_\infty^{X(e_0)X(e_\infty)} \in \mathcal{C}(\mathcal{M} \cup f)$, we have $0 \in Vf^{\sigma_f(V)} \cdot e_0^+ f^- e_\infty^{X(e_0)X(e_\infty)} = V(e_0) \boxplus -\sigma_f(V) \boxplus 0$, and so $\sigma_f(V) = V(e_0)$. Together with Equation (6.2), this shows

$$\sigma_f(V) = \begin{cases} 0 & \text{if } (V\backslash e_0) \cdot (X\backslash e_0) = \{0\}, \\ -\epsilon X(e_0) & \text{if } (V\backslash e_0) \cdot (X\backslash e_0) = \{\epsilon\}. \end{cases} \quad (6.3)$$

Finally, we show that $-\epsilon X(e_0)$ is $\sigma_f(V)$ for some vertex V. Since $(Y\backslash\{e_\infty, e_0\}) \cdot (X\backslash\{e_\infty, e_0\})) \neq \{0\}$ there is an e that's in the support of both Y and Z. Since Y is in the convex hull of the vertices of $C_{Y'}$ and $Y(e) \neq 0$, there is a vertex V_0 such that $V_0(e) = Y(e)$. The conclusion follows by applying Equation (6.3) to V_0. □

Proof of Theorem 6.73: As in our rank 3 example, we let $X \in \mathcal{C}(\mathcal{M})$ with $X(e_0) \neq 0$. If $X(e_\infty) = 0$ then we can extend $\mathcal{M}^*$ by an element g parallel to e_∞, so assume $X(e_\infty) \neq 0$. Let $\hat{X}$ be the unique signed cocircuit with $\operatorname{supp}(\hat{X}) \cap \operatorname{supp}(X) = \{e_\infty, e_0\}$ and $\hat{X}(e_\infty) = +$. Let σ_f be the localization of an extension $\mathcal{M} \cup f$ with $\hat{X}f^0 \in \mathcal{C}^*(\mathcal{M} \cup f)$ and f parallel to e_0 in $(\mathcal{M} \cup f)/e_\infty$.

Let $\mathcal{M}_1$ be the extension of $\mathcal{M}$ by a coloop g. Thus $\mathcal{C}^*(\mathcal{M}) = \{Xg^0 : X \in \mathcal{C}^*(\mathcal{M})\} \cup \{g^+, g^-\}$.

Let $\mathcal{M}_2 = \mathcal{M}_1 \cup g_\infty$ be the lexicographic extension $\mathcal{M}[g^- e_\infty^+]$. Thus

$$\mathcal{C}^*(\mathcal{M}_2) = \{g^+ g_\infty^-, g^- g_\infty^+\} \cup \{Xg^0 g_\infty^{X(e_\infty)} : X \in \mathcal{C}^*(\mathcal{M})\}$$
$$\cup \{Xg^{X(e_\infty)} g_\infty^0 : X \in \mathcal{C}^*(\mathcal{M})\}.$$

Define $\sigma : \mathcal{C}^*(\mathcal{M}_2) \to \{0, +, -\}$ by

- $\sigma(g^+ g_\infty^-) = X(e_\infty)X(e_0)$,
- $\sigma(g^- g_\infty^+) = -X(e_\infty)X(e_0)$, and
- $\sigma(Xg^0 g_\infty^{X(e_\infty)}) = X(e_0)$ and $\sigma(Xg^{X(e_\infty)} g_\infty^0) = \sigma_f(X)$ for each $X \in \mathcal{C}^*(\mathcal{M})$.

To see that this is a localization, we verify that it induces a localization on each $\mathcal{M}_2/A$ with A a corank 2 flat. The corank 2 flats have four types:

1. If A is a subset of E, then A has corank 1 in $\mathcal{M}$. Let Y be the unique signed cocircuit of $\mathcal{M}$ with $Y^0 = A$ and $Y(g_\infty) = +$. A topological representation of $\mathcal{M}_2/A$ together with the function induced by σ is shown in Figure 6.29.

Single-Element Extensions

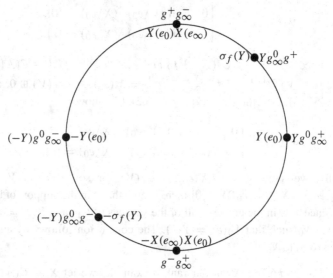

Figure 6.29 Case 1 of Theorem 6.73. Signed cocircuits are labeled outside the circle, and the values of σ are written inside the circle.

We now have two cases:

1. If $Y(e_0) = -X(e_0)X(e_\infty)$ then since $X(e_0)X(e_\infty) \neq 0$ we see that the function shown in Figure 6.29 is a localization.
2. If $Y(e_0) \in \{0, X(e_0)X(e_\infty)\}$, then since $Yf^{\sigma_f(Y)} \in \mathcal{C}^*(\mathcal{M} \cup f)$ and, by Lemma 6.74, $e_0^+ f^- e_\infty^{X(e_0)X(e_\infty)} \in \mathcal{C}(\mathcal{M} \cup f)$, we have

$$0 \in (Yf^{\sigma_f(Y)}) \cdot (e_0^+ f^- e_\infty^{X(e_0)X(e_\infty)})$$
$$= Y(e_0) \boxplus -\sigma_f(Y) \boxplus X(e_0)X(e_\infty)$$
$$= -\sigma_f(Y) \boxplus X(e_0)X(e_\infty).$$

Thus $\sigma_f(Y) = X(e_0)X(e_\infty)$, and so the function shown in Figure 6.29 is a localization.

2. If $A = A' \cup \{g\}$ or $A = A' \cup \{g_\infty\}$ for some $A' \subset E$ of corank 2 in $\mathcal{M}$ then $\mathcal{M}_2/A = \mathcal{M}'/A$, where $\mathcal{M}'$ is obtained from $\mathcal{M}$ by extending an element parallel to e_∞. The function on $\mathcal{M}_2/A$ induced by σ is the obvious tweak of the localization on $\mathcal{M}/A'$ induced by σ_f, hence is a localization.
3. If $A = A' \cup \{g, g_\infty\}$ for some $A' \subset E$ then $e_\infty \in A'$ and $\mathcal{M}_2/A = \mathcal{M}/A'$. The function on $\mathcal{M}_2/A$ induced by σ coincides with the localization on $\mathcal{M}/A'$ induced by σ_f, hence is a localization.

Thus the Las Vergnas condition tells us that σ is a localization. Let $\mathcal{M}_3$ be the extension of $\mathcal{M}_2$ by an element g_0 with this localization.

Finally, let $\mathcal{M}_4$ be obtained from $\mathcal{M}_3 \backslash \{e_\infty, e_0\}$ by renaming g_∞ as e_∞ and g_0 as e_0.

As in our example we see that $\mathcal{M}_4/g = \mathcal{M}$ and $\text{rank}(\mathcal{M}_4) = \text{rank}(\mathcal{M})+1$, and so $(\mathcal{M}_4)^*$ is a proper extension of $\mathcal{M}^*$. Also $g^+ e_\infty^- e_0^{X(e_0)X(e_\infty)} \in \mathcal{C}^*(\mathcal{M}_4) = \mathcal{C}(\mathcal{M}^* \cup g)$, and so g is parallel to e_∞ in $(\mathcal{M}^* \cup g)/e_0$.

To see that $Xg^0 \in \mathcal{C}^*(\mathcal{M}^* \cup g) = \mathcal{C}(\mathcal{M}_4)$, we will show that $Xg^0 \perp \mathcal{C}^*(\mathcal{M}_4)$. The signed cocircuits of $\mathcal{M}_4$ have four (not disjoint) types.

1. *Signed cocircuits whose 0 set is contained in* $E\backslash\{e_0, e_\infty\}$. The two poles $\pm g^+ e_\infty^- e_0^{X(e_0)X(e_\infty)}$ are clearly orthogonal to Xg^0.
2. *Signed cocircuits whose 0 set contains* g. Consider a signed cocircuit Yg^0 with $Y \in \mathcal{C}^*(\mathcal{M})$. Then since $X \perp Y$ we have $Xg^0 \perp Yg^0$.
3. *Signed ocircuits whose 0 set contains* e_∞. Consider a signed cocircuit $Z = (Y\backslash\{e_0, e_\infty\})e_\infty^0 e_0^{\sigma_f(Y)} g^{Y(e_\infty)}$ with $Y \in \mathcal{C}^*(\mathcal{M})$.

   1. If $(X\backslash\{e_\infty, e_0\}) \cdot (Y\backslash\{e_\infty, e_0\}) = \{0, +, -\}$ then $Xg^0 \cdot Z = \{0, +, -\}$.
   2. If $(X\backslash\{e_\infty, e_0\}) \cdot (Y\backslash\{e_\infty, e_0\}) = \{0\}$ then $Y = \pm \hat{X}$. Thus $\sigma_f(Y) = 0$, and $Xg^0 \cdot Z = \{0\}$.
   3. If $(X\backslash\{e_\infty, e_0\}) \cdot (Y\backslash\{e_\infty, e_0\}) = \{\epsilon\} \neq \{0\}$ then by Lemma 6.74 we have $\sigma_f(Y) = -\epsilon X(e_0)$, and so $Xg^0 \cdot Z = \epsilon \boxplus X(e_0)\sigma_f(Y) = \epsilon \boxplus -\epsilon = \{0, +, -\}$.

4. *Signed cocircuits whose 0 set contains* e_0. We need only consider such signed cocircuits that don't fall into one of the other categories. By Proposition 6.4, these have the form $(Z'_1 \circ Z'_2)e_0^0$, where Z'_1 and Z'_2 are obtained from conformal signed cocircuits Z_1 and Z_2 of $\mathcal{M}_2$ by deleting e_∞ and e_0 and renaming g_∞ as e_∞. Further $\sigma(Z_1) = -\sigma(Z_2) \neq 0$, so $Z'_1 e_0^\beta, Z'_2 e_0^{-\beta} \in \mathcal{C}^*(\mathcal{M}_4)$ for some $\beta \in \{+, -\}$. Since $Z'_1 e_0^\beta \perp Xg^0$ and $Z'_2 e_0^{-\beta} \perp Xg^0$ and $X(e_0) \neq 0$, either $Z'_1 \cdot (X\backslash e_0)g^0 = \{0, +, -\}$ or $Z'_2 \cdot (X\backslash e_0)g^0 = \{0, +, -\}$. Thus by conformality of Z'_1 and Z'_2 we have $Y \cdot Xg^0 = \{0, +, -\}$.

$\square$

Exercises

6.1 Let χ be a chirotope for $\mathcal{M}$, and let $\mathcal{M} \cup f = \mathcal{M}[e_1^{s_1}, \ldots, e_k^{s_k}]$. Prove that χ extends to a chirotope χ' for $\mathcal{M} \cup f$ as follows: For each $x_2, \ldots, x_r$ elements of $\mathcal{M}$,

- if $\chi(e_i, x_2, \ldots, x_r) = 0$ for all $i \in [k]$ then define $\chi'(f, x_2, \ldots, x_r) = 0$,
- otherwise, let i be the smallest value such that $\chi(e_i, x_2, \ldots, x_r) \neq 0$, and define $\chi'(f, x_2, \ldots, x_r) = s_i \chi(e_i, x_2, \ldots, x_r)$.

6.2 Let $\mathcal{M}$ be an oriented matroid, $e_1, \ldots, e_k$ be elements of $\mathcal{M}$, and $s_2, \ldots, s_k \in \{+, -\}$. Let $\mathcal{M} \cup f = \mathcal{M}[e_1^+, e_2^{s_2}, \ldots, e_k^{s_k}]$, and let $\mathcal{M}'$ be obtained from $(\mathcal{M} \cup f) \backslash e_1$ by renaming f as e_1. Prove that $\mathcal{M}' \rightsquigarrow \mathcal{M}$.

Use this to deduce that every oriented matroid is the weak map image of a uniform oriented matroid of the same rank.

6.3 Prove that a strong map $\mathcal{M} \to \mathcal{N}$ with rank($\mathcal{M}$) = rank($\mathcal{N}$) + 1 has a framing.

6.4 Let a be a nonloop in $\mathcal{M}$, and let A be a subset of E containing a. Let $(\mathcal{M}/A) \cup f$ be an extension of $\mathcal{M}/A$ by an element f, and let $\mathcal{N}$ be the extension of $\mathcal{M}$ by an element f parallel to a. Prove that there is an extension $\mathcal{M} \cup f$ of $\mathcal{M}$ such that $(\mathcal{M}/A) \cup f = (\mathcal{M} \cup f)/A$ and $\mathcal{M} \cup f \rightsquigarrow \mathcal{N}$.

6.5 Prove Theorem 2.68 by induction, using our results on localization. The only fact from general matroid theory you should use is that if S is a cocircuit of a rank r matroid on elements E then rank($E \backslash S$) = $r - 1$.

6.6 Use Exercise 6.5 to prove Theorem 2.72. The only additional fact from general matroid theory you should use is the cryptomorphism between bases and cocircuits of a matroid: If $\mathcal{B}$ is the set of bases of a matroid, then the set of cocircuits of that matroid is $\{E \backslash (B \backslash \{b\}) : b \in B \in \mathcal{B}\}$.

6.7 Use the Generalized Levi Enlargement Lemma (Theorem 5.40) to show that every rank 3 oriented matroid has an adjoint.

6.8 Find topological representations of several adjoints of the non-Pappus oriented matroid. (Your adjoints should not be isomorphic, even up to reorientation.)

6.9 The *Hahn–Banach separation theorem* says that if C_1 and C_2 are closed convex subsets of affine space and at least one of C_1, C_2 is compact then C_1 and C_2 are disjoint if and only if there is a hyperplane separating them. There are two straightforward potential analogs for oriented matroids: This exercise will show that one of them holds for all oriented matroids while the other holds only for oriented matroids satisfying the generalized Euclidean property.

1. In this analog, our "convex sets" are convex sets in $\mathcal{V}^*(\mathcal{M})$ – analogous to a closed convex set being an intersection of closed half-spaces.

Let $S, T \subset \mathcal{V}^*(\mathcal{M})$. We say that an extension $\mathcal{M} \cup f$ separates S and T if there is an $\epsilon \in \{+, -\}$ such that

- for each $Z \in S$, Zf^ϵ is the unique extension of Z to a covector of $\mathcal{M} \cup f$, and
- for each $Z \in T$, $Zf^{-\epsilon}$ is the unique extension of Z to a covector of $\mathcal{M} \cup f$.

Let $(\mathcal{M}, e_\infty)$ be an affine oriented matroid. Let $C(X)$ and $C(Y)$ be convex sets in $\mathcal{V}^*(\mathcal{M})$ such that $C(X)$ is contained in the affine space of $(\mathcal{M}, e_\infty)$ and $Y(e_\infty) = +$. Prove that $C(X)$ and $C(Y)$ are disjoint if and only if there is an extension $\mathcal{M} \cup f$ separating $C(X)$ and $C(Y)$.

2. (Bachem and Wanka 1988) In this analog, our "convex sets" are the convex hulls $\widehat{\text{conv}}(A)$ of sets of elements, introduced in Section 2.6 – analogous to a convex set being the convex hull of a set of points in affine space. Recall from Chapter 1 that the covectors of the oriented matroid of a vector arrangement ($\mathbf{v}_e : e \in E$) describe how hyperplanes separate pairs of subsets of $\{\mathbf{v}_e : e \in E\}$.

Let $\mathcal{M}$ be an oriented matroid on elements E satisfying the generalized Euclidean property. Let $A, B \subseteq E$ such that $\mathcal{M}(A \cup B)$ is acyclic. Prove that the following are equivalent:

1. There is no $X \in \mathcal{V}^*(\mathcal{M})$ such that $X(A) = \{+\}$ and $X(B) = \{-\}$.
2. There is an extension $\mathcal{M} \cup f$ of $\mathcal{M}$ in which $f \in \widehat{\text{conv}}(A)$ and $f \in \widehat{\text{conv}}(B)$.

6.10 (Bachem and Wanka 1988) This exercise will prove an oriented matroid generalization of *Radon's theorem*, which says that a set of n points in affine $n - 1$-space must have a partition into two sets with intersecting convex hulls.

Let $\mathcal{M}$ be a rank r oriented matroid satisfying the generalized Euclidean property. Let A be a set of elements such that $|A| \geq r + 1$ and $\mathcal{M}(A)$ is acyclic. Prove that there is a partition $A = A_1 \cup A_2$ of A and an extension $\mathcal{M} \cup f$ of $\mathcal{M}$ in which $f \in \widehat{\text{conv}}(A_1)$ and $f \in \widehat{\text{conv}}(A_2)$.

6.11 (Bachem and Wanka 1988) This exercise will prove an oriented matroid generalization of *Helly's theorem*, which says that for a collection $C_1, \ldots, C_n$ of convex sets in affine $r - 1$-space with $n > r$, if $\bigcap_{j \neq i} C_j \neq \emptyset$ for each $i \in [n]$ then $\bigcap_{j=1}^n C_j \neq \emptyset$.

Let $\mathcal{M}$ be a rank r oriented matroid on elements E satisfying the generalized Euclidean property. Let $C_1, \ldots, C_n$ be convex subsets of E (i.e., $\widehat{\text{conv}}(C_i) = C_i$ for each i). Assume that $n > r$, and that for each $i \in [n]$ we have $\bigcap_{j < i} C_j \cap \bigcap_{j > i} C_j \neq \emptyset$. Prove that there is an extension $\mathcal{M} \cup g$ of $\mathcal{M}$ in which $g \in \bigcap_{j=1}^n \widehat{\text{conv}}(C_j)$.

7
The Universality Theorem

The results of Section 2.7.2 suggest that determining whether a given oriented matroid is realizable can be difficult. In fact, the problem is NP-hard (Shor 1991). In this chapter we'll consider a still harder issue: the space of all realizations of a given oriented matroid.

We first establish exactly what we mean by "space of realizations": There are minor differences depending on whether we mean vector, hyperplane, or subspace realizations. We then settle into considering a relatively tidy little space that encodes the topology of all of these spaces.

Questions about the topology of this space actually much predate oriented matroids: The question goes back at least to Ringel's 1956 paper on pseudoline arrangements (Ringel 1956), in which he conjectured (in different language) that realization spaces of rank 3 realizable oriented matroids are path-connected. Few conjectures have ever been so thoroughly disproved. The *Universality Theorem* (Theorem 7.26) tells us that the homotopy type of the realization space of a rank 3 oriented matroid can be "arbitrarily ugly."

7.1 Realization Spaces

We introduced the vector realization space, the space Vreal($\mathcal{M}$) of all vector realizations of $\mathcal{M}$ in $\mathbb{R}^r$, in Section 5.2.2. Now let's look at some other realization spaces.

Definition 7.1 Let $\mathcal{M}$ be a rank r oriented matroid on elements $[n]$.

1. The *signed hyperplane realization space* Hreal($\mathcal{M}$) is the set of all realizations of $\mathcal{M}$ by signed hyperplane arrangements in $\mathbb{R}^r$.

   We get a topology on Hreal($\mathcal{M}$) by identifying a signed hyperplane $\mathbf{v}^\perp$ with the ray $\mathbb{R}_{>0}\mathbf{v}$ of vectors whose normal is that signed hyperplane.

7.1 Realization Spaces 215

Let Diag_n^+ be the set of all $n \times n$ diagonal matrices with all diagonal entries positive. Then an element of the quotient space $\text{Mat}(r,n)/\text{Diag}_n^+$ is an arrangement of rays $(\mathbb{R}_{>0}\mathbf{v}_1, \ldots, \mathbb{R}_{>0}\mathbf{v}_n)$, which we identify with the hyperplane arrangement $(\mathbf{v}_1^\perp, \ldots, \mathbf{v}_n^\perp)$. We topologize $\text{Hreal}(\mathcal{M})$ as a subspace of $\text{Mat}(r,n)/\text{Diag}_n^+$.

2. The *Grassmannian realization space* $\text{Greal}(\mathcal{M})$ is the set of all realizations of $\mathcal{M}$ by subspaces of $\mathbb{R}^n$. $\text{Greal}(\mathcal{M})$ is topologized as a subspace of $G(r, \mathbb{R}^n)$. The topology on $G(r, \mathbb{R}^n)$ has two good descriptions. The first is as a quotient space, via the identification of $G(r, \mathbb{R}^n)$ with the left quotient $GL_r \backslash \text{Mat}(r, n)$ given in Section 1.4. The second is as a subspace of projective space, via the Plücker embedding (Section 1.4.3).

Let $\{b_1, \ldots, b_r\}$ be a basis of $\mathcal{M}$. We have a homeomorphism

$$\text{Vreal}(\mathcal{M}) \to \text{Greal}(\mathcal{M}) \times GL_r$$
$$M \to (\text{row}(M), M_{b_1, \ldots, b_r}).$$

Assuming that $\mathcal{M}$ has no loops, we have a homeomorphism

$$\text{Vreal}(\mathcal{M}) \to \text{Hreal}(\mathcal{M}) \times (\mathbb{R}_{>0})^n$$
$$(\mathbf{v}_1, \ldots, \mathbf{v}_n) \to ((\mathbb{R}_{>0}\mathbf{v}_1, \ldots, \mathbb{R}_{>0}\mathbf{v}_n), (\|\mathbf{v}_1\|, \ldots, \|\mathbf{v}_n\|)).$$

Thus the topology of any one of these spaces determines the topology of the other two, $\text{Vreal}(\mathcal{M})$ is homotopically equivalent to $\text{Hreal}(\mathcal{M})$, and $\text{Greal}(\mathcal{M})$ is topologically simpler than $\text{Vreal}(\mathcal{M})$ and $\text{Hreal}(\mathcal{M})$. The literature is inconsistent about exactly what space is meant by "the realization space of $\mathcal{M}$," but it is generally a space that is homotopy equivalent to $\text{Greal}(\mathcal{M})$. As we shall see in Section 7.2, this leads us to the conclusion that for many $\mathcal{M}$ "the realization space of $\mathcal{M}$" is contractible.

A major goal in this chapter is to create oriented matroids for which $\text{Greal}(\mathcal{M})$ is *not* contractible. We can typically analyze the topology of a realization space most easily by identifying the realization space under consideration with a homeomorphic subspace of $\text{Vreal}(\mathcal{M})$. For instance, by identifying a signed hyperplane with its unit normal vector we get an identification of $\text{Hreal}(\mathcal{M})$ with the subspace of $\text{Vreal}(\mathcal{M})$ in which each nonzero vector has length 1. And, assuming that $[r]$ is a basis of $\mathcal{M}$, by identifying an element $\text{row}(I|A)$ of $\text{Greal}(\mathcal{M})$ with $(I|A)$, we identify $\text{Greal}(\mathcal{M})$ with the set of all vector realizations of $\mathcal{M}$ in which $[r]$ is realized as the coordinate basis.

There is an easy way to quotient $\text{Greal}(\mathcal{M})$ to a smaller space with the same homotopy type.

Definition 7.2 Assume $[r]$ is a basis for $\mathcal{M}$. The *realization space* $\mathcal{R}(\mathcal{M})$ of $\mathcal{M}$ is the space of vector realizations $(I|A)$ of $\mathcal{M}$ such that the last nonzero entry in each column of A has absolute value 1.

Assuming that $\mathcal{M}$ has no loops, we have a homeomorphism

$$\mathcal{R}(\mathcal{M}) \times (\mathbb{R}_{>o})^{n-r} \to \text{Greal}(\mathcal{M}),$$
$$((I|v_{r+1},\ldots,v_n),(c_{r+1},\ldots,c_n)) \to (I|c_{r+1}v_{r+1},\ldots,c_nv_n).$$

For many oriented matroids there is an even smaller convenient space homotopy equivalent to $\mathcal{R}(\mathcal{M})$.

Definition 7.3 A *projective basis* for $\mathcal{M}$ is an unsigned circuit of $\mathcal{M}$ with $r+1$ elements.

Let $\mathcal{M}$ be an oriented matroid such that $[r + 1]$ is a projective basis. The *projective realization space* $\mathcal{RP}(\mathcal{M})$ of $\mathcal{M}$ is the space of vector realizations $(I|v|A)$ of $\mathcal{M}$ such that each entry of v has absolute value 1 and the last nonzero entry in each column of A has absolute value 1.

Problem 7.4 Assume $[r+1]$ is a projective basis of $\mathcal{M}$ and $\mathcal{M}$ has no loops. Find a homeomorphism $\text{Greal}(\mathcal{M}) \to \mathcal{RP}(\mathcal{M}) \times (\mathbb{R}_{>0})^{n-1}$.

Clearly an oriented matroid must be connected in order to have a projective basis, but this condition is not sufficient. For example, the unsigned circuits of the rank 4 oriented matroid given by

$$\begin{pmatrix} 1 & 0 & 0 & 0 & 1 & 0 & 0 \\ 0 & 1 & 0 & 0 & 0 & 1 & 0 \\ 0 & 0 & 1 & 0 & 0 & 0 & 1 \\ 0 & 0 & 0 & 1 & 1 & 1 & 1 \end{pmatrix}$$

all have size 3 or 4.

All of our constructions in this chapter will be of rank 3 oriented matroids on $[n]$ in which $\{1,2,3\}^+4^-$ is a signed circuit, so that the fourth column of each element of $\mathcal{RP}(\mathcal{M})$ is $(1,1,1)^\top$, and in which the last nonzero entry in each element of $\mathcal{RP}(\mathcal{M})$ is 1. For each such $\mathcal{M}$ we'll depict an element of $\mathcal{RP}(\mathcal{M})$ by an arrangement of points and lines in the plane $\mathbb{A} := \{x \in \mathbb{R}^3 : x_3 = 1\}$. If the last coordinate of a vector is 1, then we can depict it as a point in $\mathbb{A}$. If the last coordinate of a vector is 0, then we treat it as a point "at infinity." For a vector v at infinity and a vector w not at infinity, the plane spanned by $\{v,w\}$ intersects $\mathbb{A}$ in the line through w with direction given by v. Thus v is associated to a parallelism class of lines in in $\mathbb{A}$, and our condition that the last nonzero component of v is 1 lets us recover v from the parallelism class. We will always draw $\mathbb{A}$ so e_1 is a direction vector for horizontal lines, pointing

7.1 Realization Spaces

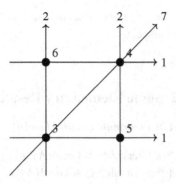

Figure 7.1 An oriented matroid $\mathcal{M}_0$ with $\mathcal{RP}(\mathcal{M}_0)$ a single point.

to the right, and $\mathbf{e}_2$ is a direction vector for vertical lines, pointing up. The condition that the last nonzero entry in each column is 1 tells us that each element appears either as a point in $\mathbb{A}$, as a direction vector pointing right, or as a direction vector pointing into the upper half-space.

Example 7.5 Figure 7.1 shows our depiction of an oriented matroid $\mathcal{M}_0$ on elements [7]. This oriented matroid has [4] as a projective basis, and $\mathcal{RP}(\mathcal{M}_0)$ has only one element, the arrangement

$$\begin{pmatrix} 1 & 0 & 0 & 1 & 1 & 0 & 1 \\ 0 & 1 & 0 & 1 & 0 & 1 & 1 \\ 0 & 0 & 1 & 1 & 1 & 1 & 0 \end{pmatrix}.$$

This is because the first four columns, together with the dependencies dictated by $\mathcal{M}_0$, dictate the remaining columns. (This will be discussed further in Section 7.2.) The realization space $\mathcal{R}(\mathcal{M}_0)$ is the set of all arrangements of the form

$$\begin{pmatrix} 1 & 0 & 0 & a & a & 0 & \frac{a}{b} \\ 0 & 1 & 0 & b & 0 & b & 1 \\ 0 & 0 & 1 & 1 & 1 & 1 & 0 \end{pmatrix},$$

with $a, b > 0$.

Remark 7.6 The hyperplane arrangement interpretation of $\mathcal{RP}(\mathcal{M})$ is the set of all unsigned hyperplane arrangements that could be signed to give a realization of $\mathcal{M}$, up to change of coordinates. The subspace interpretation of $\mathcal{RP}(\mathcal{M})$ is the set of all subspace realizations of $\mathcal{M}$ modulo *homothety*. Homothety is the equivalence relation on subspaces of $\mathbb{R}^r$ in which two subspaces are equivalent if one can be taken to the other by scaling the coordinate axes. More formally, row(M_1) is homothetic to row(M_2) if $M_1 = M_2 D$ for some $D \in \text{Diag}_n^+$. See Gel'fand and MacPherson (1982) and Gel'fand et al. (1987) for more on this perspective.

7.2 Some Elementary Results

Proposition 7.7 *Let $\mathcal{M}$ be an oriented matroid on $[n]$.*

1. *If e is a loop of $\mathcal{M}$ then $\text{Greal}(\mathcal{M}) \cong \text{Greal}(\mathcal{M}\backslash e)$.*
2. *If e is a coloop of $\mathcal{M}$ then $\text{Greal}(\mathcal{M}) \cong \text{Greal}(\mathcal{M}\backslash e)$.*
3. *If a and b are parallel or antiparallel in $\mathcal{M}$ then $\text{Greal}(\mathcal{M}\backslash a) \cong \text{Greal}(\mathcal{M})$.*
4. *$\text{Greal}(\mathcal{M})$ is homeomorphic to $\text{Greal}(\mathcal{M}_A)$ for each reorientation $\mathcal{M}_A$ of $\mathcal{M}$.*
5. *The map $V \to V^\perp$ gives a homeomorphism from $\text{Greal}(\mathcal{M})$ to $\text{Greal}(\mathcal{M}^*)$.*

The proof is straightforward. Whenever $\mathcal{R}(\mathcal{M}\backslash e)$ and $\mathcal{RP}(\mathcal{M}\backslash e)$ are defined, the conclusions about Greal also hold for $\mathcal{R}$ and $\mathcal{RP}$.

Corollary 7.8 *For every oriented matroid $\mathcal{M}$ there is a simple $\mathcal{M}'$ satisfying the conditions given at the beginning of the chapter such that $\text{Greal}(\mathcal{M}) \cong \text{Greal}(\mathcal{M}')$.*

Another easy proposition tells us about realization spaces of oriented matroids with few elements.

Proposition 7.9 *Let $\mathcal{M}$ be an oriented matroid on $[n]$ for which $[r+1]$ is a projective basis.*

1. *If $n = r + 1$ then $\mathcal{RP}(\mathcal{M})$ is a point.*
2. *Let $\mathcal{M}$ be simple and $n = r + 2$. Let $k + 2$ be the size of the smallest unsigned circuit containing $r + 2$. Then $\mathcal{RP}(\mathcal{M}) \cong (\mathbb{R}_{>0})^k$.*

Remark 7.10 Of course, the conclusion of the last part of Proposition 7.9 could have been stated more simply as $\mathcal{RP}(\mathcal{M}) \cong \mathbb{R}^k$. But, as we have seen already, in the context of realization spaces the homeomorphisms are typically seen most naturally in terms of $\mathbb{R}_{>0}$ rather than $\mathbb{R}$.

Corollary 7.11 *If $\mathcal{M}$ has rank at most 2 or corank at most 2 then $\mathcal{R}(\mathcal{M}) \cong (\mathbb{R}_{>0})^k$ for some k.*

Now we describe a route by which many oriented matroids can be shown to have contractible realization space. Assume that $\mathcal{M}$ is simple with elements $[n]$ and that $[r]$ is a basis. Let $e \in [n]\backslash[r]$. There is an obvious continuous map $\mathcal{R}(\mathcal{M}) \to \mathcal{R}(\mathcal{M}\backslash e)$, called the *forgetful map*.

7.2 Some Elementary Results

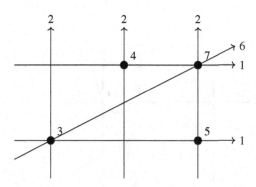

Figure 7.2 An oriented matroid not reducible by 7.

Definition 7.12 (Bokowski and Sturmfels 1989) We say e is *fixed* by $\mathcal{M}\backslash e$ if this map is a bijection. We say $\mathcal{M}$ is *reducible* by e if this map is surjective.

If $[r + 1]$ is a projective basis for $\mathcal{M}$ and $e \in [n]\backslash[r + 1]$ then there is also a forgetful map $\mathcal{RP}(\mathcal{M}) \to \mathcal{RP}(\mathcal{M}\backslash e)$, and this map is bijective resp. surjective if and only if $\mathcal{R}(\mathcal{M}) \to \mathcal{R}(\mathcal{M}\backslash e)$ is bijective resp. surjective. Our results on fixed and reducible elements apply to both projective realization space and the realization space.

For instance, in Example 7.5,

1. 7 is fixed by $\mathcal{M}_0\backslash 7$, and so $\mathcal{RP}(\mathcal{M}_0) \cong \mathcal{RP}(\mathcal{M}_0\backslash 7)$;
2. 6 is fixed by $\mathcal{M}_0\backslash\{6,7\}$, and so $\mathcal{RP}(\mathcal{M}_0\backslash 6) \cong \mathcal{RP}(\mathcal{M}_0\backslash\{6,7\})$;
3. 5 is fixed by $\mathcal{M}_0\backslash\{5,6,7\}$, and so $\mathcal{RP}(\mathcal{M}_0\backslash 5) \cong \mathcal{RP}(\mathcal{M}_0\backslash\{5,6,7\})$;
4. $\mathcal{RP}(\mathcal{M}_0\backslash\{5,6,7\})$ is a point.

This gives another explanation of why $\mathcal{RP}(\mathcal{M}_0)$ is a point.

To see an example of a nonrealizable element, notice that in the oriented matroid $\mathcal{M}$ given by Figure 7.2 $\mathcal{M}$ is not reducible by 7.

Definition 7.13 Let $S \subseteq [n]\backslash[r]$.

A **fixing sequence** from $\mathcal{M}$ to $\mathcal{M}\backslash S$ is an ordering $s_1, \ldots, s_k$ of S such that i is fixed by $\mathcal{M}\backslash\{s_j : j \geq i\}$ for each i.

A **reduction sequence** from $\mathcal{M}$ to $\mathcal{M}\backslash S$ is an ordering $s_1, \ldots, s_k$ of S such that i is reducible by $\mathcal{M}\backslash\{s_j : j \geq i\}$ for each i.

Obviously, if there is a fixing sequence from $\mathcal{M}$ to $\mathcal{M}\backslash S$ then $\mathcal{R}(\mathcal{M}) \cong \mathcal{R}(\mathcal{M}\backslash S)$, and the following proposition tells us that if there is a reduction sequence from $\mathcal{M}$ to $\mathcal{M}\backslash S$ then $\mathcal{R}(\mathcal{M}) \simeq \mathcal{R}(\mathcal{M}\backslash S)$

220 The Universality Theorem

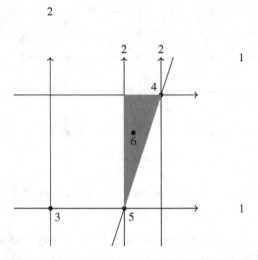

Figure 7.3 Illustration for Proposition 7.14.

Proposition 7.14 *If $\mathcal{M}$ is reducible by e then the map $\mathcal{RP}(\mathcal{M}) \to \mathcal{RP}(\mathcal{M}\backslash e)$ is a homotopy equivalence.*

The idea of the proof is mostly clear from an example. Consider the oriented matroid $\mathcal{M}$ realized by the arrangement in Figure 7.3. Let M_0 be the specific realization of $\mathcal{M}\backslash 6$ shown in the figure. Then the preimage of M under $\mathcal{RP}(\mathcal{M}) \to \mathcal{RP}(\mathcal{M}\backslash 6)$ is the set of all $(M|\mathbf{v})$ with $\mathbf{v}$ in the interior of the shaded triangle. In the same way, the preimage of each $M \in \mathcal{RP}(\mathcal{M}\backslash 6)$ is the interior of a triangle, with sides given by the lines corresponding to $\{1,3\}$, $\{2,5\}$, and $\{4,5\}$. Let $\mathbf{w}_M$ be the center of mass of this triangle. Then

$$v \colon \mathcal{RP}(\mathcal{M}\backslash 6) \to \mathcal{RP}(\mathcal{M})$$
$$M \to (M|\mathbf{w}_M)$$

is an embedding, and $\mathcal{RP}(\mathcal{M})$ retracts to the image of v by a fibrewise straight-line homotopy. That is, the map

$$\mathcal{RP}(\mathcal{M}) \times [0,1] \to \mathcal{RP}(\mathcal{M})$$
$$((M|\mathbf{v}),t) \to (M|(1-t)\mathbf{v} + t\mathbf{w}_M)$$

is a homotopy retracting $\mathcal{RP}(\mathcal{M})$ to the image of v.

For general $\mathcal{M}$ and forgetful map $\mathcal{RP}(\mathcal{M}) \to \mathcal{RP}(\mathcal{M}\backslash e)$ with $e \in [n]\backslash[r+1]$, the preimage of a realization $M = (\mathbf{v}_f : f \in [n] - \{e\})$ of $\mathcal{M}\backslash e$ is an intersection of finitely many hyperplanes and open half-spaces, together with an affine hyperplane given by the condition that the last nonzero coordinate of the vector indexed by e must have absolute value 1. Unlike in our example, this intersection need not be bounded, and so the center of mass v_e that we used in our example need not exist. To get around this, we work with a different affine hyperplane. Recall that $\mathcal{M}$ determines $\text{sign}(\mathbf{v}_e)$ for each extension $(\mathbf{v}_f : f \in [n])$ of M to a realization of $\mathcal{M}$. Let $\text{sign}(\mathbf{v}_e) = (\epsilon_1, \ldots, \epsilon_r)$, and let $\mathbb{A}' = \{\mathbf{x} \in \mathbb{R}^r : \sum_i \epsilon_i x_i = 1\}$. For each extension $(\mathbf{v}_f : f \in [n])$ of M to a realization of $\mathcal{M}$, there is a unique positive scalar multiple $\mathbf{v}'_e$ of $\mathbf{v}_e$ in $\mathbb{A}'$, and the correspondence $\mathbf{v}_e \leftrightarrow \mathbf{v}'_e$ gives a homeomorphism from the preimage of M to the interior of a convex polytope in $\mathbb{A}'$. Let $\mathbf{w}'_M$ be the center of mass of this convex polytope, and let $\mathbf{w}_M$ be its scalar multiple whose last nonzero coordinate is 1. Let $\nu(M)$ be the realization of $\mathcal{M}$ given by M and $\mathbf{w}_M$. Then, as in our example, $\nu \colon \mathcal{R}(\mathcal{M} \backslash e) \to \mathcal{R}(\mathcal{M})$ is an embedding, and $\mathcal{R}(\mathcal{M})$ retracts to the image of ν by a fibrewise straight-line homotopy.

Thus, if there is a reduction sequence from $\mathcal{M}$ to $\mathcal{M}([r])$, then $\mathcal{R}(\mathcal{M})$ is contractible.

Problem 7.15 Use Proposition 7.14 to prove that every rank 3 oriented matroid on six elements has contractible realization space.

7.3 An Oriented Matroid with Disconnected Realization Space

The earliest example of an oriented matroid whose realization space is not contractible was in Mnëv's 1986 thesis (Mnëv 1986), and the smallest example to date (rank 3 on 13 elements) is due to Tsukamoto (2013). The example we give here, due to Richter-Gebert (1996b), is perhaps the easiest to understand. We'll call this oriented matroid Ω rather than its original name Ω_{14}^+.

All of these examples have disconnected realization space.

Ω is rank 3, and its set of elements is [14]. We begin with $\Omega([11])$, given by the arrangement in Figure 7.4. From this figure one can see that [4] is a projective basis, so $\mathcal{RP}(\Omega[4])$ is a point, and 6,7,8,9,10,11 is a fixing sequence from $\Omega([11])$ to $\Omega([5])$. Also, there is a homeomorphism $h_5 \colon (0,1) \to \mathcal{RP}(\Omega([5]))$: For each $t \in (0,1)$, $h_5(t)$ is the element $(\mathbf{e}_1, \mathbf{e}_2, \mathbf{e}_3, \mathbf{v}_4, \mathbf{v}_5) \in \mathcal{RP}(\Omega([5]))$ with $\mathbf{v}_5 = (1-t)\mathbf{v}_3 + t\mathbf{v}_4$. We let $h_{11} \colon (0,1) \to \mathcal{RP}(\Omega[11])$ be the composition of h_5 with the inverse of the forgetful homeomorphism $\mathcal{RP}(\Omega[11]) \to \mathcal{RP}(\Omega[5])$.

Problem 7.16 Find the 3×9 matrix $h_{11}(t)$. The last row of your matrix will be $(0, 0, 1, 1, \ldots, 1)$. Remaining entries will be formulas in terms of t.

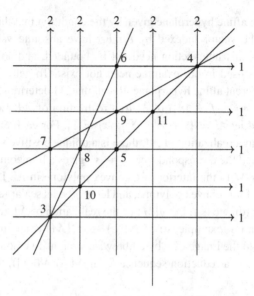

Figure 7.4 A realization of $\Omega([11])$.

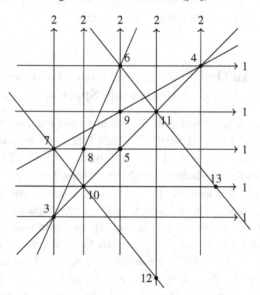

Figure 7.5 A realization of $\Omega([13])$.

Throwing in two more elements, Figure 7.5 shows a realization of $\Omega([13])$. The forgetful map $\mathcal{RP}(\Omega([13])) \to \mathcal{RP}(\Omega([11]))$ is not surjective: The only elements of $\mathcal{RP}(\Omega([11]))$ that extend to representations of $\Omega([13])$ are those with p_5 close to the midpoint of the line segment $\overline{p_3 p_4}$. Figure 7.6 shows

7.3 An Oriented Matroid with Disconnected Realization Space

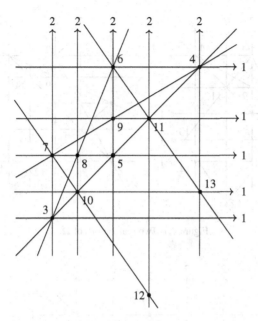

Figure 7.6 Not a realization of $\Omega([13])$.

a weak map image of $\Omega([13])$ that arises from taking p_5 too far from this midpoint: Notice that $\{2, 4, 13\}$ is dependent.

Problem 7.17 This exercise expands on Problem 7.16, where you found formulae for the entries of $h_{11}(t) = (\mathbf{e}_1, \mathbf{e}_2, \mathbf{e}_3, \mathbf{e}_1 + \mathbf{e}_2 + \mathbf{e}_3, \mathbf{v}_5, \ldots, \mathbf{v}_{11})$. Find vectors $\mathbf{v}_{12}(t)$ and $\mathbf{v}_{13}(t)$ such that the last component of each is 1 and each of the sets $\{\mathbf{v}_1, \mathbf{v}_{11}, \mathbf{v}_{12}\}$, $\{\mathbf{v}_7, \mathbf{v}_{10}, \mathbf{v}_{12}\}$, $\{\mathbf{v}_2, \mathbf{v}_4, \mathbf{v}_{13}\}$, and $\{\mathbf{v}_1, \mathbf{v}_{10}, \mathbf{v}_{13}\}$ is linearly dependent. Use your answers to show that the forgetful map $\mathcal{RP}([13]) \to \mathcal{RP}([11])$ maps $\mathcal{RP}([13])$ homeomorphically to $h_{11}(.5 - (1.5 - \sqrt{2}), .5 + (1.5 - \sqrt{2}))$.

Let $h_{13}: (.5 - (1.5 - \sqrt{2}), .5 + (1.5 - \sqrt{2})) \to \mathcal{RP}([13])$ be the homeomorphism resulting from Exercise 7.17.

Finally, Figure 7.7 shows two elements of $\mathcal{RP}(\Omega)$: one with $\mathbf{v}_5$ slightly below the midpoint of $\overline{\mathbf{v}_3, \mathbf{v}_4}$ and one with $\mathbf{v}_5$ slightly above the midpoint. There is no element of $\mathcal{RP}(\Omega)$ with $\mathbf{v}_5$ at the midpoint; Figure 7.8 shows the only projective arrangement that extends $h_{13}(.5)$ by an element $\mathbf{v}_{14}$ with $\{\mathbf{v}_1, \mathbf{v}_3, \mathbf{v}_{14}\}$ and $\{\mathbf{v}_2, \mathbf{v}_4, \mathbf{v}_{14}\}$ dependent. The resulting oriented matroid is a weak map image of Ω in which $\{12, 13, 14\}$ is dependent.

Problem 7.18 Expand your answer to Problem 7.17 to find a vector $\mathbf{v}_{14}(t)$ whose last component is 1 with each of the sets $\{\mathbf{v}_1, \mathbf{v}_3, \mathbf{v}_{14}\}$ and $\{\mathbf{v}_2, \mathbf{v}_4, \mathbf{v}_{14}\}$

224 The Universality Theorem

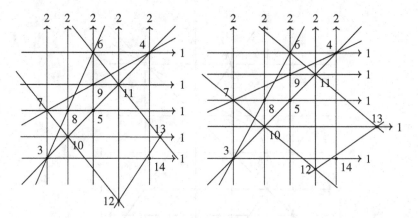

Figure 7.7 Two realizations of Ω.

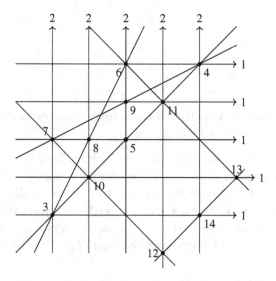

Figure 7.8 Not a realization of Ω.

linearly dependent. Show that the determinant $|\mathbf{v}_{12}(t)\ \mathbf{v}_{13}(t)\ \mathbf{v}_{14}(t)|$ is $-8t^2 + 8t - 2$ and explain why this shows that the forgetful map $\mathcal{RP}(\Omega) \to \mathcal{RP}(\Omega([13]))$ maps $\mathcal{RP}(\Omega)$ homeomorphically to $h_{13}((.5 - (1.5 - \sqrt{2}), .5 + (1.5 - \sqrt{2})) - \{.5\})$.

As a consequence of Problem 7.18 we have the following.

Proposition 7.19 (Richter-Gebert 1996b) *$\mathcal{RP}(\Omega)$ is homeomorphic to the disjoint union of two open intervals.*

Problem 7.20 Extend Ω by one element to get an oriented matroid for which the closure in $\mathbb{R}^{3\times 15}$ of the projective realization space is disconnected.

Remark 7.21 Another interesting property of Ω is its combinatorial symmetry. There is a nonidentity isomorphism $\iota\colon [14] \to [14]$ from Ω to itself, but there is no realization of Ω that also realizes this symmetry. That is, for every realization $(\mathbf{v}_i : i \in [14])$ of Ω, there is no $A \in GL_3$ such that $A\mathbf{v}_i = \mathbf{v}_{\iota(i)}$ for each i. Put another way, there is a symmetric pseudoline representation of Ω but no symmetric realization. Another example of an oriented matroid with this property is given in Shor (1991).

7.4 Semialgebraic Sets

A *basic primary semialgebraic set* is the solution set to a system of finitely many polynomial equalities and strict inequalities. The set is *defined over* $\mathbb{Z}$ if the coefficients in the polynomials can be taken to be integers. That is, $P \subseteq \mathbb{R}^n$ is a basic primary semialgebraic set defined over $\mathbb{Z}$ if there are polynomials $f_1, \ldots, f_k, g_1, \ldots, g_l \in \mathbb{Z}[x_1, \ldots, x_n]$ such that P is the set of all $\mathbf{x} \in \mathbb{R}^n$ satisfying all of the constraints $f_i(\mathbf{x}) = 0$ and $g_i(\mathbf{x}) > 0$.

Proposition 7.22 *Every realization space is a basic primary semialgebraic set defined over* $\mathbb{Z}$.

For purposes of this proof, "realization space" can mean any of our notions of realization space identified with a subset of Vreal($\mathcal{M}$).

Proof: This follows immediately from the chirotope characterization of oriented matroids. Let $\mathcal{M}$ be an oriented matroid and χ be the chirotope for $\mathcal{M}$ with $\chi(1, 2, \ldots, r) = +$. Each value $\chi(i_1, \ldots, i_r) \in \{-, 0, +\}$ is a polynomial equality (if $\chi(i_1, \ldots, i_r) = 0$) or strict inequality that must be satisfied by the entries in every realization of $\mathcal{M}$. The additional conditions on the form of our vector realizations are conditions that specified entries must be 0, 1, or -1. □

What's more surprising is a near-converse: Up to homotopy, *every basic primary semialgebraic set defined over* $\mathbb{Z}$ *is homotopy equivalent to the realization space of a rank 3 oriented matroid*. In fact, the homotopy can be taken to be of a particularly simple type, which we now describe.

Definition 7.23 Let P and Q be basic primary semialgebraic sets.
We say P is **rationally equivalent** to Q, written $P \cong_r Q$, if there is a homeomorphism $f\colon P \to Q$ such that both f and f^{-1} are rational functions with rational coefficients.

Assume $P \subseteq \mathbb{R}^{n+k}$ and $Q \subseteq \mathbb{R}^n$. Assume that projection onto the first n coordinates induces a map $\pi: P \to Q$. Let $\mathbf{x} = (x_1, \ldots, x_n)$ and $\mathbf{y} = (y_1, \ldots, y_k)$ be vectors of variables. We say π is a **stable projection** if the following conditions are satisfied:

1. There are finite sets of polynomials $\{\Phi_i : i \in I\}, \{\Psi_j : j \in J\} \subset \mathbb{Q}[\mathbf{x}, \mathbf{y}]$ such that each polynomial has degree 1 in variables $\mathbf{y}$ and

$$P = \{(\mathbf{x}, \mathbf{y}) : \mathbf{x} \in Q, \forall i \; \Phi_i(\mathbf{x}, \mathbf{y}) > 0, \forall j \; \Psi_j(\mathbf{x}, \mathbf{y}) = 0\}.$$

2. π has a continuous section. That is, there is a continuous map $\iota: Q \to P$ such that $\pi \circ \iota$ is the identity map.

We write $P \simeq_p Q$ if there is a stable projection from P to Q.

Stable equivalence, denoted $\simeq_s$, is the equivalence relation on semialgebraic sets generated by $\cong_r$ and $\simeq_p$.

A stable projection is a homotopy equivalence. This can be seen by the same argument as we used for Proposition 7.14. The first condition π tells us that preimages under π are convex sets, and so as in Proposition 7.14 there is a fiberwise straight-line homotopy retracting P to $\iota(Q)$. Thus if $P \simeq_s Q$ then P and Q are homotopy equivalent.

Remark 7.24 The definitions of "stable projection" and "stable equivalence" are matters of some debate. The idea is that a stable equivalence should preserve both homotopy type and algebraic complexity. Various sources either skip over exact definitions or (intentionally or unintentionally) give definitions that fail to guarantee that a stable equivalence is a homotopy equivalence. See Boege (n.d.) and Verkama (2023) for a discussion of possible correct definitions.

The following flows from the discussion in Section 7.1.

Proposition 7.25 *For every $\mathcal{M}$ we have* $\mathrm{Vreal}(\mathcal{M}) \cong_r \mathrm{Greal}(\mathcal{M}) \times GL_r \simeq_s \mathrm{Hreal}(\mathcal{M})$, *for all rank r $\mathcal{M}$ on elements $[n]$ with $[r]$ a basis, we have* $\mathrm{Greal}(\mathcal{M}) \simeq_s \mathcal{R}(\mathcal{M})$, *and for all rank r $\mathcal{M}$ on elements $[n]$ with $[r+1]$ a projective basis, we have* $\mathcal{R}(\mathcal{M}) \simeq_s \mathcal{RP}(\mathcal{M})$.

Thus the following theorem tells us about all of our notions of realization space.

Theorem 7.26 (The Universality Theorem; Mnëv 1988) *Every basic primary semialgebraic set defined over $\mathbb{Z}$ is stably equivalent to $\mathcal{RP}(\mathcal{M})$ for some rank 3 oriented matroid $\mathcal{M}$.*

7.4.1 Shor Semialgebraic Sets

Our proof of the Universality Theorem will loosely follow Shor's presentation (Shor 1991). As a first step, we show that the polynomials defining a basic semialgebraic set can be chosen to have a few very simple types.

Definition 7.27 A **Shor semialgebraic set** is the solution set in $\mathbb{R}^{n+1}$ of a system of conditions consisting of

- the inequalities $1 = x_0 < x_1 < \cdots < x_n$,
- a finite number of equalities of the form $x_a + x_b = x_c$ with $0 \le a \le b < c \le n$, and
- a finite number of equalities of the form $x_d x_e = x_f$ with $1 \le d \le e < c \le f$.

The purpose of this section is to prove the following.

Proposition 7.28 *Every basic primary semialgebraic set defined over $\mathbb{Z}$ is stably equivalent to a Shor semialgebraic set.*

The proof is a bit of a slog. Even worse, no oriented matroids are involved. The casual reader who is willing to assume Proposition 7.28 may move on to Section 7.5.

We prove Proposition 7.28 by a sequence of steps that transform a given basic primary semialgebraic set defined over $\mathbb{Z}$ into one satisfying progressively more of the conditions in Definition 7.27.

Let $I_=$ and $I_>$ be finite subsets of $\mathbb{Z}[x_1, \ldots, x_n]$ and P be the basic primary semialgebraic set consisting of all $\mathbf{x}$ such that $f(\mathbf{x}) = 0$ for each $f \in I_=$ and $g(\mathbf{x}) > 0$ for each $g \in I_>$. We first describe how $P \subseteq \mathbb{R}$ can be shown to be rationally equivalent to the solution set of a system given by set $I'_=$ and $I'_>$ in which each element of $I'_= \cup I'_>$ has at most three terms. For example, let $f_0(\mathbf{x}) \in I_=$, and assume f_0 is the sum of five terms, $f_0(\mathbf{x}) = m_1(\mathbf{x}) + m_2(\mathbf{x}) + m_3(\mathbf{x}) + m_4(\mathbf{x}) + m_5(\mathbf{x})$. We introduce new variables y_1, y_2 and conditions relating these variables to $\mathbf{x}$:

$$y_1 = m_1(\mathbf{x}) + m_2(\mathbf{x}),$$
$$y_2 = y_1 + m_3(\mathbf{x}),$$
$$0 = y_2 + m_4(\mathbf{x}) + m_5(\mathbf{x}).$$

Let $P' \subseteq \mathbb{R}^{n+2}$ be the solution set to the system of conditions we get by replacing the condition $f_0(\mathbf{x}) = 0$ with the above three equalities. The map $\mathbf{x} \mapsto (\mathbf{x}, m_1(\mathbf{x}) + m_2(\mathbf{x}), m_1(\mathbf{x}) + m_2(\mathbf{x}) + m_3(\mathbf{x}))$ from P to P' shows $P \cong_r P'$. In the same way, a defining condition in which the polynomial has t terms can be replaced by $t-2$ polynomial conditions, involving $t-3$ new variables, with each polynomial having at most three terms.

In the remainder of this discussion we'll carry out more complicated variations of this argument. To prepare, let's repeat our $f_0(\mathbf{x}) = m_1(\mathbf{x}) + m_2(\mathbf{x}) + m_3(\mathbf{x}) + m_4(\mathbf{x}) + m_5(\mathbf{x})$ example in a framework that lends itself to generalization. Rather than defining P' in terms of variables $x_1, \ldots, x_n, y_1, y_2$, we work with a vector of variables $\mathbf{V} = (V_{x_1}, \ldots, V_{x_n}, V_{m_1+m_2}, V_{m_1+m_2+m_3})$. For each $h(\mathbf{x}) \in \mathbb{Z}[x_1, \ldots, x_n]$, define $\hat{h}(\mathbf{V}) \in \mathbb{Z}[V_{x_1}, \ldots, V_{x_n}, V_{m_1+m_2}, V_{m_1+m_2+m_3}]$ to be the polynomial obtained from h by replacing each x_i with V_{x_i}. Let $\rho: \mathbb{R}^n \to \mathbb{R}^{n+2}$ be the map $\rho(\mathbf{x}) = (x_1, \ldots, x_n, m_1(\mathbf{x}) + m_2(\mathbf{x}), m_1(\mathbf{x}) + m_2(\mathbf{x}) + m_3(\mathbf{x}))$. Then ρ induces a rational equivalence from P to $P' := \rho(P)$, and P' is the solution set to the system of equations given by

$$\hat{f}(\mathbf{V}) = 0 \qquad \text{for each } f \in I_= - \{f_0\}.$$
$$\hat{g}(\mathbf{V}) > 0 \qquad \text{for each } g \in I_>.$$
$$V_{m_1+m_2} = \widehat{m_1}(\mathbf{V}) + \widehat{m_2}(\mathbf{V})$$
$$V_{m_1+m_2+m_3} = V_{m_1+m_2} + \widehat{m_3}(\mathbf{V})$$
$$0 = V_{m_1+m_2+m_3} + \widehat{m_4}(\mathbf{V}) + \widehat{m_5}(\mathbf{V}).$$

Let's look at what we did on a more abstract and general level. We began with a semialgebraic set P defined by conditions $f(\mathbf{x}) = 0$ for each $f \in I_=$ and $g(\mathbf{x}) > 0$ for each $g \in I_>$. We wanted to replace the defining conditions with conditions that are simpler in a particular way. We defined variables $V_{h_1}, \ldots, V_{h_{n'}}$, where each h_i is a polynomial in $\mathbf{x}$, and then defined $\rho: \mathbb{R}^n \to \mathbb{R}^{n'}$ by $\rho(\mathbf{x}) = (h_1(\mathbf{x}), \ldots, h_{n'}(\mathbf{x}))$. We defined $P' = \rho(P)$. The polynomials $h_1, \ldots, h_{n'}$ were chosen so that they have the following properties:

1. The set $\{h_1, \ldots, h_{n'}\}$ generates $\mathbb{Z}[x_1, \ldots, x_n]$; in particular, for each $k \in [n]$ there is a polynomial equality $p_k(h_{j_1}, \ldots, h_{j_n}) = x_k$. Thus ρ has a rational left inverse $\sigma: \mathbb{R}^{n'} \to \mathbb{R}^n$ defined by $\sigma(\mathbf{y}) = (p_1(\mathbf{y}), \ldots, p_n(\mathbf{y}))$, and so $P \cong_r P'$.
2. Consider the ideal of the algebraic variety $\rho(\mathbb{R}^n)$. By definition, this is $\{p \in \mathbb{R}[V_{h_1}, \ldots, V_{h_{n'}}] : \forall \mathbf{x} \in \mathbb{R}^n \; p(h_1(\mathbf{x}), \ldots, h_{n'}(\mathbf{x})) = 0\}$, so it's given by relations between the h_i. We chose our h_i so that this ideal is generated by a finite set J whose elements are all simple in our desired way. In our example, $J = \{V_{m_1+m_2} - \widehat{m_1}(\mathbf{V}) - \widehat{m_2}(\mathbf{V}), V_{m_1+m_2+m_3} - V_{m_1+m_2} - \widehat{m_3}(\mathbf{V})\}$.
3. For each $f \in I_=$ there is a $\tilde{f} \in \mathbb{Z}[V_{h_1}, \ldots, V_{h_{n'}}]$ such that $f(\mathbf{x}) = 0$ if and only if $\tilde{f}(\rho(\mathbf{x})) = 0$. Further, $\tilde{f}$ is simple in our desired way. In our example, $\tilde{f}_0 = V_{m_1+m_2+m_3} + \widehat{m_4}(\mathbf{V}) + \widehat{m_5}(\mathbf{V})$.
4. For each $g \in I_>$ there is a $\tilde{g} \in \mathbb{Z}[V_{h_1}, \ldots, V_{h_{n'}}]$ such that $g(\mathbf{x}) > 0$ if and only if $\tilde{g}(\rho(\mathbf{x})) > 0$. Further, $\tilde{g}$ is simple in our desired way.

7.4 Semialgebraic Sets

Let $\tilde{I}_= = \{\tilde{f} : f \in I_=\}$ and $\tilde{I}_> = \{\tilde{g} : g \in I_>\}$. Thus

$$P' = \{\mathbf{V} \in \rho(\mathbb{R}^n) : \forall \tilde{f} \in \tilde{I}_=\ \tilde{f}(\mathbf{V}) = 0, \forall \tilde{g} \in \tilde{I}_>\ \tilde{g}(\mathbf{V}) > 0\}$$
$$= \{\mathbf{V} \in \mathbb{R}^{n'} : \forall j \in J\ j(\mathbf{V}) = 0, \forall \tilde{f} \in \tilde{I}_=\ \tilde{f}(\mathbf{V}) = 0, \forall \tilde{g} \in \tilde{I}_>\ \tilde{g}(\mathbf{V}) > 0\}.$$

Because the conditions coming from elements of J can be read off from our subscripts h_i, we will call these conditions *subscript conditions*. The conditions $\tilde{f} = 0$ for $f \in I_=$ and $\tilde{g} > 0$ for $g \in I_>$ will be called *translated conditions*.

Using this framework, we go back to our efforts to get from P to a Shor semialgebraic set. Assume that we have dealt with all defining conditions for P involving more than three monomials, obtaining a semialgebraic set $P_1 \cong_r P$ in $\mathbb{R}^{n_1}$. For simplicity, we rename the variables we called V_{h_i} above so that the defining polynomials for P_1 are in variables $x_1, \ldots, x_{n_1}$.

We next introduce further variables and conditions to get a basic primary semialgebraic set $P_2 \cong_r P_1$, defined in terms of variables V_h, in which each equality in our conditions has one of the forms $V_a = 1$, $V_a + V_b = V_c$, or $V_a V_b = V_c$ and each inequality has the form $V_a < V_b$.

For example, assume that $3x_1^2 - 2x_2x_3 = 0$ is one of our defining conditions. We introduce new variables $V_1, V_2, V_3, V_{x_1}, V_{x_2}, V_{x_3}, V_{3x_1}, V_{3x_1^2}, V_{2x_2}$ and conditions relating these variables to **x**. The subscript conditions are

$$V_1 = 1$$
$$V_2 = V_1 + V_1$$
$$V_3 = V_2 + V_1$$
$$V_{3x_1} = V_3 V_{x_1}$$
$$V_{3x_1^2} = V_{3x_1} V_{x_1}$$
$$V_{2x_2} = V_2 V_{x_2}.$$

We replace the defining condition $3x_1^2 - 2x_2x_3 = 0$ for P_1 with the translated condition $V_{3x_1^2} = V_{2x_2} V_{x_3}$. In the same way, we can introduce variables and conditions for each of our equality conditions. Also, an inequality $0 < g(\mathbf{x})$ can be replaced with $x_a < g(\mathbf{x}) + x_a$, and by introducing new variables as above we can get this to the form $V_{x_a} < V_h$.

Once we have done this for all of our conditions on P_1, we have a collection of n_2 variables, for some n_2, and conditions on these variables whose solution set in $\mathbb{R}^{n_2}$ is the desired semialgebraic set $P_2 \cong_r P_1$. Then once again we rename variables V_h so that our defining conditions for P_2 are in variables $x_1, \ldots, x_{n_2}$.

To proceed further, we wish to introduce a variable w that is related to our previous variables by $w > -x_a + 1$ for each a and $w > 1$. By the same rigmarole as in the preceding examples, we can introduce new variables defined in terms of w and the x_a to put these conditions in the form $V_w > V_{x_b}$. Let P_3 be the solution set to the union of the conditions on P_2 and these new conditions. The map $P_3 \to P_2$ that forgets V_w and these new variables and sends each V_{x_a} to x_a is a stable projection: We get a section of this map by taking each $\mathbf{x}$ to the point in its preimage with $V_w = \max(-x_1 + 2, \ldots, -x_{n_2} + 2, 2)$. Once again, we rename variables so that our defining conditions for P_3 are in variables $x_1, \ldots, x_{n_3}, w$; the variables V_{x_a} with $a \in [n_2]$ will be renamed x_a. We write $(x_1, \ldots, x_{n_3}, w)$ as $(\mathbf{x}, w)$, and an element of P_3 as $(\mathbf{p}, q)$.

Our next step is to get a P_4 that is rationally equivalent to P_3 and is defined to be the set of all $\mathbf{V}$ satisfying

- $V_1 = 1$,
- some conditions of the form $V_{h_a} < V_{h_b}$,
- some conditions of the form $V_{h_a} + V_{h_b} = V_{h_c}$,
- some conditions of the form $V_{h_a} V_{h_b} = V_{h_c}$, and
- $V_{h_a} > 1$ for all $h_a \neq 1$.

We first introduce variables V_1, V_w, V_{w^2}, V_{w+w^2}, and for each $a \in [n_2]$, V_{x_a+w}. The subscript conditions are

$$V_1 = 1$$
$$V_w V_w = V_{w^2}$$
$$V_w + V_{w^2} = V_{w+w^2}.$$

The translated conditions coming from $w > 1$ and $w > -x_a + 1$ are $V_w > 1$ and $V_{x_a+w} > 1$ for all $a \in [n_2]$. It follows from the preceding conditions that

$$V_{w^2} > 1$$

and

$$V_{w+w^2} > 2. \tag{7.1}$$

In the following discussion the reader should note that each new variable $V_{h(\mathbf{x},w)}$ we introduce is a sum of variables that are constrained to be greater than 1, and hence satisfies $V_{h(\mathbf{x},w)} > 1$.

Each condition $x_a = 1$ on P_3 translates to a condition $V_{x_a+w} = V_1 + V_w$ on P_4, and each condition $x_a < x_b$ on P_3 translates to a condition $V_{x_a+w} < V_{x_b+w}$.

We restate each condition $x_a + x_b = x_c$ on P_3 as $(x_a + w) + (x_b + w) = x_c + 2w = (x_c + w) + w$. Inspired by this, we introduce a variable V_{x_c+2w},

7.4 Semialgebraic Sets

subscript condition $V_{x_c+2w} = V_{x_c+w} + V_w$, and translated condition $V_{x_a+w} + V_{x_b+w} = V_{x_c+2w}$.

We restate each condition $x_a x_b = x_c$ on P_3 as

$$(x_a + w)(x_b + w) = x_c + x_a w + x_b w + w^2,$$
$$(x_a + w)(x_b + w) + w^2 = x_c + (x_a + w)w + (x_b + w)w.$$

We use the last line to get equivalent conditions using our previously defined variables V_{x_a+w}, V_{x_b+w}, V_{x_c+w}, V_w, and V_{w^2}. We introduce variables and subscript conditions

$$V_{(x_a+w)w} = V_{x_a+w} V_w,$$
$$V_{(x_b+w)w} = V_{x_b+w} V_w,$$
$$V_{(x_a+w)w+(x_b+w)w} = V_{(x_a+w)w} + V_{(x_b+w)w},$$
$$V_{(x_a+w)(x_b+w)} = V_{x_a+w} V_{x_b+w},$$
$$V_{(x_a+w)(x_b+w)+w^2} = V_{(x_a+w)(x_b+w)} + V_{w^2}$$

and translated condition

$$V_{(x_a+w)(x_b+w)+w^2} = V_{x_c} + V_{(x_a+w)w+(x_b+w)w}.$$

Ordering our variables $V_1, V_{h_2}, \ldots, V_{h_{n_4}}$, we get a map $\rho : P_3 \to \mathbb{R}^{n_4}$ whose image is our $P_4 \cong_r P_3$.

Once again, we rename our variables $x_1, \ldots, x_{n_4}$. As in our definition of P_3, we introduce a new variable v, this time with the condition that $v > x_a$ for each a, and call the solution set to this expanded set of conditions P_5. Inequality (7.1) tells us that every element $(\mathbf{x}, v)$ of P_5 satisfies $v > 2$. Forgetting v gives us a stable projection $P_5 \to P_4$; one section of this map is $\iota(\mathbf{x}) = (\mathbf{x}, \max(x_1, \ldots, x_{n_4}) + 1)$.

At last, we define a Shor semialgebraic set $P_6 \cong_r P_5$. The only remaining hurdle is to get P_6 to satisfy $1 < x_1 < \cdots < x_{n_6}$. The construction of P_6 has the same flavor as that for P_4. For each $a \in [n_4]$ we introduce a variable $V_{x_a+v^a}$. We also need many variables of the form V_{v^a}. Exactly how many such variables would be tedious to say: Suffice it to say that we introduce variables $V_v, V_{v^2}, \ldots, V_{v^r}$, where r is large enough to make the following discussion work, and subscript conditions $V_{v^{a+1}} = V_{v^a} V_v$. The idea is that, just as each x_a gets its own associated power v^a of v, each condition on P_4 gets a set of associated powers of v, with no power associated to more than one condition or variable. Then the total ordering on our variables follows from the fact that for each $(\mathbf{x}, v) \in P_5$ and each i, j, k, l with $i < j$, since $1 < x_k < v$, $1 < x_l$, and $v > 2$, we have

$$x_k + v^i < 2v^i < v^j < x_l + v^j.$$

In particular,
$$x_1 + v^1 < x_2 + v^2 < \cdots < x_{n_4} + v^{n_4}$$
and so elements of P_6 satisfy the translated conditions
$$V_{x_1+v^1} < V_{x_2+v^2} < \cdots < V_{x_{n_4}+v^{n_4}}.$$

For each condition $x_a < x_b$ on P_4, we choose a power v^α not yet associated to anything (in particular, $\alpha > n_4$) and introduce variables $V_{v^\alpha-v^a}, V_{v^\alpha-v^b}, V_{x_a+v^\alpha}$,
$V_{x_b+v^\alpha}$ and subscript conditions

$$V_{v^\alpha} = V_{v^\alpha-v^a} + V_{v^a},$$
$$V_{v^\alpha} = V_{v^\alpha-v^b} + V_{v^b},$$
$$V_{x_a+v^a} + V_{v^\alpha-v^a} = V_{x_a+v^\alpha},$$
$$V_{x_b+v^b} + V_{v^\alpha-v^b} = V_{x_b+v^\alpha}.$$

Since $x_a < x_b$, if $a < b$ we have
$$v^\alpha - v^b < v^\alpha - v^a < v^\alpha < x_a + v^\alpha < x_b + v^\alpha,$$
and if $b < a$ we have
$$v^\alpha - v^a < v^\alpha - v^b < v^\alpha < x_a + v^\alpha < x_b + v^\alpha.$$

This leads to the set of coordinates $\{V_{v^\alpha-v^b}, V_{v^\alpha-v^a}, V_{v^\alpha}, V_{x_a+v^\alpha}, V_{x_b+v^\alpha}\}$ of each element of P_6 being totally ordered. If another condition $x_{a'} < x_{b'}$ is associated to a power $\beta > \alpha$, then in the same way we see that for each element of P_6 the coordinates associated to this condition are strictly larger than the coordinates we introduced for $x_a < x_b$.

For each condition $x_a + x_b = x_c$, we choose powers $v^\alpha, v^\beta, v^\gamma$ not yet associated to anything, with $\alpha < \beta < \gamma$. Rewriting $x_a + x_b = x_c$ as $(x_a + v^\alpha) + (x_b + v^\beta) + (v^\gamma - v^\alpha - v^\beta) = x_c + v^\gamma$ inspires us to introduce variables $V_{v^\alpha-v^a}, V_{v^\beta-v^b}, V_{v^\gamma-v^c}, V_{x_a+v^\alpha}, V_{x_b+v^\beta}, V_{x_c+v^\gamma}$ related by subscript conditions

$$V_{v^\alpha} = V_{v^\alpha-v^a} + V_{v^a}, \tag{7.2}$$
$$V_{v^\beta} = V_{v^\beta-v^b} + V_{v^b}, \tag{7.3}$$
$$V_{v^\gamma} = V_{v^\gamma-v^c} + V_{v^c}, \tag{7.4}$$
$$V_{x_a+v^a} + V_{v^\alpha-v^a} = V_{x_a+v^\alpha}, \tag{7.5}$$
$$V_{x_b+v^b} + V_{v^\beta-v^b} = V_{x_b+v^\beta}, \tag{7.6}$$
$$V_{x_c+v^c} + V_{v^\gamma-v^c} = V_{x_c+v^\gamma}, \tag{7.7}$$

7.4 Semialgebraic Sets

then introduce variables $V_{v^\gamma - v^\alpha}, V_{v^\gamma - v^\alpha - v^\beta}$ related by subscript conditions

$$V_{v^\gamma - v^\alpha} + V_{v^\alpha} = V_{v^\gamma},$$
$$V_{v^\gamma - v^\alpha - v^\beta} + V_{v^\beta} = V_{v^\gamma - v^\alpha},$$

and finally introduce variables $V_{x_a + x_b + v^\alpha + v^\beta}, V_{v^\gamma - v^\alpha}$ related by subscript conditions

$$V_{v^\gamma - v^\alpha} + V_{v^\alpha} = V_{v^\gamma},$$
$$V_{v^\gamma - v^\alpha - v^\beta} + V_{v^\beta} = V_{v^\gamma - v^\alpha},$$
$$V_{x_a + v^\alpha} + V_{x_b + v^\beta} = V_{x_a + x_b + v^\alpha + v^\beta}$$

and translated condition

$$V_{x_a + x_b + v^\alpha + v^\beta} + V_{v^\gamma - v^\alpha - v^\beta} = V_{x_c + v^\gamma}.$$

This formidable list of new variables is totally ordered, since for each $(\mathbf{x}, v) \in P_5$,

$$v^\alpha - v^a < x_a + v^a < v^\beta - v^b < x_b + v^\beta < x_a + x_b + v^\alpha + v^\beta < v^\gamma - v^\alpha - v^\beta$$
$$< v^\gamma - v^\alpha < v^\gamma - v^c < v^\gamma - v^a < x_c + v^\gamma.$$

Our last hurdle is multiplication. For each condition $x_a x_b = x_c$, we choose powers $v^\alpha, v^\beta, v^\gamma$ not yet associated to anything, with $\alpha < \beta < \gamma$ and $\alpha + \beta = \gamma$. Rewriting $x_a x_b = x_c$ as $(x_a + v^\alpha)(x_b + v^\beta) + 2v^\gamma = (x_c + v^\gamma) + v^\beta(x_a + v^\alpha) + v^\alpha(x_b + v^\beta) + v^\gamma$ inspires us to once again introduce the variables and conditions of Equations (7.2)–(7.7), as well as the following variables and conditions:

$$V_{2v^\gamma} = V_{v^\gamma} + V_{v^\gamma},$$
$$V_{x_a x_b + v^\beta x_a + v^\alpha x_b + v^\gamma} = V_{x_a + v^\alpha} V_{x_b + v^\beta},$$
$$V_{x_a x_b + v^\beta x_a + v^\alpha x_b + 3v^\gamma} = V_{x_a x_b + v^\beta x_a + v^\alpha x_b + v^\gamma} + V_{2v^\gamma},$$
$$V_{v^\beta x_a + v^\gamma} = V_{v^\beta} V_{x_a + v^\alpha},$$
$$V_{v^\alpha x_b + v^\gamma} = V_{v^\alpha} V_{x_b + v^\beta},$$
$$V_{v^\alpha x_b + v^\beta x_a + 2v^\gamma} = V_{v^\beta x_a + v^\gamma} + V_{v^\alpha x_b + v^\gamma},$$
$$V_{x_a x_b + v^\beta x_a + v^\alpha x_b + 3v^\gamma} = V_{v^\alpha x_b + v^\beta x_a + 2v^\gamma} + V_{x_c + v^\gamma}.$$

With more tedious checking we can see that the resulting P_6 satisfies a list of inequalities $V_{h_1} < \cdots < V_{h_{n_6}}$ involving all of the variables.

7.5 Proof of the Universality Theorem

Our aim in this section is to prove the following.

Proposition 7.29 *For each Shor semialgebraic set P, there is a rank 3 oriented matroid $\mathcal{M}$ such that $P \simeq_s \mathcal{RP}(\mathcal{M})$.*

Together with Proposition 7.28, this proves Universality (Theorem 7.26).

Let $P \subset \mathbb{R}^n$ be the set of all $\mathbf{x} \in \mathbb{R}^n$ satisfying $1 = x_0 < x_1 < \cdots < x_n$ and a finite set R of additional conditions, each of the form $x_a + x_b = x_c$ or $x_a x_b = x_c$. Order R somehow, and let R_i be the set consisting of the first i elements of R. Let P_0 be the set of all $\mathbf{x} \in \mathbb{R}^n$ satisfying $1 = x_0 < x_1 < \cdots < x_n$, and for $i > 0$ let P_i be the subset of P_0 of $\mathbf{x}$ satisfying the conditions in R_i. We will inductively construct oriented matroids $\mathcal{M}_i$ such that $\mathcal{RP}(\mathcal{M}_i) \simeq_s P_i$.

Let $E_0 = \{1, 2, 3, 4, e_1, \ldots, e_n\}$, ordered this way. Let $\mathcal{M}_0$ be the oriented matroid on E_0 given by the arrangement in Figure 7.9.

Thus elements of $\mathcal{RP}(\mathcal{M}_0)$ have the form

$$\begin{pmatrix} 1 & 0 & 0 & 1 & 1 & x_2 & \ldots, & x_n \\ 0 & 1 & 0 & 1 & 0 & 0 & \ldots & 0 \\ 0 & 0 & 1 & 1 & 1 & 1 & \ldots & 1 \end{pmatrix},$$

with $1 < x_2 < \cdots < x_n$.

For every rank 3 $\mathcal{M}$ with $\mathcal{M}_0$ as a minor, we have a rational map $\xi : \mathcal{RP}(\mathcal{M}) \to P_0$ given by entries in the first row. Our inductive argument will construct each $\mathcal{M}_i$ with $i > 0$ as an extension of $\mathcal{M}_{i-1}$ in such a way that the image of $\xi_i : \mathcal{RP}(\mathcal{M}_i) \to P_0$ is P_i. Let $\phi_i : \mathcal{RP}(\mathcal{M}_i) \to \mathcal{RP}(\mathcal{M}_{i-1})$ denote the forgetful map. Then ξ_i factors as

$$\mathcal{RP}(\mathcal{M}_i) \xrightarrow{\phi_i} \mathrm{im}(\phi_i) \xrightarrow{\xi_{i-1}} P_i.$$

Our construction of $\mathcal{M}_i$ will make it clear that the first map is a stable projection, and the induction hypothesis will guarantee that the second map is a stable homotopy equivalence.

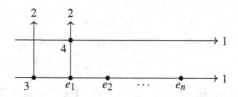

Figure 7.9 A realization of $\mathcal{M}_0$.

7.5 Proof of the Universality Theorem

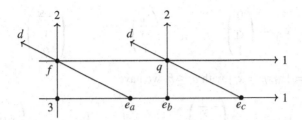

Figure 7.10 Constructing addition geometrically.

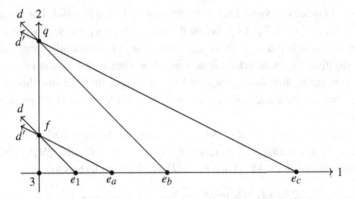

Figure 7.11 Constructing multiplication geometrically.

The construction is based on two elementary observations about arrangements, illustrated in Figures 7.10 and 7.11. Assume that in each of the arrangements shown the points labeled e_a, e_b, e_c have first coordinates x_a, x_b, x_c. Then by congruence of triangles we see that in the arrangement of Figure 7.10 $x_a + x_b = x_c$, and by similarity of triangles we see that in the arrangement of Figure 7.11 $x_a x_b = x_c$. It follows that if $\mathcal{M}$ is an extension of both $\mathcal{M}_0$ and the oriented matroid of Figure 7.10 then $\xi(\mathcal{RP}(\mathcal{M})) \subseteq \{\mathbf{x} \in \mathbb{R}^n : 1 = x_1 < x_2 < \cdots < x_n, x_a + x_b = x_c\}$, and if $\mathcal{M}$ is an extension of both $\mathcal{M}_0$ and the oriented matroid of Figure 7.11 then $\xi(\mathcal{RP}(\mathcal{M})) \subseteq \{\mathbf{x} \in \mathbb{R}^n : 1 = x_1 < x_2 < \cdots < x_n, x_a x_b = x_c\}$.

To construct $\mathcal{M}_1$, we extend $\mathcal{M}_0$ by elements f_1, d_1, q_1 (if $R_1 = \{x_a + x_b = x_c\}$) or by elements f_1, d_1, d'_1, q_1 (if $R_1 = \{x_a x_b = x_c\}$) to reflect the appropriate equality. Let E_i be the set of new elements. Examining Figures 7.10 and 7.11, we see that, for each $(\mathbf{v}_e : e \in E_0)$ that extends as desired, the set of extensions $(\mathbf{v}_e : e \in E_0 \cup E_1)$ with $(\mathbf{v}_e : e \in \{1, 2, 3, e_a, e_b, e_c\} \cup E_1)$ as desired is parametrized by a single number y_1. If $R_1 = \{x_a + x_b = x_c\}$ with $a \leq b$, we have

$$\mathbf{v}_{f_1} = \begin{pmatrix} 0 \\ y_1 \\ 1 \end{pmatrix} \quad \mathbf{v}_{q_1} = \begin{pmatrix} x_b \\ y_1 \\ 1 \end{pmatrix} \quad \mathbf{v}_{d_1} = \begin{pmatrix} -\frac{x_a}{y_1} \\ 1 \\ 0 \end{pmatrix},$$

and if $R_1 = \{x_a x_b = x_c\}$ with $a \leq b$, we have

$$\mathbf{v}_{f_1} = \begin{pmatrix} 0 \\ y_1 \\ 1 \end{pmatrix} \quad \mathbf{v}_{d_1} = \begin{pmatrix} -\frac{1}{y_1} \\ 1 \\ 0 \end{pmatrix} \quad \mathbf{v}_{d_1'} = \begin{pmatrix} -\frac{x_a}{y_1} \\ 1 \\ 0 \end{pmatrix} \quad \mathbf{v}_{q_1} = \begin{pmatrix} 0 \\ \frac{y_1(x_c - x_b)}{x_a - 1} \\ 1 \end{pmatrix}.$$

Not all choices of y_1 will result in the same oriented matroid: For instance, small values of y_1 will put $\mathbf{v}_{f_1}$ below the line through $\mathbf{v}_4$ and $\mathbf{v}_{e_1}$, while larger values of y_1 will put $\mathbf{v}_{f_1}$ above this line. But all sufficiently large values of y_1 do give the same oriented matroid. Specifically, all values of y_1 such that the resulting line through $\mathbf{v}_{e_a}$ and $\mathbf{v}_{f_1}$ is steeper than the line through $\mathbf{v}_4$ and $\mathbf{v}_{e_1}$ induce the same oriented matroid $\mathcal{M}_1$, while no smaller values of y_1 induce $\mathcal{M}_1$.

For a given realization M of $\mathcal{M}_0$, let $c(M)$ be the lower bound of the values y_1 inducing a realization of $\mathcal{M}_1$, and for each $y_1 > c(M)$, let $\langle M, y_1 \rangle$ be the induced realization of $\mathcal{M}_1$. Then $\xi_1: \mathcal{RP}(\mathcal{M}_1) \to P_0$ factors as

$$\mathcal{RP}(\mathcal{M}_1) \to \text{im}(\phi_1) \times \mathbb{R}_{>0} \to \text{im}(\phi_1) \to P_0.$$

The first map sends a realization $\langle M, y_1 \rangle$ to the pair $(M, y_1 - c(M))$. The map $\mathcal{RP}(\mathcal{M}_1) \to \text{im}(\phi_1) \times \mathbb{R}_{>0}$ is a rational equivalence and the map $\text{im}(\phi_1) \times \mathbb{R}_{>0} \to \text{im}(\phi_1)$ is a stable projection. The map $\text{im}(\phi_1) \to P_0$ is a rational homeomorphism to its image. Our observations on Figure 7.11 tell us that this image is a subset of P_1. Conversely, every element of P_1 is $\xi_0(M)$ for some $M \in \mathcal{RP}(\mathcal{M}_0)$ that extends to $\langle M, 1 \rangle \in \mathcal{RP}(\mathcal{M}_1)$. Thus the image of $\xi_1: \mathcal{RP}(\mathcal{M}_1) \to P_0$ is exactly P_1. The map $M \to \langle M, 1 \rangle$ from $\text{im}(\phi_1)$ to $\mathcal{RP}(\mathcal{M}_1)$ gives us the cross-section required in the definition of stable projection.

Now assume we have $\mathcal{M}_{i-1}$ on elements $\bigcup_{j=0}^{i-1} E_j$ as desired. We want to extend $\mathcal{M}_{i-1}$ by $E_i := \{f_i, d_i, q_i\}$ (if $R_i = \{x_a + x_b = x_c\}$) or by $E_i := \{f_i, d_i, d_i', q_i\}$ (if $R_i = \{x_a x_b = x_c\}$). Consider an element $M = (\mathbf{v}_e : e \in \bigcup_{j=0}^{i-1} E_j)$ of $\mathcal{RP}(\mathcal{M}_{i-1})$. By choosing a parameter y sufficiently large and defining vectors $\mathbf{v}_e$ for each $e \in E_i$ as we did for $\mathcal{M}_1$, we obtain an arrangement in which $\mathbf{v}_{f_i}$ and $\mathbf{v}_{q_i}$ are above each line in $\mathbb{A}$ spanned by two columns of M, and the direction vector in $\mathbb{A}$ given by $\mathbf{v}_{d_i'}$ is closer to vertical than that of any column of M at infinity except $\mathbf{e}_2$ itself. For each such y, the resulting arrangement has the same oriented matroid $\mathcal{M}_i$. Let $c(M)$ be the lower bound on the set of suitable y for a given M. Then $M \to c(M)$ is a rational function, and so, as in our argument for $\mathcal{M}_1$, we see $\mathcal{RP}(\mathcal{M}_i) \cong_r$ $\{(M, y): \mathcal{M} \in \text{im}(\phi_i), y > c(M)\} \simeq_p \text{im}(\phi_i)$. The induction hypothesis tells us that ξ_{i-1} restricts to a map from $\text{im}(\phi_i)$ to $\text{im}(\xi_i)$ that is a composition of rational equivalences and stable projections. By construction $\text{im}(\xi_i) \subseteq P_i$. Conversely, given an element $(1, x_1, \ldots, x_n)$ of P_i, let M_0 be the corresponding realization of $\mathcal{M}_0$. A choice of $0 << y_1 << \cdots << y_i$ constructs an extension M of M_0 realizing $\mathcal{M}_i$ with $\xi_i(M) = (1, x_1, \ldots, x_n)$. Putting this together, we have a stable equivalence $\mathcal{RP}(\mathcal{M}_i) \simeq_s P_i$.

Remark 7.30 On the level of projective geometry, the constructions of Figures 7.10 and 7.11 are known as the *von Staudt constructions*. The von Staudt constructions are typically thought of as geometric tools for showing that, for a projective line L in a projective plane, a choice of three distinct points on L to label 0, 1, and ∞ determines a unique coordinatization of $L - \{\infty\}$.

7.6 History

Ringel (1956) studied the combinatorics of pseudoline arrangements in the plane, which amounts to the study of reorientation classes of rank 3 oriented matroids, and asked essentially whether the realization spaces of rank 3 oriented matroids are path-connected. Goodman and Pollack pointed out the significance of Ringel's conjecture for oriented matroids around 1980. The stronger conjecture that realization spaces are contractible gained prominence via Gel'fand and Serganova's introduction of *thin Schubert cells* (Gel'fand and Serganova 1987, Gel'fand et al. 1987, Gel'fand and MacPherson 1982).

During the Cold War these conjectures were pursued independently in the Soviet Union and in the West. Mnëv, in Leningrad (now St. Petersburg), proved Universality in his PhD thesis (Mnëv 1986). Among the specific examples in his thesis of oriented matroids with disconnected extension space were a rank 3 oriented matroid on 15 elements and a uniform rank 3 oriented matroid on 19 elements (Mnëv 1986, pp. 103 and 105). While his work got attention within the Soviet Union, his thesis was never published as a journal article, and the geopolitics of the time prevented the news from spreading to the West informally. Unaware of his work, various Western mathematicians pursued specific counterexamples to Ringel's conjecture, which appeared in rapid succession in 1988–89: a rank 3 nonuniform example on 42 elements (White 1989), followed (in the next issue of the same journal!) by a rank 3 nonuniform example with 17 elements, and a construction to produce, from a rank 3 oriented matroid $\mathcal{M}$, a uniform rank 3 oriented matroid $\widehat{\mathcal{M}}$ whose realization space has at least as many connected components as that of $\mathcal{M}$ (Jaggi and Mani-Levitska 1988, Bokowski and Sturmfels 1989, Jaggi et al. 1989). Only in 1988 did researchers in the West become aware of the work being done in Leningrad, including the Universality Theorem and Suvorov's construction of a rank 3 nonuniform example on 14 elements, both of which had their first English announcement in the same 1988 collection (Mnëv 1988, Suvorov 1988).

As noted in Section 7.3, the smallest known example of an oriented matroid whose realization space is not contractible is Tsukamoto's rank 3 example on 13 elements. A result of Richter-Gebert (1988) shows that every rank 3 oriented matroid with at most nine elements has contractible realization space.

7.7 Universality and the Plücker Embedding

We define orthants in $\mathbb{RP}^{N-1}$ to be quotients of orthants in $\mathbb{R}^N$.

Notation 7.31 For $X \in \{0, +, -\}^N$, $\mathcal{O}_{\pm X}$ denotes the orthant $\{\mathbb{R}\mathbf{x} : \text{sign}(\mathbf{x}) = X\}$ in $\mathbb{RP}^{N-1}$.

$\mathbb{R}^{\binom{n}{r}}$ denotes the set of vectors $\mathbf{x} = (x_{i_1,\ldots,i_r} : 1 \leq i_1 < \cdots < i_r \leq n)$ with components in $\mathbb{R}$. $\mathbb{RP}^{\binom{n}{r}-1}$ denotes the space of lines $\mathbb{R}\mathbf{x}$ with $\mathbf{x} \in \mathbb{R}^{\binom{n}{r}}$.

Let $\chi : [n]^r \to \{0, +, -\}$ be an alternating function. We denote its restriction to increasing sequences by $\hat{\chi}$.

Since an alternating function χ is determined by $\hat{\chi}$, we see a bijection $\pm\chi \leftrightarrow \mathcal{O}_{\pm\hat{\chi}}$ between pairs of nonzero alternating functions $[n]^r \to \{0, +, -\}$ and orthants in $\mathbb{RP}^{\binom{n}{r}-1}$.

Another perspective on realization spaces comes from the Plücker embedding (Section 1.4.3). Recall that $P : G(r, \mathbb{R}^n) \to \mathbb{P}(\bigwedge^n \mathbb{R}^m) \cong \mathbb{RP}^{\binom{m}{n}-1}$ takes a subspace V to a line $\mathbb{R}\mathbf{v}$ such that $\text{sign}(\mathbf{v}) = \hat{\chi}$ for a chirotope χ of the oriented matroid corresponding to V. The following proposition just restates this in terms of the language we've just introduced.

Proposition 7.32 *For a rank r oriented matroid $\mathcal{M}$ on elements $[n]$ with chirotopes $\pm\chi$,*

$$\text{Greal}(\mathcal{M}) = P^{-1}(\mathcal{O}_{\pm\hat{\chi}}).$$

From Proposition 7.32, the Universality Theorem, and Exercise 7.4, we get the following.

Corollary 7.33 *The intersection of the Plücker embedding of the Grassmannian with an orthant in projective space can have the stable homotopy type of any basic primary semialgebraic set.*

We'll discuss the partition of $G(r, \mathbb{R}^n)$ into sets $\text{Greal}(\mathcal{M})$ more in Chapter 10.

There is a surprising amount to say about the orthant $\mathcal{O}_{\pm(+,+,...,+)}$ and its faces. Define a **positively oriented matroid** to be an oriented matroid on $[n]$ with a chirotope whose value on every increasing sequence is in $\{0,+\}$. For instance, the alternating function $\chi : [n]^r \to \{0, +, -\}$ with value $+$ on each increasing sequence is indeed a chirotope, whose oriented matroid $\mathcal{M}^{n,r}_{\text{alt}}$ was shown to be realizable in Exercise 1.8. Da Silva (1987) conjectured that every positively oriented matroid is realizable. Postnikov (2006) dodged this conjecture, defining a **positroid** to be a realizable positively oriented matroid and showing that the realization space of a positroid is always an open ball. He defined the **totally nonnegative part** $G(r, \mathbb{R}^n)_{\geq 0}$ of $G(r, \mathbb{R}^n)$ to be the set of elements of $G(r, \mathbb{R}^n)$ whose corresponding oriented matroids are positroids and conjectured that the set of closures $\overline{\text{Greal}(\mathcal{M})}$ of Grassmannian realization spaces of positroids is a regular cell decomposition of $G(r, \mathbb{R}^n)_{\geq 0}$. In particular, by applying Proposition 5.21 one can see that this set of closures would have a unique maximum $\text{Greal}(\mathcal{M}^{n,r}_{\text{alt}})$, and so Postnikov's conjecture would imply that $G(r, \mathbb{R}^n)_{\geq 0}$ itself is a closed ball.

Da Silva's conjecture was proved in Ardila et al. (2017), so positively oriented matroids coincide with positroids. $G(r, \mathbb{R}^n)_{\geq 0}$ was shown to be a closed ball in 2021 (Galashin et al. 2022), and the set of closures $\overline{\text{Greal}(\mathcal{M})}$ of Grassmannian realization spaces of positroids was shown to be a CW decomposition of $G(r, \mathbb{R}^n)_{\geq 0}$ (Postnikov et al. 2009). As of this writing Postnikov's full conjecture remains open.

In short, all issues of realizability and Universality melt away when we restrict to positively oriented matroids and the totally nonnegative part of the Grassmannian.

Positroids have intriguing connections to cluster algebras (Scott 2006) and to theoretical physics (cf. Arkani-Hamed et al. 2014, 2016).

7.8 Applications

Recall that the *combinatorial type* of a convex polytope is the isomorphism class of its face poset. Let P be a combinatorial type of d-dimensional convex polytope with n vertices, together with an identification of its vertex set with

[n]. A **realization** of P is a convex polytope in affine d-dimensional space and an isomorphism from its face poset to P. We identify a realization of P with its vertex set, so that the set of all realizations of P is a subset of $\mathbb{A}^n$. The quotient of this set by affine transformations is the **realization space** $\mathcal{R}(P)$ of P.

Problem 7.34 Prove that $\mathcal{R}(P)$ is a basic primary semialgebraic set defined over $\mathbb{Z}$.

In Section 8.4 we'll use Universality to prove the following.

Theorem 7.35 (Vershik 1988, Bokowski and Sturmfels 1989) *For every basic primary semialgebraic set S defined over $\mathbb{Z}$ there is a combinatorial type of convex polytope P such that $\mathcal{R}(P)$ is stably homotopy equivalent to S.*

The convex polytopes arising in the proof we'll give can be very high-dimensional. Steinitz's Theorem shows that realization spaces of convex polytopes of dimensions less than 4 are contractible. Richter-Gebert expanded on the techniques described in this chapter to prove that *realization spaces of four-dimensional polytopes are universal.*

Theorem 7.36 (Richter-Gebert 1996a) *For every basic primary semialgebraic set S defined over $\mathbb{Z}$ there is a combinatorial type of four-dimensional convex polytope P such that $\mathcal{R}(P)$ is stably homotopy equivalent to S.*

A scheme-theoretic version of Universality has found application in algebraic geometry. We will not explore this here: See Kapovich and Millson (1998), Lafforgue (2003), Belkale and Brosnan (2003), Vakil (2006), Lee and Vakil (2013).

The flavor of results applying Universality are sometimes called "Murphy's Law": They show that some class of objects that one might hope to have simple structure actually have arbitrarily complicated structure. To quote Vakil (2006), *"How bad can the deformation space of an object be?"* The answer seems to be: *"Unless there is some a priori reason otherwise, the deformation space may be as bad as possible."* And to lightly paraphrase Mnëv (n.d.), *Universality is useless, in the sense that it states that a class of nice-looking problems is universal and therefore useless. However, it has a way of showing up in remote corners of natural geometric moduli problems and making them useless too, understandably provoking some uneasy feelings in the correspondent community.*

Exercises

7.1 Prove that if $\overline{\text{Greal}(\mathcal{M})} \cap \text{Greal}(\mathcal{N}) \neq \emptyset$ then $\mathcal{M} \rightsquigarrow \mathcal{N}$.

7.2 Let $j \in [n]$, and let $\pi_j \colon \mathbb{R}^n \to \mathbb{R}^{n-1}$ be the map that forgets the component indexed by j.

Let $\mathcal{M}$ be an oriented matroid on elements $[n]$. Consider the composition

$$\text{Greal}(\mathcal{M}) \to \text{Greal}(\mathcal{M}^*) \to \text{Greal}(\mathcal{M}^* \backslash j) \to \text{Greal}(\mathcal{M}/j),$$

where the first and third maps are the canonical bijections $V \to V^\perp$ and the second map is the forgetful map. Show that this composition takes a space V to $\pi_j(V \cap \{\mathbf{x} \in \mathbb{R}^n : x_j = 0\})$.

Conclude that this composition need not be surjective.

7.3 Find an oriented matroid $\mathcal{M}$ and an element e such that $\mathcal{R}(\mathcal{M})$ is of lower dimension than $\mathcal{R}(\mathcal{M}\backslash e)$.

7.4 Find a rank 3 oriented matroid $\mathcal{M}$ with an element e such that $\mathcal{R}(\mathcal{M}\backslash e)$ has more components than $\mathcal{R}(\mathcal{M})$.

8

Oriented Matroid Polytopes

In Chapters 4 and 6 we considered *convex sets of covectors*, and in Chapter 6 we considered *polytopal sets of covectors* – collections of covectors that behave like vertex sets of convex polytopes. This chapter will explore a more widely studied connection between convex polytopes and oriented matroids, *oriented matroid polytopes*. These are oriented matroids that behave like the oriented matroids associated to vertex sets of convex polytopes in affine space. (They also behave like the oriented matroids associated to sets of hyperplanes defining the facets of convex polytopes.)

By now we know what to ask when we hear that an oriented matroid concept "behaves like" a geometric concept: *How well does it behave, exactly?* Or more pessimistically: *How does it not behave properly?* One answer to the pessimistic question that we'll explore is *polarity*: We'll see that questions on the existence of oriented matroid polars are closely related to questions on the existence of adjoints.

In Section 8.4 we'll look at a simple but powerful technique, closely related to Perles's construction in Section 1.6. For every oriented matroid $\mathcal{M}$, we'll find an oriented matroid polytope $\Lambda(\mathcal{M})$ with a positive tope T such that $\mathcal{V}^*(\mathcal{M})$ embeds into $\mathcal{V}^*(\Lambda(\mathcal{M}))_{\leq T}$. In our analogy with convex polytopes, this makes the entire covector set of $\mathcal{M}$ analogous to a subset of the face poset of a convex polytope. Section 8.4 will explore some of the many interesting examples arising from this construction.

One of the most fruitful connections between oriented matroids and convex polytopes is Gale diagrams, which we considered already in Section 1.6. We will touch on it again briefly in Section 8.4.1.

8.1 Convex Polytopes

We first summarize the facts about convex polytopes motivating our oriented matroid discussion. Many of these facts are "obvious" based on intuition in two and three dimensions but nontrivial to prove in general. The diligent reader who questions the assertions made here can find proofs in Ziegler (1995). The less diligent reader may go ahead and treat them as obvious.

Every convex polytope can be described in two ways: as the convex hull of a finite set of points, and as a bounded nonempty intersection of finitely many closed half-spaces. We will view convex polytopes as lying in an affine space $\mathbb{A}$, and so the arrangements to have in mind are point arrangements in affine space and signed affine hyperplane arrangements. A face of a convex polytope P is a subset P' of P such that, for some signed hyperplane $\mathcal{H} = (H, H^+, H^-)$, we have $P \cap H^- = \emptyset$ and $P \cap H^0 = P'$. Such an $\mathcal{H}$ is called a *defining signed hyperplane* for P'. In particular, P itself is the unique maximal face of P (with the degenerate signed hyerplane as its defining signed hyperplane), and $\emptyset$ is the unique minimal face. A *vertex* is a zero-dimensional face, and a *facet* of a d-dimensional convex polytope is a $(d-1)$-dimensional face. Each convex polytope is the convex hull of its vertex set. Assuming P spans $\mathbb{A}$, each facet has a unique defining hyperplane, and P is the intersection of the positive closed half-spaces of the signed hyperplanes defining facets. The set of vertices of P is the unique minimum among subsets of $\mathbb{A}$ with convex hull P, and the set of facet-defining signed hyperplanes is the unique minimum among sets of signed hyperplanes in $\mathbb{A}$ whose positive closed half-spaces have intersection P.

The poset $\mathcal{F}(P)$ of faces of P is a lattice, called the *face lattice* of P. It is both atomic (every face is a convex hull of vertices of P) and coatomic (every face is an intersection of facets of P). Every face of P is a convex hull of a unique set of vertices of P, and it is often convenient to identify the face by this set. Every face is also the intersection of a set of facets of P, but this set may not be unique. For instance, if P is a square pyramid and P' is its apex, then the intersection of any three of the facets incident with P' is P'. It is often convenient to identify a face by the set of all facet-defining hyperplanes containing that face.

The preceding discussion included several instances in which a statement about vertices corresponded neatly with a statement about facets. These correspondences are consequences of a natural duality for polytopes, which we call *polarity* and make precise in the following proposition.

Proposition 8.1 *Let* $\mathbb{A} = \{\mathbf{x} \in \mathbb{R}^{d+1} : x_{d+1} = 1\}$. *Let P be a d-dimensional convex polytope in $\mathbb{A}$ with $(0, \ldots, 0, 1)$ in its interior, and let*

$$P^\square = \{\mathbf{w} \in \mathbb{A} : \mathbf{w} \cdot \mathbf{v} \geq 0 \text{ for all } \mathbf{v} \in P\}.$$

Then $P^\square$ is a convex polytope, and we have a poset isomorphism

$$\mathcal{F}(P)^{\mathrm{op}} \to \mathcal{F}(P^\square)$$
$$F \to \{\mathbf{w} \in P^\square : \mathbf{w} \cdot \mathbf{v} = 0 \text{ for all } \mathbf{v} \in F\}.$$

In particular,

- *the vertices of $P^\square$ are the vectors $\mathbf{w}$ such that $\mathbf{w}^\perp$ is a facet-defining signed hyperplane for P, and*
- *the facets of $P^\square$ are the signed hyperplanes $\mathbf{v}^\perp$ such that $\mathbf{v}$ is a vertex of P.*

The $P^\square$ of Proposition 8.1 is closely related to the *polar* $P^\triangle$ defined, for instance, in Ziegler (1995). For a d-dimensional convex polytope P in $\mathbb{R}^d$, $P^\triangle$ is defined to be $\{\mathbf{a} \in (\mathbb{R}^d)^* : \mathbf{ax} \leq 1 \text{ for all } \mathbf{x} \in P\}$. If $\mathbf{0}$ is in the interior of P then (cf. corollaries 2.13 and 2.14 in Ziegler 1995) $P^\triangle$ is a convex polytope and $\mathcal{F}(P^\triangle) \cong \mathcal{F}(P)^{\mathrm{op}}$. Proposition 8.1 follows from these corollaries: Given P satisfying the hypotheses of Proposition 8.1, let $\widehat{P} = \{\mathbf{x} : (\mathbf{x}, 1) \in P\}$. For each $\mathbf{w} = (\mathbf{y}, 1) \in P^\square$, let $\mathbf{w}^* \in (\mathbb{R}^d)^*$ be the function $\mathbf{w}^*\mathbf{x} = \mathbf{y} \cdot \mathbf{x}$, and let $\widehat{P^\square} = \{\mathbf{w}^* : \mathbf{w} \in P^\square\}$. Then $\widehat{P}$ satisfies the hypotheses of corollaries 2.13 and 2.14 in Ziegler (1995), and $\widehat{P^\square} = -\widehat{P}^\triangle$. Proposition 8.1 then follows from these corollaries and straightforward isomorphisms $\mathcal{F}(\widehat{P}) \cong \mathcal{F}(P)$ and $\mathcal{F}(\widehat{P}^\triangle) \cong \mathcal{F}(-\widehat{P}^\triangle) \cong \mathcal{F}(\widehat{P^\square})$.

Definition 8.2 For P satisfying the hypotheses of Proposition 8.1, we call $P^\square$ the **affine polar** of P.

Now we can preview the topics of this chapter.

In Section 8.2 we generalize the notion of convex polytopes to acyclic oriented matroids from two perspectives. The *vertex perspective*, or *Las Vergnas perspective*, generalizes convex polytopes by viewing an oriented matroid as analogous to an affine point arrangement. The *facet perspective*, or *Edmonds–Mandel perspective*, reaches the same goal by viewing an oriented matroid as analogous to a signed hyperplane arrangement. Proposition 8.1 suggests that thinking of a single oriented matroid $\mathcal{M}$ from both of these perspectives should give us a generalization of a pair P, $P^\square$ of convex polytopes. Indeed, the two perspectives lead to generalizations of face posets that are opposite to each other: From the Edmonds–Mandel perspective the face poset is $\mathcal{V}^*(\mathcal{M})_{\leq E^+}$, while from the Las Vergnas perspective it's the opposite poset.

Section 8.3 considers a more subtle question on polarity. Suppose we stick to just one of our two perspectives. Given an oriented matroid $\mathcal{M}$, analogous to a convex polytope P, is there an oriented matroid $\mathcal{M}^\square$ analogous to $P^\square$? We'll see that if $\mathcal{M}$ has an adjoint, then a suitable choice of adjoint is such a "polar oriented matroid."

All of these comparisons of face posets of convex polytopes with $\mathcal{V}^*(\mathcal{M})_{\leq E^+}$ suggests a question: Is $\mathcal{V}^*(\mathcal{M})_{\leq E^+}$ always isomorphic to the face poset of a convex polytope? Put another way: We know that $\mathcal{M}$ need not be realizable by an arrangement of signed hyperplanes, but perhaps we can realize a single tope of $\mathcal{M}$, together with its faces, as an intersection of closed half-spaces? The answer is No: Section 8.4 will introduce the *Lawrence construction*, which produces counterexamples to many conjectures, including this one.

8.2 Face Lattices and Oriented Matroid Polytopes

Consider a finite affine point arrangement $\mathcal{A}_{\text{pt}} = (p_e : e \in E_{\text{pt}})$, its associated oriented matroid $\mathcal{M}_{\text{pt}}$, and its convex hull P. Thus $\mathcal{M}_{\text{pt}}$ is acyclic, and P is a convex polytope. For each $F \subseteq E_{\text{pt}}$, $(p_e : e \in F)$ is the subarrangement of elements of $\mathcal{A}_{\text{pt}}$ in a face of P if and only if $(E_{\text{pt}} \setminus F)^+ \in \mathcal{V}^*(\mathcal{M}_{\text{pt}})$. This gives an order-reversing correspondence between faces of P and positive covectors of $\mathcal{V}^*(\mathcal{M}_{\text{pt}})$, motivating the following definition.

Definition 8.3 (Las Vergnas 1975b, 1980) Let $\mathcal{M}$ be an acyclic oriented matroid on elements E. A subset F of E is a **face** of $\mathcal{M}$ if $(E \setminus F)^+ \in \mathcal{V}^*(\mathcal{M})$.

The **Las Vergnas face lattice** $\text{LV}(\mathcal{M})$ of $\mathcal{M}$ is the poset of faces, ordered by inclusion.

The Las Vergnas face lattice is indeed a lattice, since $\mathcal{V}^*(\mathcal{M}) \cup \{\hat{1}\}$ is a lattice (Proposition 3.28) and $\text{LV}(\mathcal{M})$ is an interval in $\mathcal{V}^*(\mathcal{M})^{\text{op}}$.

Example 8.4 Figure 8.1 shows an affine point arrangement whose oriented matroid has faces $[6]$, $\{1,2,3\}$, $\{3,4\}$, $\{4,5\}$, $\{1,5\}$, $\{1\}$, $\{3\}$, $\{4\}$, $\{5\}$, $\emptyset$.

The Las Vergnas face lattice will play a prominent role in Chapter 9.

Now consider a finite arrangement $\mathcal{A}_{\text{hyp}} = (\mathcal{H}_e : e \in E_{\text{hyp}})$ of signed affine hyperplanes whose closed positive sides intersect in a convex polytope Q. Let $\mathcal{M}_{\text{hyp}}$ be the oriented matroid of $\mathcal{A}_{\text{hyp}}$. For each $F \subseteq E_{\text{hyp}}$, $(\mathcal{H}_e : e \in F)$ is the subarrangement of elements $\mathcal{H}_e$ of $\mathcal{A}_{\text{hyp}}$ with H_e containing a face of Q if and only if $(E \setminus F)^+ \in \mathcal{V}^*(\mathcal{M}_{\text{hyp}})$. This gives an order-preserving

246 Oriented Matroid Polytopes

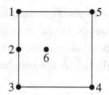

Figure 8.1 Illustration for Example 8.4.

correspondence between faces of Q and positive covectors of $\mathcal{V}^*(\mathcal{M}_{\text{hyp}})$, motivating the following definition.

Definition 8.5 (Mandel 1982) Let $\mathcal{M}$ be an acyclic oriented matroid on elements E. The **Edmonds–Mandel face lattice** $\mathrm{EM}(\mathcal{M})$ of $\mathcal{M}$ is $\mathcal{V}^*(\mathcal{M})_{\leq E^+}$.

The Edmonds–Mandel face lattice is anti-isomorphic to the Las Vergnas face lattice, by the order-reversing map taking an element A^+ of $EM(\mathcal{M})$ to the face $E\backslash A$.

Clearly the Las Vergnas face lattice associated to an affine point arrangement is unchanged when we delete a point that is in the convex hull of the remaining points, and the Edmonds–Mandel face lattice associated to an affine signed hyperplane arrangement is unchanged when we delete a signed hyperplane that is not facet-defining. The following extends this to nonrealizable oriented matroids.

Recall that the convex hull $\widehat{\mathrm{conv}}(A)$ of a set A of elements of $\mathcal{M}$ is $A \cup \{f \in E : \hat{A}^+\{f\}^- \in \mathcal{V}(\mathcal{M}) \text{ for some } \hat{A} \subseteq A\}$.

Proposition 8.6 *Let $\mathcal{M}$ be an acyclic oriented matroid on elements E, and let $e \in E$. The following are equivalent:*

1. *e is in the convex hull of the remaining elements.*
2. *$(E\backslash e)^+ \notin \mathcal{V}^*(\mathcal{M})$.*

If these conditions hold then $\mathrm{LV}(\mathcal{M}) \cong \mathrm{LV}(\mathcal{M}\backslash e)$ *by the correspondence* $F \leftrightarrow F\backslash\{e\}$ *and* $\mathrm{EM}(\mathcal{M}) \cong \mathrm{EM}(\mathcal{M}\backslash e)$ *by the correspondence* $A^+ \leftrightarrow (A\backslash\{e\})^+$.

Proof: The first condition implies the second by orthogonality of vectors and covectors, and the second implies the first by the Farkas property.

Assuming the two conditions hold, the map $\mathrm{EM}(\mathcal{M}) \to \mathrm{EM}(\mathcal{M}\backslash e)$ taking A^+ to $(A\backslash\{e\})^+$ is surjective by the definition of deletion and injective by the second condition. The conclusion for $\mathrm{LV}(\mathcal{M})$ follows immediately. □

Definition 8.7 An **oriented matroid polytope** is an acyclic oriented matroid in which no element is in the convex hull of the remaining elements.

Corollary 8.8 *An oriented matroid $\mathcal{M}$ on elements E is an oriented matroid polytope if and only if $(E\setminus\{e\})^+ \in \mathcal{V}^*(\mathcal{M})$ for each $e \in E$.*

Note that an oriented matroid polytope is simple, and for every acyclic oriented matroid we obtain an oriented matroid polytope with the same face lattices by deleting elements. A realizable oriented matroid is an oriented matroid polytope if and only if it is acyclic and some (and also all) of its realizations by affine point arrangements constitute the vertices of a convex polytope.

Las Vergnas initiated the study of oriented matroid polytopes from the perspective of abstracting vertex sets of convex polytopes in Las Vergnas (1975b, 1980), while the hyperplane perspective first appeared in Mandel's thesis (Mandel 1982).

8.3 Polarity and Adjoints

Proposition 8.1 showed that a convex polytope P appropriately situated in an affine subspace of $\mathbb{R}^n$ has a well-defined affine polar $P^\square$, satisfying $\mathcal{F}(P^\square) = \mathcal{F}(P)^{\mathrm{op}}$, and two associated oriented matroid polytopes $\mathcal{M}_{\mathrm{pt}}$ and $\mathcal{M}_{\mathrm{hyp}}$, satisfying $\mathcal{F}(P) = \mathrm{LV}(\mathcal{M}_{\mathrm{pt}})$ and $\mathcal{F}(P^\square) = \mathrm{LV}(\mathcal{M}_{\mathrm{hyp}}) \cong \mathrm{EM}(\mathcal{M}_{\mathrm{pt}})$. Thus elements of $\mathcal{M}_{\mathrm{hyp}}$ are in bijection with positive signed cocircuits of $\mathcal{M}_{\mathrm{pt}}$. The relationship between $\mathcal{M}_{\mathrm{pt}}$ and $\mathcal{M}_{\mathrm{hyp}}$ is not canonical: Neither oriented matroid determines the other.

We want to define polars for oriented matroid polytopes so that a polar $\mathcal{M}^\square$ of $\mathcal{M}$ is an oriented matroid polytope such that $\mathrm{EM}(\mathcal{M}^\square)^{\mathrm{op}} \cong \mathrm{EM}(\mathcal{M})$, or in other words $\mathrm{LV}(\mathcal{M}^\square) \cong \mathrm{EM}(\mathcal{M})$. Since the elements of $\mathcal{M}^\square$ must correspond to positive cocircuits of $\mathcal{M}$, we might as well take the elements of $\mathcal{M}^\square$ to be the set of positive cocircuits of $\mathcal{M}$, which we will denote by $\mathcal{C}_+^*$. Thus $\mathcal{C}_+^*$, viewed as a set of sign vectors, is the set of atoms of $\mathrm{EM}(\mathcal{M})$, and $\mathcal{C}_+^*$, viewed as a set of elements of a polar $\mathcal{M}^\square$ of $\mathcal{M}$, should be the set of atoms of $\mathrm{LV}(\mathcal{M}^\square)$.

Definition 8.9 A **polar** to an oriented matroid polytope $\mathcal{M}$ on elements E is an oriented matroid polytope $\mathcal{M}^\square$ on elements $\mathcal{C}_+^*$ such that there is a poset isomorphism $\mathrm{LV}(\mathcal{M}^\square) \to \mathrm{EM}(\mathcal{M})$ taking each atom $\{X\}$ of $\mathrm{LV}(\mathcal{M}^\square)$ to the minimal element X of $\mathrm{EM}(\mathcal{M})$.

Proposition 8.10 *If $\mathcal{M}^\square$ is a polar to $\mathcal{M}$ then the poset isomorphism $f : \mathrm{EM}(\mathcal{M}) \to \mathrm{LV}(\mathcal{M}^\square)$ takes each Y to $\{X \in \mathcal{C}_+^* : X \leq Y\}$.*

Proof: Since $\{X \in \mathcal{C}_+^* : X \leq Y\}$ is the set of atoms of $\mathrm{EM}(\mathcal{M})_{\leq Y}$ and f is a poset isomorphism, we must have that $\{\{X\} : X \in \mathcal{C}_+^*, X \leq Y\}$ is the set of atoms of $\mathrm{LV}(\mathcal{M}^\square)_{\leq f(Y)}$. But each element of the Las Vergnas face lattice is just the union of the atoms below it. □

Proposition 8.11 *If $\mathcal{M}^\square$ is a polar to $\mathcal{M}$ and $A \subseteq \mathcal{C}_+^*$ then A^+ is a signed cocircuit of $\mathcal{M}^\square$ if and only if there is an element e of $\mathcal{M}$ such that $A = \{X \in \mathcal{C}_+^* : X(e) = +\}$.*

Proof: The set of facets of E^+ in $\mathcal{M}$ is $\{(E\backslash\{e\})^+ : e \in E\}$, and the set of signed cocircuits less than or equal to $(E\backslash\{e\})^+$ is $\{X \in \mathcal{C}^* : X(e) = 0\}$.

A^+ is a signed cocircuit of $\mathcal{M}^\square$ if and only if $\mathcal{C}_+^*\backslash A$ is a maximal element of $LV(\mathcal{M}^\square)\backslash\{E\}$, thus if and only if there is a facet Y of E^+ in $\mathcal{V}^*(\mathcal{M})$ such that $\mathcal{C}_+^*\backslash A = \{X \in \mathcal{C}^* : X \leq Y\}$. By the remarks in the previous paragraph, we see that there is such a Y if and only if there is an e such that $\mathcal{C}_+^*\backslash A = \{X \in \mathcal{C}_+^* : X(e) = 0\}$, that is, $A = \{X \in \mathcal{C}_+^* : X(e) = +\}$. □

The quest for a polar to an oriented matroid polytope $\mathcal{M}$ is reminiscent of the quest for an adjoint to a simple $\mathcal{M}$. For instance, the role of elements (rank 1 flats) and cocircuits (corank 1 flats) of $\mathcal{M}$ is reversed in an adjoint to $\mathcal{M}$. A polar to $\mathcal{M}$ is similar but weaker: Elements of $\mathcal{M}$ correspond to corank 1 flats in a polar, but we only need some of the signed cocircuits of $\mathcal{M}$ (the positive ones) to correspond to elements of the polar. So our first result on the existence of polars (Proposition 8.13) is no surprise: We'll see that if an oriented matroid polytope $\mathcal{M}$ has an adjoint then $\mathcal{M}$ has a polar.

Proposition 8.11 gives us a tidy way to describe the positive signed cocircuits of a dual $\mathcal{M}^\square$. Let the elements of $\mathcal{M}$ be $\{e_1, \ldots, e_n\}$, and let $\mathcal{C}_+^* = \{X_1, \ldots, X_m\}$. Consider the $m \times n$ matrix N of signs with columns indexed by $\{e_1, \ldots, e_n\}$ and with X_i as its ith row. Then Proposition 8.11 tells us that the columns of N are the positive cocircuits of $\mathcal{M}^\square$. (This matrix is the submatrix $[\mathcal{M}](\mathcal{C}_+^*)$ of the matrix discussed in Section 6.5.)

Example 8.12 Let $\mathcal{M}$ be the oriented matroid polytope given by the vertices of the convex polytope shown on the left in Figure 8.2. If we list the elements of $\mathcal{C}_+^*$ in the appropriate order, we get the matrix

$$N = \begin{pmatrix} 0 & 0 & 0 & + & + \\ 0 & 0 & + & 0 & + \\ 0 & + & 0 & 0 & + \\ + & 0 & 0 & + & 0 \\ + & 0 & + & 0 & 0 \\ + & + & 0 & 0 & 0 \end{pmatrix}.$$

8.3 Polarity and Adjoints

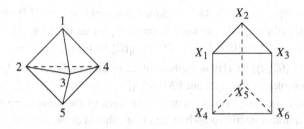

Figure 8.2 Illustration for Example 8.12.

As expected, the six columns of the matrix give us the six facets of the polar polytope on the right.

Recall from Section 6.5 that if $\mathcal{M}$ has an adjoint and we have a preferred element of each pair $\pm X$ of signed cocircuits of $\mathcal{M}$, then there is an adjoint $\mathcal{M}_{\text{adj}}$ of $\mathcal{M}$ whose elements are these preferred signed cocircuits. In particular, if $\mathcal{M}$ has an adjoint then it has an adjoint whose set of elements contains $\mathcal{C}_+^*$.

Proposition 8.13 (Bachem and Kern 1986, Cordovil 1987) *Let $\mathcal{M}$ be an oriented matroid polytope that has an adjoint. Let $\mathcal{M}_{\text{adj}}$ be an adjoint of $\mathcal{M}$ whose set of elements contains $\mathcal{C}_+^*$. Then $\mathcal{M}_{\text{adj}}(\mathcal{C}_+^*)$ is a polar of $\mathcal{M}$.*

Notation 8.14 If $\{X_e : e \in S\}$ is a set of conformal sign vectors then $\circ_{e \in S} X_e$ denotes the composition of all elements of this set.

Proof: Recall that each element e of $\mathcal{M}$ gives us a signed cocircuit Y_e of $\mathcal{M}_{\text{adj}}$ defined by $Y_e(Z) = Z(e)$. (These are the columns of the matrix $[\mathcal{M}]$ of Section 6.5.) Let Y_e' be the restriction of Y_e to $\mathcal{C}_+^*$. Thus Y_e' is a positive covector of $\mathcal{M}_{\text{adj}}(\mathcal{C}_+^*)$, and the composition $\circ_{e \in E} Y_e'$ is $(\mathcal{C}_+^*)^+$. Thus $\mathcal{M}_{\text{adj}}(\mathcal{C}_+^*)$ is acyclic. Additionally, for each $X \in \mathcal{C}_+^*$ we have a covector $\circ_{e \in X^0} Y_e'$ of $\mathcal{M}_{\text{adj}}(\mathcal{C}_+^*)$ that is 0 only on X. Thus by Corollary 8.8 $\mathcal{M}_{\text{adj}}(\mathcal{C}_+^*)$ is an oriented matroid polytope.

Let Z be a positive covector of $\mathcal{M}$, and let $\{X_1, \ldots, X_k\} = \mathcal{C}^*(\mathcal{M})_{\leq Z}$. Then $\circ_{e \in Z^0} Y_e'$ is a positive covector of $\mathcal{M}_{\text{adj}}(\mathcal{C}_+^*)$ that is 0 only on $\{X_1, \ldots, X_k\}$. Thus we have an injective poset map

$$\text{EM}(\mathcal{M}) \hookrightarrow \text{LV}(\mathcal{M}_{\text{adj}}(\mathcal{C}_+^*))$$
$$Z \to (\circ_{e \in Z^0} Y_e')^0 = \{X_1, \ldots, X_k\}.$$

This map takes $\hat{0} = \mathbf{0}$ to $\hat{0} = \emptyset$ and $\hat{1} = E^+$ to $\hat{1} = \mathcal{C}_+^*$. Consider the restriction $\text{EM}(\mathcal{M}) \setminus \{\hat{0}, \hat{1}\} \hookrightarrow \text{LV}(\mathcal{M}_{\text{adj}}(\mathcal{C}_+^*)) \setminus \{\hat{0}, \hat{1}\}$. The resulting map of order complexes is an injective map. Since $\text{EM}(\mathcal{M}) \setminus \{\hat{0}, \hat{1}\} = (\mathcal{V}^*(\mathcal{M}) \setminus \{\mathbf{0}\})_{\leq E^+}$, the Topological Representation Theorem tells us that

$\|\Delta(\text{EM}(\mathcal{M})\setminus\{\hat{0},\hat{1}\})\|$ is a sphere of dimension rank$(\mathcal{M}) - 1$. Likewise, since $\text{LV}(\mathcal{M}_{\text{adj}}(\mathcal{C}_+^*))\setminus\{\hat{0},\hat{1}\}$ has the same order complex as $\text{EM}(\mathcal{M}_{\text{adj}}(\mathcal{C}_+^*))\setminus\{\hat{0},\hat{1}\})$ and $\text{EM}(\mathcal{M}_{\text{adj}}(\mathcal{C}_+^*))\setminus\{\hat{0},\hat{1}\} = (\mathcal{V}^*(\mathcal{M}_{\text{adj}}(\mathcal{C}_+^*))\setminus\{0\})_{\leq(\mathcal{C}_+^*)^+}$, we have that $\|\Delta(\text{LV}(\mathcal{M}_{\text{adj}}(\mathcal{C}_+^*))\setminus\{\hat{0},\hat{1}\})\|$ is a sphere of dimension rank$(\mathcal{M}) - 1$. So the map of order complexes arising from $\text{EM}(\mathcal{M})\setminus\{\hat{0},\hat{1}\} \hookrightarrow \text{LV}(\mathcal{M}_{\text{adj}}(\mathcal{C}_+^*))\setminus\{\hat{0},\hat{1}\}$ is a continuous injective map between two spheres of the same dimension. But every continuous injective map between two spheres of the same dimension is bijective. Thus our map $\text{EM}(\mathcal{M}) \to \text{LV}(\mathcal{M}_{\text{adj}}(\mathcal{C}_+^*))$ is a poset isomorphism.
□

Remark 8.15 The use of topology shortens the proof of Proposition 8.13, but to see a purely combinatorial proof see the cited papers.

Recall from Section 6.5 that a topological representation of an adjoint of $\mathcal{M}$ gives a representation of $\mathcal{M}$ by a pseudoconfiguration of points. If $\mathcal{M}$ is an oriented matroid polytope then the topological representation of an adjoint satisfying the conditions of Proposition 8.13 restricts to a topological representation $(\mathcal{S}_X : X \in \mathcal{C}_+^*)$ of $\mathcal{M}_{\text{adj}}(\mathcal{C}_+^*)$. Proposition 8.13 tells us that the cell corresponding to the positive tope has as 0-cells the Y_e with $e \in E$, and that the poset of proper faces of this cell is indeed the opposite of the poset $\mathcal{V}^*(\mathcal{M})\setminus\{0\}_{<E^+}$. So the transition from a representation of an oriented matroid polytope $\mathcal{M}$ by a pseudosphere arrangement to a representation of $\mathcal{M}$ by a pseudoconfiguration of points is also a transition from $\mathcal{M}$ to a polar $\mathcal{M}^\square$.

Polars of oriented matroid polytopes are typically not unique. For instance, in Section 6.5 we saw that the oriented matroid $\mathcal{M}_{\text{hex}}$ given by the vertices of a convex hexagon has two different adjoints, which give two different polars.

In Section 8.4 we'll find examples of oriented matroid polytopes that do not have polars.

8.4 The Lawrence Construction

The Lawrence construction associates to an arbitrary oriented matroid $\mathcal{M}$ an oriented matroid polytope $\Lambda(\mathcal{M})$ whose face lattice encodes all of $\mathcal{V}^*(\mathcal{M})$. Lawrence didn't publish the construction himself, but he passed it on to others, for whom it has been a fruitful route between oriented matroids and convex polytopes. The construction first appeared in print in Munson's thesis (Munson 1981, see also Billera and Munson 1984a, 1984b, 1984c, Sturmfels 1988, Bayer and Sturmfels 1990). A variation has been important in the study of triangulations of oriented matroids (Anderson 1999, Santos 2002): This will be examined further in Section 9.6.

8.4 The Lawrence Construction

For an oriented matroid $\mathcal{M}$ on elements E, $\Lambda(\mathcal{M})$ is defined to be the dual of the oriented matroid obtained from $\mathcal{M}^*$ by adding an antiparallel $\tilde{e}$ to each element e of E. Thus if $\mathcal{M}$ is rank r then $\Lambda(\mathcal{M})$ is rank $r + |E|$ with elements $E \cup \tilde{E}$. (Here $\tilde{E}$ is $\{\tilde{e} : e \in E\}$. More generally, for every $S \subseteq E$ we denote $\{\tilde{e} : e \in S\}$ by $\tilde{S}$.) Since $\{e, \tilde{e}\}^+ \in \mathcal{V}^*(\Lambda(\mathcal{M}))$ for every e, $(E \cup \tilde{E})^+$ is a tope of $\Lambda(\mathcal{M})$, so $\Lambda(\mathcal{M})$ is acyclic, and its positive cocircuits are easy to see.

Proposition 8.16 *If $\mathcal{M}$ has no coloops then*

$$\mathcal{C}^*(\Lambda(\mathcal{M}))_{\leq (E \cup \tilde{E})^+} = \{\{e, \tilde{e}\}^+ : e \in E\} \cup \{A^+ \tilde{B}^+ : A^+ B^- \in \mathcal{C}^*(\mathcal{M})\}.$$

(The restriction to coloopless oriented matroids is a convenience: If e is a coloop of $\mathcal{M}$ then $\{e\}^+ \in \mathcal{V}^*(\Lambda(\mathcal{M}))$, and so the element $\{e, \tilde{e}\}^+$ of $\mathcal{V}^*(\mathcal{M}) \setminus \{0\}$ is not minimal.)

Proof: Certainly each $\{e, \tilde{e}\}^+$ is in $\mathcal{C}^*(\Lambda(\mathcal{M}))_{\leq (E \cup \tilde{E})^+}$, and if $A^+ B^- \in \mathcal{C}^*(\mathcal{M})$ then $A^+ B^- \in \mathcal{C}((\Lambda(\mathcal{M}))^*)$, and by elimination with the signed circuits $\{b, \tilde{b}\}^+$ with $b \in B$ we get $A^+ \tilde{B}^+ \in \mathcal{C}((\Lambda(\mathcal{M}))^*)$. Conversely, if a positive cocircuit of $\Lambda(\mathcal{M})$ has a pair $\{e, \tilde{e}\}$ in its support then by Incomparability it must be $\{e, \tilde{e}\}^+$. Otherwise it has support $C \cup \tilde{D}$ with C and D disjoint, and so by elimination with the signed cocircuits $\{d, \tilde{d}\}^+$ with $d \in D$ we get a signed cocircuit $A^+ B^-$ with $A \cup B = C \cup D$. Thus $A^+ B^- \in \mathcal{C}(\Lambda(\mathcal{M})^*)$, and the signed circuits of $\Lambda(\mathcal{M})^*$ with support in E are exactly the signed cocircuits of $\mathcal{M}$. □

Corollary 8.17 *If $\mathcal{M}$ has no loops then $\Lambda(\mathcal{M})$ is an oriented matroid polytope.*

Proof: We apply Corollary 8.8. For each $e \in E$ we have that $\{e, \tilde{e}\}^+ \in \mathcal{V}^*(\Lambda(\mathcal{M}))$, and for each $X \in \mathcal{V}^*(\mathcal{M})$, we have that $X \tilde{E}^0$ and $\tilde{X} E^0$ are elements of $\mathcal{V}^*(\Lambda(\mathcal{M}))$. Thus by taking compositions we see that

$$\text{coat}(\text{EM}(\Lambda(\mathcal{M}))) = \{((E \cup \tilde{E}) \setminus \{e\})^+ : e \in E\} \cup \{((E \cup \tilde{E}) \setminus \{\tilde{e}\})^+ : e \in E\}.$$

□

Proposition 8.16 tells us that $\text{EM}(\Lambda(\mathcal{M}))$ determines $\mathcal{M}$. The following theorem tells us that even the isomorphism type of the poset $\text{EM}(\Lambda(\mathcal{M}))$ determines $\mathcal{M}$, up to isomorphism.

Theorem 8.18 *Let $\mathcal{M}$ be an oriented matroid on elements E. If $\mathcal{N}$ is an acyclic oriented matroid on elements $E \cup \tilde{E}$ and $EM(\mathcal{N})$ is isomorphic as a poset to $EM(\Lambda(\mathcal{M}))$ then $\mathcal{N}$ is isomorphic to $\Lambda(\mathcal{M})$.*

To streamline the proof, we prove this only for simple $\mathcal{M}$.

Proof: For each $e \in E \cup \tilde{E}$ let X_e denote the coatom $((E \cup \tilde{E}) \setminus \{e\})^+$ of $\text{EM}(\Lambda(\mathcal{M}))$.

Let $F\colon \mathrm{EM}(\Lambda(\mathcal{M})) \to \mathrm{EM}(\mathcal{N})$ be an isomorphism. Then F restricts to a bijection $\mathrm{coat}(\mathrm{EM}(\Lambda(\mathcal{M}))) \to \mathrm{coat}(\mathrm{EM}(\mathcal{N}))$. The number of coatoms of $\mathrm{EM}(\Lambda(\mathcal{M}))$ is $2|E|$, and so the number of coatoms of $\mathrm{EM}(\mathcal{N})$ is as well. A coatom of $\mathrm{EM}(\mathcal{N})$ must have the form $((E \cup \tilde{E})\setminus P)^+$ with P a parallelism class of nonloops in $\mathcal{N}$. The only way $\mathcal{N}$ can have $2|E|$ such coatoms is if $\mathcal{N}$ is simple and $\mathrm{coat}(\mathcal{N}) = \{X_e : e \in E \cup \tilde{E}\} = \mathrm{coat}(\mathcal{M})$. So the bijection F induces a bijection $\hat{F}\colon E \cup \tilde{E} \to E \cup \tilde{E}$, where $F(X_e) = X_{\hat{F}(e)}$. By renaming elements of $\mathcal{N}$ we may assume $\hat{F}$ is the identity.

For each $e \in E$, $EM(\Lambda(\mathcal{M}))$ has an element $\{e, \tilde{e}\}^+$ that's less than or equal to every coatom except X_e and $X_{\tilde{e}}$. The corresponding element of $EM(\mathcal{N})$ must also have support $\{e, \tilde{e}\}$, so it must be $\{e, \tilde{e}\}^+$.

Thus, for each $e \in E$, e and $\tilde{e}$ are antiparallel in $\mathcal{N}^*$, so $\mathcal{N}^*$ is obtained from $\mathcal{N}^*\setminus \tilde{E}$ by adding an antiparallel copy $\tilde{e}$ of each element e. Since $\mathcal{N}^*\setminus \tilde{E} = (\mathcal{N}/\tilde{E})^*$, we see $\mathcal{N} = \Lambda(\mathcal{N}/\tilde{E})$. Also, for each $A^+B^- \in \{0, +, -\}^E$,

$$A^+B^- \in \mathcal{V}^*(\mathcal{M}) \Leftrightarrow (A \cup \tilde{B})^+ \in \mathrm{EM}(\Lambda(\mathcal{M}))$$
$$\Leftrightarrow (A \cup \tilde{B})^+ \in \mathrm{EM}(\mathcal{N})$$
$$\Leftrightarrow A^+B^- \in \mathcal{V}^*(\mathcal{N}/\tilde{E}).$$

Thus $\mathcal{M} \cong \mathcal{N}/\tilde{E}$, and so $\Lambda(\mathcal{M}) \cong \Lambda(\mathcal{N}/\tilde{E}) = \mathcal{N}$. □

8.4.1 Some Examples

Already in our discussion of Gale diagrams (Section 1.6) we implicitly used the Lawrence construction to find a counterexample. To fit the example there into our current discussion, we note a corollary to Corollary 8.17.

Corollary 8.19 *Let $\mathcal{M}$ be a realizable oriented matroid with no loops. Let $\mathcal{A}$ be a realization of $\Lambda(\mathcal{M})$ as an affine point arrangement. Then the elements of $\mathcal{A}$ are the vertices of a convex polytope.*

If $\mathcal{M}$ is realizable and either $\mathcal{M}$ or $\mathcal{M}^*$ is low rank, we can draw a realization of it to study properties of the convex polytope given by $\Lambda(\mathcal{M})$: In Section 1.6 we got a counterexample by drawing $\mathcal{M}^*$ (and we never gave much thought to the rank 8 oriented matroid $\mathcal{M}$).

Now let's use the Lawrence construction to get another counterexample. In Section 4.5 we saw that, in each topological representation of $\mathcal{M}$, the boundary of each cell can be shelled in much the same way that the boundary of a convex polytope can be shelled. This raises a question: Does each cell have the combinatorial type of a convex polytope? In other words, for each $X \in \mathcal{V}^*(\mathcal{M})$ is there a convex polytope P whose face poset is isomorphic to $\mathcal{V}^*(\mathcal{M})_{\leq X}$?

8.4 The Lawrence Construction

We can see an example of an $\mathcal{M}$ whose positive tope does *not* have the combinatorial type of a convex polytope as a corollary to Theorem 8.18.

Corollary 8.20 *Let $\mathcal{M}$ be an oriented matroid with no loops. $\mathcal{M}$ is realizable if and only if there is a convex polytope whose poset of faces is isomorphic to* $\text{EM}(\Lambda(\mathcal{M}))$.

Proof: If $\mathcal{M}$ is realizable then $\Lambda(\mathcal{M})$ is also realizable, and, in every realization of $\Lambda(\mathcal{M})$ by signed hyperplanes, the intersection of the positive cone with an appropriate affine hyperplane will be a convex polytope whose poset of faces is isomorphic to $\text{EM}(\Lambda(\mathcal{M}))$.

Conversely, given a convex polytope P in an affine hyperplane whose poset of faces is $\text{EM}(\Lambda(\mathcal{M}))$, consider the linear hyperplane arrangement determined by the facets of P, with each hyperplane oriented toward P. Theorem 8.18 says that this arrangement (with appropriate labels on the hyperplanes) must realize $\Lambda(\mathcal{M})$. But $\Lambda(\mathcal{M})$ is realizable if and only if $\mathcal{M}$ is realizable. □

So, for instance, the Lawrence construction of the non-Pappus oriented matroid is a rank 12 oriented matroid polytope whose positive tope does not have the combinatorial type of a convex polytope.

We can use Corollary 8.20 to prove Theorem 7.35, that realizations of convex polytopes are universal.

Lemma 8.21 $\mathcal{R}(\mathcal{M})$ *is stably homotopy equivalent to* $\mathcal{R}(\Lambda(\mathcal{M}))$ *for all* $\mathcal{M}$.

Proof: For every oriented matroid $\mathcal{N}$, we have

- $\mathcal{R}(\mathcal{N}) \simeq_s \text{Greal}(\mathcal{N})$, by Proposition 7.25,
- $\text{Greal}(\mathcal{N}) \cong_r \text{Greal}(\mathcal{N}^*)$, by the homeomorphism $W \leftrightarrow W^\perp$, and
- if f is an element of $\mathcal{N}$ and $\mathcal{N} \cup \tilde{f}$ is an extension by an element antiparallel to f then $\mathcal{R}(\mathcal{N}) \cong_r \mathbb{R}(\mathcal{N} \cup \tilde{f})$.

□

Proof of Theorem 7.35: Let S be a basic primary semialgebraic set. By the Universality Theorem (Theorem 7.26), there is an oriented matroid $\mathcal{M}$ with $\mathcal{R}(\mathcal{M}) \simeq_s S$, and so by Lemma 8.21 $\mathcal{R}(\Lambda(\mathcal{M})) \simeq_s S$, Consider $LV(\Lambda(\mathcal{M}))$. Corollary 8.20 tells us this is a combinatorial type of convex polytope, and Theorem 8.18 tells us that its realization space is $\mathcal{R}(\Lambda(\mathcal{M}))$. □

To kill off yet another conjecture, let's return to polars. Proposition 8.13 showed that every oriented matroid polytope with an adjoint has a polar. Does every oriented matroid polytope have a polar? Put another way, is the set of all isomorphism types of posets that arise as Edmonds–Mandel face lattices the same as the set of types for Las Vergnas face lattices?

Proposition 8.22 (Billera and Munson 1984b) *Let $\mathcal{M}$ be an oriented matroid polytope without loops or coloops. If $\mathcal{M}$ does not have an adjoint then $\Lambda(\mathcal{M})$ does not have a polar.*

Proof: Recall that an adjoint for $\mathcal{M}$ has one element for every pair $\{X, -X\}$ of signed cocircuits of $\mathcal{M}$. We define an oriented matroid to be an **extended adjoint** of $\mathcal{M}$ if it's obtained from an adjoint of $\mathcal{M}$ by adding an antiparallel $-X$ to each element X. Thus an extended adjoint for $\mathcal{M}$ is an oriented matroid on elements $\mathcal{C}^*(\mathcal{M})$ such that for each element e of $\mathcal{M}$ there is a $W_e \in \mathcal{C}^*(\mathcal{M}_{\text{adj}})$ given by $W_e(X) = X(e)$. We'll show that if $\Lambda(\mathcal{M})$ does have a polar then we can get an extended adjoint of $\mathcal{M}$ isomorphic to a contraction of this polar.

We assume $\mathcal{M}$ is rank r on elements $[n]$, and we number the signed cocircuits of $\mathcal{M}$ as $X_1, \ldots, X_m$.

Let $(\Lambda(\mathcal{M}))^\square$ be a polar for $\Lambda(\mathcal{M})$. Thus its set of elements is the set of positive cocircuits of $\Lambda(\mathcal{M})$. Recall that these positive cocircuits come in two flavors: For each $i \in [n]$ we have the signed cocircuit $Z_i := \{i, \tilde{i}\}^+$, and for each $X_i = A^+ B^-$ we have the signed cocircuit $X_i' := (A \cup \tilde{B})^+$.

The contraction $\mathcal{M}' := (\Lambda(\mathcal{M}))^\square / \{Z_1, \ldots, Z_n\}$ is an oriented matroid on elements $\{X_1', \ldots, X_m'\}$: We'll show that it's isomorphic to an adjoint of $\mathcal{M}$, by an isomorphism taking each X_i' to X_i.

Recall from Section 8.3 the description of the positive cocircuits Y_e of a polar as columns of a matrix: In our case, the matrix is

$$\begin{array}{c} \\ Z_1 \\ Z_2 \\ \vdots \\ Z_n \\ X_1' \\ X_2' \\ \vdots \\ X_m' \end{array} \begin{pmatrix} Y_1 & Y_2 & \cdots & Y_n & Y_{1'} & Y_{2'} & \cdots & Y_{n'} \\ + & 0 & & & + & 0 & & \\ 0 & + & & & 0 & + & & \\ & & \ddots & & & & \ddots & \\ & & & + & & & & + \\ a_{11} & a_{12} & \cdots & a_{1n} & a_{11}' & a_{12}' & \cdots & a_{1n}' \\ a_{21} & a_{22} & \cdots & a_{2n} & a_{21}' & a_{22}' & \cdots & a_{2n}' \\ \vdots & \vdots & & \vdots & \vdots & \vdots & & \vdots \\ a_{m1} & a_{m2} & \cdots & a_{mn} & a_{m1}' & a_{m2}' & \cdots & a_{mn}' \end{pmatrix}.$$

Here

$$a_{ij} = X_i'(j) = \begin{cases} + & X_i(j) = +, \\ 0 & X_i(j) \neq + \end{cases}$$

for each $i \in [n]$, and

$$a_{ij}' = X_i'(j) = \begin{cases} + & X_i(j) = -, \\ 0 & X_i(j) \neq -. \end{cases}$$

8.4 The Lawrence Construction

The rank of $\Lambda(\mathcal{M})$ is $n + r$, and so

$$\text{rank}(\mathcal{M}') = n + r - \text{rank}_{(\Lambda(\mathcal{M}))^\square}(\{Z_1, \ldots, Z_n\}).$$

Assume by way of contradiction that $\{Z_1, \ldots, Z_n\}$ is dependent in $(\Lambda(\mathcal{M}))^\square$: Let Z be a signed circuit of $(\Lambda(\mathcal{M}))^\square$ with support contained in $\{Z_1, \ldots, Z_n\}$, and let Z_i be in the support of Z. Then Z is orthogonal to the signed cocircuit Y_i, but $\text{supp}(Z) \cap \text{supp}(Y_i) = \{Z_i\}$, and so $0 \notin Z \cdot Y_i$, a contradiction. We conclude that the rank of $\mathcal{M}'$ is r.

Eliminating Z_e between the signed cocircuits Y_e and $-Y'_e$ for each e, we get a covector of $(\Lambda(\mathcal{M}))^\square$ with support contained in $\{X'_1, \ldots, X'_m\}$: The restriction of this covector to $\{X'_1, \ldots, X'_m\}$ is a covector W'_e of $\mathcal{M}'$. From our description of the entries a_{ij} and a'_{ij} we see that for each $j \in [m]$, if $X_i(e) = +$ then $a_{ie} = +$ and $a'_{ie} = -$, and so $W'_e(X'_i) = +$, and likewise if $X_i(e) = -$ then $W_e(X'_i) = -$, and if $X_i(e) = 0$ then $W'_e(X_i) = 0$. Thus $W'_e(X'_i) = X_i(e)$.

It remains to see that $W'_e \in \mathcal{C}^*(\mathcal{M}')$ for each $e \in [n]$. Since $\mathcal{M}$ has no loops, each e is in a basis $\{e = e_1, \ldots, e_r\}$ of $\mathcal{M}$. For each $j \in [r]$ there is a signed cocircuit $X_{\alpha(j)}$ of $\mathcal{M}$ such that $X_{\alpha(j)}(e_j) = +$ and $X_{\alpha(j)}(e_i) = 0$ for each $i \in [r] \setminus \{j\}$. Thus $W'_{e_j}(X_{\alpha(j)}) = +$ and $W'_{e_j}(X_{\alpha(i)}) = 0$ for each $i \in [r] \setminus \{j\}$. We conclude that

$$0 < W'_{e_1} < W'_{e_1} \circ W'_{e_2} < \cdots < W'_{e_1} \circ W'_{e_2} \circ \cdots \circ W'_{e_r}.$$

This is a chain of $r+1$ elements in the rank r poset $\mathcal{V}^*(\mathcal{M}')$, and so the smallest nonzero element $W'_{e_1} = W'_e$ is indeed a signed cocircuit. □

8.4.2 Some Geometric Insight

Notice that $\{X\tilde{E}^0 : X \in \mathcal{V}^*(\mathcal{M})\} \subset \mathcal{V}^*(\Lambda(\mathcal{M}))$, and so $\mathcal{M} = \Lambda(\mathcal{M})/\tilde{E}$. For instance, if $\mathcal{M}$ is realizable then the vertex set of a convex polytope realizing $\Lambda(\mathcal{M})$ quotients to a realization of $\mathcal{M}$.

To give a little bit of insight on going from a realization of $\mathcal{M}$ to a realization of $\Lambda(\mathcal{M})$, let's think about the dual operation of extending by an antiparallel element. Suppose $\mathcal{A} = (\mathbf{v}_e : e \in E)$ is a realization of $\mathcal{M}$, $f \in E$, and $\mathcal{M}^* \cup \tilde{f}$ is an extension of $\mathcal{M}^*$ by an element $\tilde{f}$ antiparallel to f. How do we go from $\mathcal{A}$ to a realization of $(\mathcal{M}^* \cup \tilde{f})^*$?

$(\mathcal{M}^* \cup \tilde{f})^*$ is a lifting of $\mathcal{M}$ – that is, $(\mathcal{M}^* \cup \tilde{f})^*/\tilde{f} = \mathcal{M}$. In particular, if $\mathcal{A}$ spans a vector space V then a realization of $(\mathcal{M}^* \cup \tilde{f})^*$ should be in a vector space V' whose dimension is one greater than the dimension of V. We'll construct a realization $\mathcal{A}' = (\mathbf{v}'_e : e \in E \cup \{\tilde{f}\})$ of $(\mathcal{M}^* \cup \tilde{f})^*$ in a vector space $V' \supset V$.

256   Oriented Matroid Polytopes

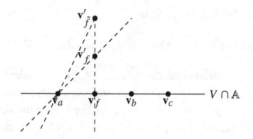

Figure 8.3 Realizing a lifting $(\mathcal{M}^* \cup \tilde{f})^*$.

Because $\{f, \tilde{f}\}^+$ is a covector of $(\mathcal{M}^* \cup \tilde{f})^*$, we must construct $\mathcal{A}' = (\mathbf{v}'_e : e \in E \cup \{\tilde{f}\})$ such that there is a signed hyperplane $\mathcal{H}$ in V' with $\{\mathbf{w}_e : e \in E\backslash\{f\}\} \subset H$ and $\{f, \tilde{f}\} \subset H^+$. If we define $\mathbf{v}'_e = \mathbf{v}_e$ for each $e \notin \{f, \tilde{f}\}$ then V is our H. Since f is a coloop of $(\mathcal{M}^* \cup \tilde{f})^*\backslash\{f\}$, we can choose $\mathbf{v}'_f$ to be any element of $V'\backslash V$.

Let R be the ray in V' beginning at $\mathbf{v}_f$ and going through $\mathbf{v}_f'$. Choose $\mathbf{v}'_{\tilde{f}}$ to be a point on this ray beyond $\mathbf{v}'_f$. (See Figure 8.3 for an example depicted in affine space.) It's straightforward to verify that the $\mathcal{A}'$ we've constructed realizes $(\mathcal{M}^* \cup \tilde{f})^*$: Just confirm that it has the correct signed cocircuits.

$\mathcal{C}^*((\mathcal{M}^* \cup \tilde{f})^*) = \{A^+B^- : A^+B^- \in \mathcal{C}^*(\mathcal{M}) \text{ and } (A \cup B) \subseteq E\backslash\{f\}\}$
$\cup \{\pm A^+ B^- : A^+ B^- \in \mathcal{C}^*(\mathcal{M}) \text{ and } f \in A\}$
$\cup \{\pm(A\backslash\{f\})^+(B \cup \tilde{f})^- : A^+B^- \in \mathcal{C}^*(\mathcal{M}) \text{ and } f \in A\}$
$\cup \{\pm\{f, \tilde{f}\}^+\}.$

Each element of the first set on the right-hand side arises from our original arrangement $\mathcal{A}$ by way of a signed hyperplane $\mathcal{H}$ in V with $\mathbf{v}_f \in H$; we get a corresponding $\mathcal{H}'$ in V' by letting H' be the span of $H \cup \{\mathbf{v}'_f\}$ and signing appropriately. (In Figure 8.3, the vertical line is such an H'.) In a similar way we get signed hyperplanes demonstrating each element of the second and third sets on the left-hand side: The diagonal lines in our figure show examples. For the final set $\{\pm\{f, \tilde{f}\}^+\}$, V is the corresponding hyperplane.

This demonstrates the geometric sense for extending $\mathcal{M}^*$ by a single $\tilde{f}$ parallel to an element f: To carry out the Lawrence construction one repeats this process over, and over, and over ...

Exercises

8.1 (Bayer and Sturmfels 1990) Prove that if $\mathcal{M}$ is realized by a matrix M then $\Lambda(\mathcal{M})$ is realized by

$$\begin{pmatrix} M & 0 \\ I & I \end{pmatrix}.$$

8.2 Find a nonrealizable oriented matroid polytope $\mathcal{M}$ for which $P(\mathcal{M})$ is the face poset of a convex polytope.

9

Subdivisions and Triangulations

A *polytopal decomposition* of a set S of points in affine space is a regular cell complex whose elements are convex polytopes, whose vertex set is S and whose underlying space is the convex hull of S. If the decomposition is a simplicial complex, we call the decomposition a *triangulation* of S. (Some sources only require the vertex set to be a subset of S rather than all of S: We'll call a simplicial complex satisfying this weaker definition a *partial triangulation* of S.) The following problem shows that the set of abstract simplicial complexes arising from triangulations of S depends only on the oriented matroid of S.

Problem 9.1 Let $P = (p_e : e \in E)$ and $Q = (q_e : e \in E)$ be affine point arrangements with realizing oriented matroids $\mathcal{M}_P$ and $\mathcal{M}_Q$. Let $\mathcal{G}$ be a triangulation of P. Let $f : \text{conv}(P) \to \text{conv}(Q)$ be the map that is linear on each element of $\mathcal{G}$ and sends each p_e to q_e.

1. Prove that if f is not injective then $\mathcal{M}_P \neq \mathcal{M}_Q$.
2. Prove that if f is not surjective then there is a codimension 1 element of $\mathcal{G}$ contained in the boundary of $\text{conv}(P)$ whose interior is mapped by f into the interior of $\text{conv}(Q)$. Conclude from this that $\mathcal{M}_P \neq \mathcal{M}_Q$.
3. Conclude from the previous two parts that if $\mathcal{M}_P = \mathcal{M}_Q$ then the map from P to Q sending each p_e to q_e induces a bijection from the set of triangulations of P to the set of triangulations of Q.

The results of Problem 9.1 open hope for a notion of *triangulation of an oriented matroid*. Defining such a notion is unexpectedly tricky. Certainly a triangulation of an oriented matroid should be an abstract simplicial complex whose vertices are the elements of the oriented matroid. But how to generalize the notion of "underlying space" of a geometric simplicial complex? One property one would want from such a notion: For every single-element extension $\mathcal{M} \cup f$ of $\mathcal{M}$ in which f is in the convex hull of the remaining elements, the set of elements of the triangulation with f in their convex hull has a unique minimal element. Another is that a triangulation of an acyclic oriented matroid, like a triangulation of a set of points in affine space, should be a topological ball.

Section 9.3 will introduce two different notions for oriented matroids analogous to polytopal subdivisions. Each notion gives us a definition of oriented matroid triangulation. The notions coincide for realizable oriented matroids, but it's not known whether they coincide in general. This section will also introduce a third definition of triangulation which is less intuitive but most powerful to work with (because it's recursive in rank). In Section 9.6 we'll show the equivalence of the recursive definition with one of the other notions of triangulation. Also, in Section 9.5 we'll show that every oriented matroid subdivision can be subdivided further to a triangulation with the same topology, and so for many purposes we can restrict our attention to triangulations.

Also open is whether every subdivision (by any of these definitions) of an acyclic oriented matroid is a topological ball. In Section 9.7 we'll settle this question only for Euclidean oriented matroids.

We'll consider subdivisions of arbitrary oriented matroids, not just acyclic ones. The geometric analog to such a subdivision is a *fan*. We generalize in part out of necessity: Inductive arguments will lead us to triangulations of oriented matroids that are not necessarily acyclic. Additionally, as we'll discuss in Section 9.8, one of the original motivations for oriented matroid triangulations was focused on totally cyclic oriented matroids, envisioning such oriented matroids as analogs to coordinate charts on simplicial manifolds.

One reason to look at triangulations of oriented matroids comes from thinking of a triangulation of $\mathcal{M}$ as a "piecewise-linear (PL) realization," something weaker than a true (linear) realization. Thinking of an oriented matroid as an abstraction of a signed hyperplane arrangement brought us to the Topological Representation Theorem. We already made an unsuccessful attempt at a similar result from the vector arrangement perspective: In Chapter 6 we saw that not every oriented matroid has a representation as a pseudo-configuration of points. Triangulations offer another, weaker approach to the vector arrangement perspective. The abstract simplicial complex $\mathcal{T}$ associated to a triangulation of a point set S in affine space encodes a PL structure on the convex hull of S, in the sense that there is a canonical simplexwise-linear homeomorphism from a geometric realization $\|\mathcal{T}\|$ to conv(S). We might hope to think of a triangulation of an acyclic oriented matroid likewise as a PL space that behaves like the convex hull of a set of points, and more generally, to think of a triangulation of an arbitrary oriented matroid as a PL space that behaves like a fan whose underlying space is convex. The results of this chapter will realize this hope for Euclidean oriented matroids, but the question for general oriented matroids remains open.

9.1 Some PL Topology

In all of the topological discussions in previous chapters we worked in the category of topological spaces and continuous maps. In fact, the Topological Representation Theorem and all of the results that flow from it could have been strengthened to results about PL spaces and PL maps. We avoided this at the time to avoid yet another layer of material to cover. Now it makes more sense to work with PL topology, so we review the subject here.

For a detailed introduction to PL topology, see Hudson (1969) and Rourke and Sanderson (1982).

Definition 9.2 A **polyhedron** is a union of finitely many convex polytopes.

Two polyhedra are **PL homeomorphic** if there is a PL homeomorphism between them.

Example 9.3 If $\mathcal{T}_1$ and $\mathcal{T}_2$ are abstract simplicial complexes with geometric realizations $\|\mathcal{T}_1\|$ and $\|\mathcal{T}_2\|$ then a simplicial map from $\mathcal{T}_1$ to $\mathcal{T}_2$ induces a map from $\|\mathcal{T}_1\|$ to $\|\mathcal{T}_2\|$ that is linear on each simplex. In particular, every two geometric realizations of the same abstract simplicial complex are PL homeomorphic.

Definition 9.4 Let $d \in \mathcal{N}$. A **PL d-ball** is a polyhedron that is PL homeomorphic to a d-simplex. A **PL $(d-1)$-sphere** is a polyhedron that is PL homeomorphic to the boundary of a d-simplex.

Example 9.5 Every convex polytope is a PL ball. The boundary of a convex polytope is a PL sphere.

Surprisingly, a polyhedron may be a topological sphere but not a PL sphere. For instance, the double suspension of the Poincare homology sphere is a simplicial complex homeomorphic to S^5 but is not a PL sphere. For a concise survey of related issues, see chapter 2 of Buchstaber and Panov (2002).

Definition 9.6 A **PL cell complex** is a finite collection $\mathcal{G}$ of nonempty PL balls with the following two properties.

1. The boundary of each element of $\mathcal{G}$ is a union of elements of $\mathcal{G}$.
2. For each $\sigma, \tau \in \mathcal{G}$, either $\sigma \subseteq \tau$ or $\sigma \cap \tau$ is an element of $\mathcal{G}$ in the boundary of σ.

A **polytopal complex** is a PL cell complex whose elements are all convex polytopes.

A **vertex** of $\mathcal{G}$ is a zero-dimensional element of $\mathcal{G}$, The **underlying space** of $\mathcal{G}$ is the union of all elements of $\mathcal{G}$. $\mathcal{G}$ is called a **PL cell decomposition** of its underlying space.

The **augmented face poset** of $\mathcal{G}$ is the poset $\mathcal{G} \cup \{\hat{1}, \hat{0}\}$, where $\mathcal{G}$ is partially ordered by inclusion.

Definition 9.7 A **PL subdivision** of a PL cell complex $\mathcal{G}$ is a PL cell complex $\mathcal{G}'$ with the same underlying space such that every element of $\mathcal{G}'$ is a subset of an element of $\mathcal{G}$.

Example 9.8 Let S be a finite set of points in an affine space $\mathbb{A}$, let $\mathcal{G}_1$ be a (geometric) triangulation of S, and let $\mathcal{G}_2$ be another geometric simplicial complex whose vertex set is a subset of the convex hull of S. Then $\{\sigma \cap \tau : \sigma \in \mathcal{G}_1, \tau \in \mathcal{G}_2\}$ is a PL subdivision of $\mathcal{G}_2$.

Example 9.8 lends itself to generalization to oriented matroids. We'll revisit it in Section 9.6 in oriented matroid terms as the *superposition* of two abstract simplicial complexes.

The following result is a consequence of results in chapter 3 in Rourke and Sanderson (1982).

Proposition 9.9 *If P is the poset of cells in a shellable cell complex then P is isomorphic to the poset of cells in a PL cell decomposition of a PL ball (if P is subthin) or PL sphere (if P is thin).*

Proposition 9.9 allows us to strengthen our results from Chapter 4.

Corollary 9.10 *Let P be a poset with a recursive coatom ordering. If P is thin, then P is isomorphic to the augmented face poset of a PL cell complex whose underlying space is a PL sphere. If P is subthin, then P is isomorphic to the augmented face poset of a PL cell complex whose underlying space is a PL ball.*

With this strengthening of Theorem 4.29 in hand, the rest of proof of the Topological Representation Theorem carries through to give the following.

Corollary 9.11 (PL Topological Representation Theorem) *Let $\mathcal{M}$ be a rank r oriented matroid. There is an arrangement of signed pseudospheres in a PL $(r - 1)$-sphere S such that the resulting cell decomposition of S is a PL cell complex with augmented face poset $\mathcal{V}^*(\mathcal{M}) \cup \{\hat{1}\}$.*

Definition 9.12 Let $\mathcal{G}$ be a PL cell complex, viewed as a poset ordered by inclusion. Fix a geometric realization $\|\Delta\mathcal{G}\|$ of the order complex of $\mathcal{G}$. For each $g \in \mathcal{G}$ let

$$\sigma_g = \bigcup_{\substack{\gamma \in \Delta \mathcal{G}: \\ \max(\gamma) \leq g}} \|\gamma\|.$$

The **standard presentation** of $\mathcal{G}$ is $\{\sigma_g : g \in \mathcal{G}\}$.

Proposition 9.13 *The standard presentation of $\mathcal{G}$ is a PL cell complex, and there is a PL homeomorphism from the underlying space of $\mathcal{G}$ to the underlying space of its standard presentation, taking each cell g to σ_g.*

Proposition 9.13 allows us to reduce our considerations on PL cell complexes to the particularly nice ones that arise as standard presentations.

Proof: Let $\mathcal{G}^{(k)}$ denote the cells of $\mathcal{G}$ of dimension at most k. We recursively construct a PL homeomorphism from the underlying space of $\mathcal{G}^{(k)}$ to $\|\Delta \mathcal{G}^{(k)}\|$ sending each cell g to σ_g. When $k = 0$ the map is clear. Once we have the homeomorphism defined on the underlying space of $\mathcal{G}^{(k-1)}$, for each k-cell σ its boundary has been mapped homeomorphically to $\bigcup_{\max(\gamma) < g}^{\gamma \in \Delta \mathcal{G}:} \|\gamma\|$. In particular, the latter space is a PL $(k-1)$-sphere, and σ_g is a cone on this sphere, so is a PL k-ball. A homeomorphism between the boundaries of two PL k-balls extends to a PL homeomorphism of the balls themselves, and so we can extend our map to the underlying space of $\mathcal{G}^{(k)}$. □

Corollary 9.14 *If P_1 and P_2 are PL cell complexes and $f: P_1 \to P_2$ is a poset isomorphism then there is a PL homeomorphism of the underlying spaces taking each $\sigma \in P_1$ to $f(\sigma)$.*

Definition 9.15 An n-dimensional **PL manifold** is a polyhedron in which every point has an open neighborhood that is PL homeomorphic to an open set in $\mathbb{R}^n$. An n-dimensional **PL manifold with boundary** is a polyhedron in which every point has an open neighborhood that is PL homeomorphic to either an open set in $\mathbb{R}^n$ or an open set in $\mathbb{R}^{n-1} \times \mathbb{R}_{\geq 0}$.

Proposition 9.16 *If P is isomorphic to the face poset of a PL cell decomposition of a sphere then P^{op} is as well.*

Proof: By Proposition 9.13, $\|\Delta P\|$ is a PL sphere. A PL sphere is a PL manifold, so by basic PL topology the link of every simplex of $\|\Delta P\|$ is a PL sphere. In particular, for each $p \in P$ and each maximal chain in $P_{\leq p}$, its link in $\|\Delta P\|$ is a PL sphere. But this link is $\|\Delta P_{>p}\|$. Let $\tau_p = \bigcup_{\min(\gamma) \geq p}^{\gamma \in \bar{\Delta} P:} \|\gamma\|$. Then τ_p is a cone on a PL sphere, hence is a PL ball, and it is straightforward to verify that $\{\tau_p : p \in P\}$ is a PL cell complex with face poset P^{op}.

Since $\Delta P = \Delta P^{\mathrm{op}}$, the underlying space of this cell complex is the same as the underlying space of $\|\Delta P\|$, hence is a PL sphere. □

Proposition 9.17 *If $\mathcal{M}$ is rank r and acyclic then $\mathrm{LV}(\mathcal{M})$ and $\mathrm{EM}(\mathcal{M})$ are each isomorphic to the augmented face poset of a PL cell decomposition of an $(r-2)$-sphere.*

Proof: That $\mathrm{LV}(\mathcal{M})$ is isomorphic to the augmented face poset of a PL cell decomposition of an $(r-2)$-sphere follows from the PL Topological Representation Theorem (Corollary 9.11). This implies the result for $\mathrm{EM}(\mathcal{M})$ by Proposition 9.16. □

Definition 9.18 Let $\mathcal{G}$ be a polytopal complex and $\sigma_0 \in \mathcal{G}$. Let x be in the interior of σ_0. The **stellar subdivision** of $\mathcal{G}$ by x in σ_0 is the polytopal complex

$$\mathcal{G}\setminus\{\omega : \omega \supseteq \sigma_0\} \cup \{\mathrm{conv}(\{x\} \cup \omega) : \omega \cup \sigma_0 \in \mathcal{G}, \sigma_0 \not\subseteq \omega\}.$$

Let $\omega \in \mathcal{G}$. We define the **subdivision** sub(ω) **of** ω **in** $\mathcal{G}'$ to be

- $\{\omega\}$, if $\omega \not\supseteq \sigma_0$,
- $\{\mathrm{conv}((\omega\setminus\nu) \cup \{x\}) : \nu \subseteq \sigma_0\}$ if $\omega \supseteq \sigma_0$.

The following is straightforward to verify.

Proposition 9.19 *If $\mathcal{G}$ is a polytopal complex then the stellar subdivision of $\mathcal{G}$ by x in σ_0 is a PL subdivision of $\|\mathcal{G}\|$. For each $\omega \in \mathcal{G}$ the set of faces of elements of* sub(ω) *is a PL subdivision of ω.*

A subdivision of $\|\mathcal{G}\|$ as described in Proposition 9.19 is a **geometric stellar subdivision**.

Proposition 9.19 tells us that $\|\mathcal{G}\|$ is PL homeomorphic to every realization of every stellar subdivision of $\mathcal{G}$, and also that a choice of a point x in $\|\mathcal{G}\|\setminus\|\mathcal{G}^0\|$ gives rise to a geometric stellar subdivision.

9.2 Geometric Fans

Definition 9.20 The **oriented matroid of an arrangement of rays** $(\mathbb{R}_{\geq 0}v_e : e \in E)$ is the oriented matroid of the vector arrangement $(v_e : e \in E)$.

Definition 9.21 Given an arrangement of rays $(R_e : e \in E)$, for each $\sigma \subseteq E$ we define the **cone** of σ, denoted cone(σ), to be the set of all nonnegative linear combinations of $\bigcup_{e \in \sigma} R_e$.

A **face** of a cone C is a subset F of C such that, for some signed hyperplane (H, H^+, H^-), we have $F \subseteq H$ and $C \setminus F \subseteq H^+$. A **vertex** of C is a face consisting of a single ray. We denote the set of vertices of C by vert(C).

A cone is **acyclic** if $\{0\}$ is a face.

Definition 9.22 A **fan** in a real vector space V is a finite collection $\mathcal{G}$ of acyclic cones satisfying both of the following:

1. Every face of an element of $\mathcal{G}$ is also in $\mathcal{G}$.
2. The intersection of any two elements of $\mathcal{G}$ is a face of both.

The **underlying space** of a fan $\mathcal{G}$ in a vector space V is the union of all elements of $\mathcal{G}$.

The **face poset** of $\mathcal{G}$ is the set $\mathcal{G}$, partially ordered by inclusion.

The set of vertices in a fan is an arrangement of rays, and the **underlying oriented matroid** of the fan is the oriented matroid of this arrangement. Studying fans whose underlying oriented matroid is acyclic is essentially the same as studying polytopal cell complexes in affine space: In this case there is an affine space $\mathbb{A}$ that intersects each such vertex in a single point and intersects each cone in a convex polytope. To generalize this correspondence between fans and cell complexes of dimension to arbitrary fans, instead of intersecting with an affine space we intersect with a PL sphere around the origin.

Definition 9.23 Let $\mathcal{G}$ be a fan and P a convex polytope with $\mathbf{0}$ in its interior. The PL cell complex $\{C \cap \partial P : C \in \mathcal{G}\}$ is the **spherical complex** $S(\mathcal{G})$ associated to $\mathcal{G}$ (with respect to P).

Different choices of P give different spherical complexes, but they are all PL cell complexes with face posets isomorphic to $\mathcal{G} \setminus \{\mathbf{0}\}$, and so by Corollary 9.14 they are PL homeomorphic. For our purposes the choice of P is not important. The underlying space of $\mathcal{G}$ is a cone on the underlying space of $S(\mathcal{G})$. In particular, if $\mathcal{G}$ is a fan whose underlying space is convex, then the associated spherical complex is either a ball or a sphere. In much of this chapter we'll be generalizing fans on an arrangement of rays to objects over oriented matroids. Instead of an actual fan, we'll have a poset analogous to the poset of cones in a fan. This poset will be isomorphic to the poset of cells in a PL cell complex, and intuitively we will imagine this cell complex as the spherical complex of the fan.

Definition 9.24 Let $\mathcal{A} = (R_e : e \in E)$ be an arrangement of rays and $\mathcal{G}$ a fan whose vertex set is a subset of $\mathcal{A}$. We say $\mathcal{G}$ is a **decomposition** of $\mathcal{A}$ if the underlying space of $\mathcal{G}$ is cone(E).

Proposition 9.25 *Consider an arrangement* $\mathcal{A} = (R_e : e \in E)$ *of rays whose associated oriented matroid* $\mathcal{M}$ *has rank r and a decomposition* $\mathcal{G}$ *of* $\mathcal{A}$.

If $\mathcal{M}$ *is totally cyclic then every PL cell complex with face poset isomorphic to* $\mathcal{G}\backslash\{0\}$ *is a PL sphere.*

If $\mathcal{M}$ *is not totally cyclic then every PL cell complex with face poset isomorphic to* $\mathcal{G}\backslash\{0\}$ *is a PL ball.*

Proof: The associated spherical complex of $\mathcal{G}$ is a PL cell complex, and every PL cell complex with face poset isomorphic to $\mathcal{G}\backslash\{0\}$ is PL homeomorphic to this complex, by Proposition 9.13, and so it's enough to show that the spherical complex of $\mathcal{G}$, defined with respect to a convenient convex polytope P, is a PL sphere or ball.

If $\mathcal{M}$ is totally cyclic, then the underlying space of $\mathcal{G}$ is the entire vector space, so the underlying space of the spherical complex is the entire boundary of P, hence is a PL sphere.

If $\mathcal{M}$ is not totally cyclic, then all of our rays lie in a closed half-space $H^+ \cup H$, where (H, H^+, H^-) is a signed hyperplane. Let P be a d-simplex in $\mathbb{R}^r$ with $\mathbf{0}$ in its interior, with some facet Δ' parallel to H and contained in H^-, and with the opposite vertex $\mathbf{v}$ contained in the interior of a maximal cone of $\mathcal{G}$. Thus the underlying space of the resulting spherical complex intersects the interior of each face of P containing $\mathbf{v}$. We can get a PL homeomorphism from the underlying space U of the spherical complex to the PL ball $\partial P \cap (H \cup H^+)$ by *pseudo-radial projection*, developed in chapter 2 of Rourke and Sanderson (1982). □

The following proposition, whose proof is straightforward, gives us a characterization of the fans corresponding to simplicial spherical cell complexes.

Proposition 9.26 *Let I be a finite subset of $\mathbb{R}^n\backslash\{0\}$ and C the cone of nonnegative linear combinations of elements of I. The following are equivalent:*

1. I *is independent.*
2. $\mathbb{R}_{\geq 0}\mathbf{v} \neq \mathbb{R}_{\geq 0}\mathbf{v}'$ *for each* $\mathbf{v} \neq \mathbf{v}'$ *in* I, $\{\mathbb{R}_{\geq 0}\mathbf{v} : \mathbf{v} \in I\}$ *is the set of vertices of C, and the faces of C are exactly the cones on subsets of I.*
3. *If* $\mathbf{w} \in \mathbb{R}^n$ *such that* $\mathbf{w} \cdot \mathbf{v} > 0$ *for each* $\mathbf{v} \in I$ *and* $\mathbb{A} = \{\mathbf{x} \in \mathbb{R}^n : \mathbf{w} \cdot \mathbf{x} = 1\}$ *then* $C \cap \mathbb{A}$ *is a simplex of dimension* $|I| - 1$.

Definition 9.27 If the conditions of Proposition 9.26 hold then we say C is **simplicial**.

A fan is **simplicial** if every element is simplicial.

The **abstract simplicial complex** associated to a simplicial fan with vertex set indexed by E is the set of subsets of E indexing vertex sets of elements of $\mathcal{F}$.

Definition 9.28 A **triangulation** of an arrangement of rays is a simplicial fan that is a decomposition of the arrangement.

9.3 Oriented Matroid Fans

We want to define oriented matroid fans, decompositions, and triangulations so that for an arrangement $\mathcal{A}$ of rays with oriented matroid $\mathcal{M}$, each definition for $\mathcal{M}$ coincides with the corresponding definition for $\mathcal{A}$.

Recall from Chapter 8 that a *face* of an oriented matroid $\mathcal{M}$ on elements E is an $A \subseteq E$ such that $(E \backslash A)^+ \in \mathcal{V}^*(\mathcal{M})$.

Definition 9.29 Let $\mathcal{M}$ be an oriented matroid on elements E and σ an acyclic subset of E. A **face** of σ in $\mathcal{M}$ is a face of $\mathcal{M}(\sigma)$.

The **vertex set** of σ, denoted vert$(\sigma, \mathcal{M})$, is the set of one-dimensional faces of σ.

We want to define a *fan on an oriented matroid* $\mathcal{M}$ in such a way that the fans on a realizable $\mathcal{M}$ are exactly the face posets of fans on a realization of $\mathcal{M}$. The following proposition suggests two equivalent ways we might combinatorialize the second condition in Definition 9.22 (that the intersection of any two elements of a fan is a face of both).

Proposition 9.30 *Let $\mathcal{M}$ be an oriented matroid on elements E and σ, τ subsets of E. The following are equivalent:*

1. *There is a covector X of $\mathcal{M}$ such that $X(\sigma \cap \tau) = 0$, $X(\sigma \backslash \tau) = +$, and $X(\tau \backslash \sigma) = -$.*
2. *If Y is a signed circuit of $\mathcal{M}$ such that $Y^+ \subseteq \sigma$ and $Y^- \subseteq \tau$ then supp$(Y) \subseteq \sigma \cap \tau$.*

Further, if we have a fixed realization $(R_e : e \in E)$ of $\mathcal{M}$ then these conditions are equivalent to each of the following:

1. cone$(\{R_e : e \in \sigma\}) \cap$ cone$(\{R_e : e \in \tau\})$ *is a face of both* cone$(\{R_e : e \in \sigma\})$ *and* cone$(\{R_e : e \in \tau\})$.
2. cone$(\{R_e : e \in \sigma\}) \cap$ cone$(\{R_e : e \in \tau\}) =$ cone$(\{R_e : e \in \sigma \cap \tau\})$.

The equivalence of the first two conditions in Proposition 9.30 follows from the Farkas Property (Section 2.3.3). We leave the details of the proof to the reader.

The first part of Proposition 9.30 suggests one way to replace the second condition in Definition 9.22 in our definition of oriented matroid fans, stated

9.3 Oriented Matroid Fans

in terms of signed circuits and/or covectors. The second part of Proposition 9.30 suggests another way to replace the second condition in Definition 9.22, in terms of single-element extensions.

Definition 9.31 Let $\mathcal{M}$ be an oriented matroid on elements E and σ, τ subsets of E. We say that σ and τ are **circuit compatible** if every signed circuit Y of $\mathcal{M}$ such that $Y^+ \subseteq \sigma$ and $Y^- \subseteq \tau$ satisfies $\mathrm{supp}(Y) \in \sigma \cap \tau$.

Recall that for a set σ of elements of $\mathcal{M}$, the convex hull $\widehat{\mathrm{conv}}(\sigma)$ is defined to be $\sigma \cup \{e : \hat{\sigma}^+\{e\}^- \in \mathcal{V}(\mathcal{M}) \text{ for some } \hat{\sigma} \subseteq \sigma\}$.

Definition 9.32 Let $\mathcal{M}$ be an oriented matroid on elements E and σ, τ subsets of E. We say that σ and τ are **extension compatible** if for each extension $\mathcal{M} \cup f$ of $\mathcal{M}$, if $f \in \widehat{\mathrm{conv}}(\sigma) \cap \widehat{\mathrm{conv}}(\tau)$ then $f \in \widehat{\mathrm{conv}}(\sigma \cap \tau)$.

Circuit elimination in $\mathcal{M} \cup f$ tells us that every pair σ, τ that fails to be extension compatible also fails to be circuit compatible. A pair can be extension compatible without being circuit compatible. For instance, consider a non-Euclidean rank r oriented matroid $\mathcal{M}$ and a pair σ of rank 2 and τ of rank $r - 1$ demonstrating this non-Euclideanness. Thus there is no extension $\mathcal{M} \cup f$ by a nonloop with $\mathrm{rank}(\sigma \cup \{f\}) = 2$ and $\mathrm{rank}(\tau \cup \{f\}) = r - 1$. In particular σ and τ are extension compatible, and $\sigma \cap \tau = \emptyset$, since an extension by an element parallel to an element of $\sigma \cap \tau$ would be in the convex hull of both. Thus $|\sigma \cup \tau| = r + 1$, and so there is a signed circuit of $\mathcal{M}$ with support contained in $\sigma \cup \tau$. Since Euclideanness is preserved under reorientation, we may assume $\mathcal{M}$ is oriented so that this signed circuit is $\hat{\sigma}^+ \hat{\tau}^-$ with $\hat{\sigma} \subseteq \sigma$ and $\hat{\tau} \subseteq \tau$.

Definition 9.33 Let $\mathcal{M}$ be an oriented matroid on elements E. Let $\mathcal{F}$ be a collection of subsets of E satisfying the following properties:

1. For each $\sigma \in \mathcal{F}$, $\mathcal{M}(\sigma)$ is acyclic and has no signed circuits of the form $\hat{\sigma}^+ f^-$.
2. If $\sigma \in \mathcal{F}$ then every face of σ is an element of $\mathcal{F}$.
3. If $\sigma, \tau \in \mathcal{F}$ then $\sigma \cap \tau$ is a face of σ (possibly empty).

$\mathcal{F}$ is a **fan** on $\mathcal{M}$ if every pair of elements of $\mathcal{F}$ is extension compatible.
$\mathcal{F}$ is a **strong fan** on $\mathcal{M}$ if every pair of elements of $\mathcal{F}$ is circuit compatible.
A fan $\mathcal{F}$ on $\mathcal{M}$ is **simplicial** if every element of $\mathcal{F}$ is independent in $\mathcal{M}$.

Every strong fan is a fan. Our preceding discussion of a non-Euclidean oriented matroid and simplices σ and τ which are extension compatible but not circuit compatible points us to a weak fan that is not strong: Just take the abstract simplicial complex whose maximal elements are σ and τ.

The following proposition follows from Proposition 9.30.

Proposition 9.34 *Let $\mathcal{M}$ be realizable on elements E, and let $\mathcal{F}$ be a set of subsets of E. The following are equivalent:*

1. *$\mathcal{F}$ is a strong fan on $\mathcal{M}$.*
2. *$\mathcal{F}$ is a weak fan on $\mathcal{M}$.*
3. *For every realization $(\mathbf{v}_e : e \in E)$ of $\mathcal{M}$ there is a fan with vertex set contained in $\{\mathbb{R}_{\geq 0}\mathbf{v}_e : e \in E\}$ and face poset given by $\mathcal{F}$.*
4. *For some realization $(\mathbf{v}_e : e \in E)$ of $\mathcal{M}$ there is a fan with vertex set contained in $\{\mathbb{R}_{\geq 0}\mathbf{v}_e : e \in E\}$ and face poset given by $\mathcal{F}$.*

Proposition 9.35 *Let $\mathcal{F}$ satisfy the first three conditions to be a fan. The following are equivalent:*

1. *$\mathcal{F}$ is a fan.*
2. *For every extension $\mathcal{M} \cup f$ the set $\{\sigma \in \mathcal{F} : f \in \widehat{\mathrm{conv}}(\sigma)\}$ either is empty or has a unique minimal element.*
3. *For every extension $\mathcal{M} \cup f$ the set $\{\sigma \in \mathcal{F} : \sigma^+ f^- \in \mathcal{V}(\mathcal{M} \cup f)\}$ has at most one element.*

If these conditions hold then either both sets $\{\sigma \in \mathcal{F} : f \in \widehat{\mathrm{conv}}(\sigma)\}$ and $\{\sigma \in \mathcal{F} : \sigma^+ f^- \in \mathcal{V}(\mathcal{M} \cup f)\}$ are empty or the unique $\sigma \in \mathcal{F}$ such that $\sigma^+ f^- \in \mathcal{V}(\mathcal{M} \cup f)$ is the unique minimum $\sigma \in \mathcal{F}$ such that $f \in \widehat{\mathrm{conv}}(\sigma)$.

Proof: That (3) implies (2) implies (1) is immediate. To see (1) implies (2), if $\{\sigma \in \mathcal{F} : f \in \widehat{\mathrm{conv}}(\sigma)\} \neq \emptyset$ then the intersection of all elements of this set is the unique minimal element.

Now assume (2), and assume that $\{\sigma \in \mathcal{F} : \sigma^+ f^- \in \mathcal{V}(\mathcal{M} \cup f)\}$ is nonempty. Then $\{\sigma \in \mathcal{F} : f \in \widehat{\mathrm{conv}}(\sigma)\}$ is nonempty as well, and so it has a unique minimal element σ_0, and $\sigma_0^+ f^- \in \mathcal{V}(\mathcal{M} \cup f)$. Assume by way of contradiction that there is another $\tau \in \mathcal{F}$ such that $\tau^+ f^- \in \mathcal{V}(\mathcal{M} \cup f)$. Then σ_0 is a face of τ, so there is an $X \in \mathcal{V}^*(\mathcal{M} \cup f)$ such that $X(\sigma_0) = \{0\}$ and $X(\tau\backslash\sigma_0) = \{+\}$. Since $X \perp \sigma_0^+ f^-$ we know $X(f) = 0$, but then $X \cdot \tau^+ f^- = \{+\}$, a contradiction. Thus we have (3). □

Lemma 9.36 *Let $\mathcal{M}$ be an acyclic oriented matroid on elements E. If A is a face of $\mathcal{M}$ then the set of faces of A is the set of faces of $\mathcal{M}$ contained in A.*

Proof: If F is a face of A then there is a covector X of $\mathcal{M}$ such that $X(F) = 0$ and $X(A\backslash F) = \{0\}$. The composition $(E\backslash A)^+ \circ X = (E\backslash F)^+$ shows that F is a face of $\mathcal{M}$. Conversely, if $F \subseteq A$ is a face of $\mathcal{M}$ then the covector $(E\backslash F)^+$ of $\mathcal{M}$ restricts to a covector $(A\backslash F)^+$ of $\mathcal{M}(A)$. □

9.3 Oriented Matroid Fans

Proposition 9.37 *If σ is acyclic in $\mathcal{M}$ then the set of faces of σ is a strong fan that is isomorphic to the augmented face poset of a PL cell decomposition of a sphere of dimension* rank$(\sigma) - 2$.

Proof: The first two conditions for a fan are clear. To see the third condition, let ν and ω be faces of σ. Then $(\sigma\backslash\nu)^+$ and $(\sigma\backslash\omega)^+$ are covectors of $\mathcal{M}(\sigma)$, and so their composition $(\sigma\backslash(\nu \cap \omega))^+$ is as well, and so $\nu \cap \omega$ is a face of σ. To see circuit compatibility, again let ν and ω be faces of σ. There is a $Y_\nu \in \mathcal{V}^*(\mathcal{M})$ whose restriction to σ is $(\sigma\backslash\nu)^+$ and a $Y_\omega \in \mathcal{V}^*(\mathcal{M})$ whose restriction to σ is $(\sigma\backslash\omega)^+$. A signed circuit X with $X^+ \subseteq \nu$ and $X^- \subseteq \omega$ must be orthogonal to both of these, and so supp$(X) \subseteq \nu \cap \omega$.

The augmented face poset result follows from Proposition 9.17 applied to $\mathcal{M}(\sigma)$. □

Proposition 9.38 *Let $\mathcal{F}$ be a fan on $\mathcal{M}$. Fix a realization $\|\Delta(\mathcal{F}\backslash\{0\})\|$, and for each $f \in \mathcal{F}\backslash\{0\}$ let*

$$\sigma_f = \bigcup_{\substack{\gamma \in \Delta(\mathcal{F}\backslash\{0\}): \\ \max(\gamma) \leq f}} \|\gamma\|.$$

Then $\{\sigma_f : f \in \mathcal{F}\backslash\{0\}\}$ is a PL cell complex. If $\mathcal{M}$ is realizable then this cell complex is isomorphic to the spherical cell complex associated to the realization of $\mathcal{F}$.

Proof: For each $f \in \mathcal{F}\backslash\{0\}$ we have that $\mathcal{F}_{\leq f}$ is the set of faces of f. Thus by Proposition 9.37 $\mathcal{F}\backslash\{0\}$ is the face poset of a PL ball, and $\{\sigma_f : f \in \mathcal{F}\backslash\{0\}\}$ is its standard presentation. The second statement is clear. □

Definition 9.39 The cell complex $\{\sigma_f : f \in \mathcal{F}\backslash\{0\}\}$ is called the **realization** of $\mathcal{F}$ and is denoted $\|\mathcal{F}\|$.

Definition 9.40 Let $\mathcal{M}$ be a simple rank r oriented matroid on elements E. A **fan decomposition** of $\mathcal{M}$ is a fan $\mathcal{F}$ on $\mathcal{M}$ with vertex set E with the following properties:

1. Every maximal element of $\mathcal{F}$ has rank r.
2. Every element of $\mathcal{F}$ of rank $r - 1$ either is contained in a facet of $\mathcal{M}$ or is contained in exactly two maximal elements of $\mathcal{F}$.

A **triangulation** of $\mathcal{M}$ is a simplicial fan decomposition of $\mathcal{M}$.

A **strong fan decomposition** is a strong fan that is a fan decomposition, and a **strong triangulation** is a strong fan decomposition that is a triangulation.

A **partial fan decomposition** (resp. **partial triangulation, partial strong fan decomposition, partial strong triangulation**) is a fan satisfying all of the conditions except perhaps having vertex set E.

The unintuitive results on oriented matroid extensions in Chapter 6 make strong fan decompositions more immediately appealing than general fan decompositions. However, most of the results we'll get are on the level of the weaker notion. While we noted earlier that not all fans are strong fans, the following question is open.

Conjecture 9.41 *Every fan decomposition of $\mathcal{M}$ is a strong fan decomposition of $\mathcal{M}$.*

Let's return to the questions on topology of triangulations discussed at the beginning of this chapter. In Section 9.5 we'll see that every oriented matroid has a triangulation that behaves as we expect. For brevity, we will often refer to the topology of $\mathcal{F}$ when we mean the topology of $\|\mathcal{F}\|$ – for instance, by referring to a fan being a PL ball or sphere.

Proposition 9.42 *If $\mathcal{M}$ is totally cyclic then $\mathcal{M}$ has a strong triangulation that is a PL sphere. Otherwise $\mathcal{M}$ has a strong triangulation that is a PL ball.*

In Section 9.5 we'll see that for each subdivision of $\mathcal{M}$ there is a triangulation of $\mathcal{M}$ with the same topology. The question then is whether all triangulations behave as we expect.

Conjecture 9.43 *Let $\mathcal{M}$ be a rank r oriented matroid. If $\mathcal{M}$ is totally cyclic then every triangulation of $\mathcal{M}$ is a PL $(r-1)$-sphere. Otherwise every triangulation of $\mathcal{M}$ is a PL $(r-1)$-ball.*

In Section 9.6 we'll prove this conjecture for Euclidean oriented matroids by way of Proposition 9.42 and the following.

Theorem 9.44 *If $\mathcal{M}$ is Euclidean and $\mathcal{T}_1$ and $\mathcal{T}_2$ are triangulations of $\mathcal{M}$ then $\|\mathcal{T}_1\|$ is PL homeomorphic to $\|\mathcal{T}_2\|$.*

One key tool in the proof of Theorem 9.44 is a notion of *superposition* of triangulations of $\mathcal{M}$, which will give a common PL subdivision when $\mathcal{M}$ is Euclidean. Another is a third definition of oriented matroid triangulation.

Definition 9.45 (Anderson 1999) Let $\mathcal{M}$ be a simple oriented matroid on elements E, and let $\mathcal{T}$ be a nonempty set of subsets of E. We say $\mathcal{T}$ is a **partial recursive triangulation** of $\mathcal{M}$ if all of the following hold:

1. For each $\sigma \in \mathcal{T}$, σ is independent in $\mathcal{M}$ and $\widehat{\text{conv}}(\sigma) \cap \mathcal{T}^0 = \sigma$.
2. If $\mathcal{M}$ is rank 1 and totally cyclic then $\mathcal{T} = S^0$.
3. If $\text{rank}(\mathcal{M}) > 1$ then $\text{link}(t)$ is a partial recursive triangulation of $\mathcal{M}/t$ for each $t \in \mathcal{T}^0$.

9.4 Stellar Subdivision in an Oriented Matroid Fan

A **recursive triangulation** of $\mathcal{M}$ is a partial recursive triangulation $\mathcal{T}$ such that $\mathcal{T}^0 = E$.

In Section 9.6 we'll prove the following two propositions, giving us several perspectives on fans.

Proposition 9.46 (Anderson 1999, Santos 2002) *Let $\mathcal{M}$ be an oriented matroid on elements E and $\mathcal{T}$ a simplicial complex with $\mathcal{T}^0 \subseteq E$. The following are equivalent:*

1. $\mathcal{T}$ *is a partial triangulation of* $\mathcal{M}$.
2. $\mathcal{T}$ *is a partial recursive triangulation of* $\mathcal{M}$.
3. $\mathcal{T}$ *is a simplicial fan on* $\mathcal{M}$, *and if* $\mathcal{M} \cup f$ *is an extension of* $\mathcal{M}$ *with* $f \in \widehat{\mathrm{conv}}(E)$ *then there is a unique* $\sigma \in \mathcal{T}$ *such that* $\sigma^+ f^- \in \mathcal{V}(\mathcal{M} \cup f)$.

Proposition 9.47 (Santos 2002) *Every strong triangulation of $\mathcal{M}$ is a triangulation of $\mathcal{M}$.*

Despite being the least intuitive notion of triangulation, the recursive definition is the most useful, because it lends itself to proofs by induction on rank. In particular, we'll prove Theorem 9.44 and Proposition 9.47 using the recursive definition.

9.4 Stellar Subdivision in an Oriented Matroid Fan

Consider an oriented matroid $\mathcal{M}$ realized by an arrangement $(p_e : e \in E)$ of points in affine space and $\mathcal{F}$ a fan on $\mathcal{M}$. Let $\mathcal{G} = \{\mathrm{conv}(p_e : e \in \omega) : \omega \in \mathcal{F}\}$ be the polytopal complex corresponding to $\mathcal{F}$, $\sigma_0 \in \mathcal{G}$, and $x \in E$ with $p_x \in \mathrm{conv}(\sigma_0)$. Then we can take a geometric stellar subdivision of $\mathcal{G}$ by p_x in σ_0. This motivates the following definition.

Definition 9.48 Let $\mathcal{M}$ be an oriented matroid on elements E and $\mathcal{F}$ a fan on $\mathcal{M}$. Let $\sigma_0 \in \mathcal{F}$ and $x \in E$ such that $\sigma_0^+ x^- \in \mathcal{V}(\mathcal{M})$. The $\mathcal{M}$-**stellar subdivision** of $\mathcal{F}$ by x in σ_0 is

$$\mathcal{F}' = \{\omega \in \mathcal{F} : \omega \not\supseteq \sigma_0\} \cup \bigcup_{\substack{\omega \in \mathcal{F}: \\ \omega \supseteq \sigma_0}} \{v \cup \{x\} : v \in \mathcal{F}, \sigma_0 \not\subseteq v \subseteq \omega\}.$$

For each $\omega \in \mathcal{F}$ we define the **subdivision** sub(ω) of ω in $\mathcal{F}'$ to be

- ω, if $\omega \not\supseteq \sigma_0$,
- $\{v \cup \{x\} : v \in \mathcal{F}, \sigma_0 \not\subseteq v \subseteq \omega\}$, if $\omega \supseteq \sigma_0$.

As an analog to Proposition 9.19 we have the following.

Proposition 9.49 *Let $\mathcal{M}$ be an oriented matroid, $\mathcal{F}$ a fan on $\mathcal{M}$, and $\mathcal{F}'$ an $\mathcal{M}$-stellar subdivision of $\mathcal{F}$. Then $\mathcal{F}'$ is a fan on $\mathcal{M}$, and there is a PL homeomorphism from $\|\mathcal{F}\|$ to $\|\mathcal{F}'\|$ taking each $\|\omega\|$ to $\bigcup_{v \in \mathrm{sub}(\omega)} \|v\|$.*

Proof: $\mathcal{F}'$ is the stellar subdivision of $\mathcal{S}$ by an element x in a simplex σ_0 such that $\sigma_0^+ x^- \in \mathcal{V}(\mathcal{M})$.

To see the acyclicity condition, consider $\tau \in \mathcal{F}'\backslash\mathcal{F}$. We have that $\tau \subset \sigma \cup x$ for some $\sigma \in \mathcal{F}$ with $\sigma_0 \subseteq \sigma$. Because σ is acyclic, there is a $Y \in \mathcal{V}^*(\mathcal{M})$ such that $Y(\sigma) = \{+\}$. Since $\sigma_0^+ x^- \in \mathcal{V}(\mathcal{M})$, Y must be orthogonal to $\sigma_0^+ x^-$, and so $Y(x_0) = +$. Thus $Y(\tau) = \{+\}$, and so $\mathcal{M}(\tau)$ is indeed acyclic. Also, assume by way of contradiction there is a $\hat{\tau} \subset \tau$ and a $f \in \tau\backslash\hat{\tau}$ such that $\hat{\tau}^+ f^- \in \mathcal{V}(\mathcal{M})$. Since $\tau\backslash\{x\} \subset \sigma \in \mathcal{F}$, we know that $x \in \hat{\tau}$. Since $\sigma_0 \not\subseteq \hat{\tau}$, Proposition 9.35 tells us that $f \neq x$. Thus we can eliminate x between $\hat{\tau}^+ f^-$ and $\sigma_0^+ x^-$ to get a vector $\tilde{\tau}^+ \tilde{\sigma}^-$ with $\tilde{\tau} \cup \tilde{\sigma} \subseteq \sigma$, contradicting our result (Proposition 9.37) that the set of faces of σ is a strong fan.

To see extension compatibility, let $\mathcal{M} \cup f$ be an extension of $\mathcal{M}$ and τ and ν be elements of $\mathcal{F}'$ such that $\tau^+ f^-$ and $\nu^+ f^-$ are elements of $\mathcal{V}(\mathcal{M})$. If both τ and ν are in $\mathcal{F}$ then $\tau = \nu$, since $\mathcal{F}$ satisfies extension compatibility. So assume $\tau \notin \mathcal{F}$, so $\tau = \{x\} \cup \tau'$ for some τ' such that $\tau' \cup \sigma_0 \in \mathcal{F}$. Eliminating x between $\tau^+ f^-$ and $\sigma_0^+ x^-$ gives us $(\tau' \cup \sigma_0)^+ f^- \in \mathcal{V}(\mathcal{M})$. Thus τ' is a face of some $\omega \in \mathcal{F}$ with $\sigma_0 \subseteq \omega$ and $\sigma_0 \not\subseteq \tau'$, and ω is the unique element of $\mathcal{F}$ giving us a signed circuit of this form. We conclude $\nu \notin \mathcal{F}$, and so by the same argument as for τ we get ν' a face of ω.

Assume by way of contradiction that $\tau \neq \nu$. Eliminating f between $\tau^+ f^-$ and $f^+ \nu^-$ gives a vector $\tilde{\tau}^+ \tilde{\nu}^-$ with $\tilde{\tau} \subseteq \tau$ and $\tilde{\nu} \subseteq \nu$. If x were in neither $\tilde{\tau}$ nor $\tilde{\nu}$ we would contradict Proposition 9.37, so without loss of generality assume $x \in \tilde{\nu}$. Since τ' is a face of ω, there is a covector X of $\mathcal{M} \cup f$ such that $X(\tau') = \{0\}$ and $X(\omega\backslash\tau') = \{+\}$. Since $X \perp \sigma_0^+ x^-$ and $\sigma_0 \not\subseteq \tau'$, we have $X(x) = +$. But then $X \cdot \tilde{\tau}^+ \tilde{\nu}^- = \{-\}$, contradicting $X \in \mathcal{V}^*(\mathcal{M} \cup f)$ and $\tilde{\tau}^+ \tilde{\nu}^- \in \mathcal{V}(\mathcal{M} \cup f)$.

To construct the PL homeomorphism $h: \|\mathcal{F}\| \to \|\mathcal{F}'\|$, we first define h to be the identity on $\|\mathcal{F} \cap \mathcal{F}'\|$ and then define h on the remaining cells by induction on dimension. Since $\bigcup_{v \in \mathrm{sub}(\sigma_0)} \|v\|$ is a cone on the PL sphere $\partial\|\sigma_0\|$, it's a PL ball of the same dimension as σ_0, so our definition of h on the boundary of this ball extends to a PL homeomorphism $\|\sigma_0\| \to \bigcup_{v \in \mathrm{sub}(\sigma_0)} \|v\|$. For $\omega \supset \sigma_0$, we see $\bigcup_{v \in \mathrm{sub}(\omega)} \|v\|$ is a cone on the PL ball $\partial\|\omega\|\backslash\bigcup_{\sigma_0 \subset \nu \subset \omega} \|v\|$, and so again a homeomorphism from the boundary of $\|\omega\|$ to the boundary of this cone extends. □

We'll use stellar subdivisions in Section 9.7 to construct a common PL subdivision of two triangulations. Our immediate use for them is to reduce

9.5 Lifting Subdivisions

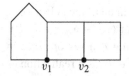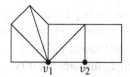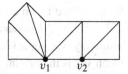

Figure 9.1 Pulling two vertices.

questions on the topology of oriented matroid fans to questions on simplicial oriented matroid fans. For each oriented matroid fan with associated PL cell complex P, we will find a simplicial fan with the same vertices whose associated PL cell complex is a PL subdivision of P.

Given a fan $\mathcal{F}$ on $\mathcal{M}$ and $v \in \mathcal{F}^0$, take an extension of $\mathcal{M}$ by a vertex x parallel to v, and let $\mathcal{F}_1$ be the stellar subdivision of $\mathcal{F}$ by x in $\{v\}$. Thus $\{x\} \in \mathcal{F}_1^0$, $v \notin \mathcal{F}_1^0$, and every element of $\|\mathcal{F}_1\|$ containing $\|\{x\}\|$ is a cone with apex $\{x\}$. Let $\mathcal{F}_2$ be the fan obtained from $\mathcal{F}_1$ by replacing x with v in every element of $\mathcal{F}_1$ that contains x. Thus $\mathcal{F}_2$ is the set

$$\{F \in \mathcal{F} : v \notin F\} \cup \{F \cup \{v\} : F \in \mathcal{F}, v \notin F \text{ and } F \cup \{v\} \subseteq G \text{ for some } G \in \mathcal{F}\}.$$

$\mathcal{F}_2$ is a fan on $\mathcal{M}$, and $\mathcal{F}_2$ can be realized as a subdivision of $\|\mathcal{F}_1\|$.

Definition 9.50 The operation of going from $\mathcal{F}$ to $\mathcal{F}_2$ as above is called **pulling** the vertex v.

By pulling enough vertices in succession, eventually we get to a fan that is simplicial, so we have the following.

Corollary 9.51 *For each fan $\mathcal{F}$ in $\mathcal{M}$ there is a simplicial fan $\mathcal{F}'$ with the same vertices such that $\|\mathcal{F}\|$ is PL homeomorphic to $\|\mathcal{F}'\|$.*

Example 9.52 The leftmost cell complex in Figure 9.1 shows a realization of a rank 3 $\mathcal{M}$ and a nonsimplicial fan $\mathcal{F}$ on $\mathcal{M}$. By pulling the vertex v_1, then the vertex v_2, we get a realization of a simplicial fan that is a subdivision of $\mathcal{F}$.

9.5 Lifting Subdivisions

Our next goal is to show that every totally cyclic oriented matroid has a triangulation that is a PL sphere and every oriented matroid that is not totally cyclic has a triangulation that is a PL ball.

Recall that *lifting* is the dual concept to extension: A lifting of $\mathcal{M}$ is $(\mathcal{M}^* \cup f)^*$ for some extension of $\mathcal{M}^*$ by a non-coloop f. Equivalently, $\hat{\mathcal{M}}$ is a lifting of $\mathcal{M}$ if there is a nonloop f of $\hat{\mathcal{M}}$ such that $\hat{\mathcal{M}}/f = \mathcal{M}$. If a lifting

274   Subdivisions and Triangulations

$\hat{\mathcal{M}}$ is realized by an arrangement $(\mathbf{w}_e : e \in E \cup \{f\})$ in an affine subspace $\mathbb{A}$ of $\mathbb{R}^{n+1}$ then $\mathcal{M}$ is realized by the vector arrangement $(\mathbf{w}_e - \mathbf{w}_f : e \in E)$: Intuitively, we get a vector realization of $\mathcal{M}$ by making $\mathbb{A}$ a vector space in which $\mathbf{w}_f$ is the origin. Turning this around, if $(\mathbf{v}_e : e \in E)$ is a realization of $\mathcal{M}$ in a vector space V then we get an acyclic lifting $\mathcal{M} \cup f$ of $\mathcal{M}$ by treating V as an affine space and the origin of V as $\mathbf{w}_f$. More precisely, we take an affine bijection η from V to an affine space $\mathbb{A}$, let $\mathbf{w}_f = \eta(\mathbf{0})$, and for each $e \in E$ let $\mathbf{w}_e = \eta(\mathbf{v}_e)$. Then $(\mathbf{w}_e : e \in E \cup \{f\})$ realizes a lift of $\mathcal{M}$.

Example 9.53 Each row in Figure 9.2 shows a realization of a rank 2 oriented matroid and the corresponding rank 3 affine point arrangement. Notice that the first two rows show realizations of the same rank 2 oriented matroid $\mathcal{M}$ but different liftings.

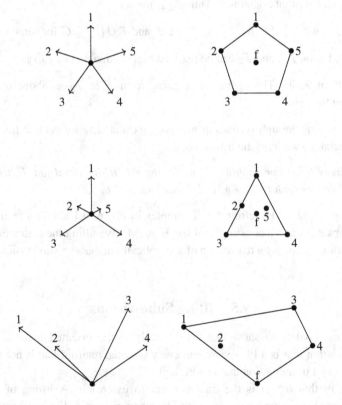

Figure 9.2 Some rank 2 vector arrangements and corresponding rank 3 affine arrangements.

9.5 Lifting Subdivisions

Proposition 9.54 *Let $\hat{\mathcal{M}}$ be an acyclic oriented matroid of rank at least 2 on elements $E \cup \{f\}$. Let $\mathcal{M} = \hat{\mathcal{M}}/f$ and $\mathcal{D}$ be the set of all faces of $\hat{\mathcal{M}}$ not containing f. Let $\mathcal{F} = \{\text{vert}(\sigma) : \sigma \in \mathcal{D}\}$. Then $\mathcal{F}$ is a partial strong fan decomposition of $\mathcal{M}$. If $\mathcal{M}$ is totally cyclic then $\mathcal{F}$ is isomorphic to the poset of cells in a PL sphere. Otherwise $\mathcal{F}$ is isomorphic to the poset of cells in a PL ball.*

Remark 9.55 The set $\mathcal{D}$ of Proposition 9.54 appeared in Santos (2002) as an example of what was there called a *subdivision*.

Lemma 9.56 *Let $\mathcal{M}$ be a simple acyclic oriented matroid, $\mathcal{D}$ the set of faces of $\mathcal{M}$, and $\mathcal{F} = \{\text{vert}(\sigma) : \sigma \in \mathcal{D}\}$.*

1. *The map $\sigma \to \text{vert}(\sigma)$ from $\mathcal{D}$ to $\mathcal{F}$ is a rank-preserving bijection with inverse map $\tau \to \widehat{\text{conv}}(\tau)$.*
2. *Each element of $\mathcal{F}$ is acyclic, and if $\tau \in \mathcal{F}$ then every face of τ is in $\mathcal{F}$.*
3. *For each $\sigma, \tau \in \mathcal{F}$ we have $\widehat{\text{conv}}(\sigma) \cap \widehat{\text{conv}}(\tau) = \widehat{\text{conv}}(\sigma \cap \tau)$.*

Proof: Each element of $\mathcal{D}$ is a flat, and for each subset S of a flat F we have $\widehat{\text{conv}}(S) \subseteq F$, so for each $\sigma \in \mathcal{D}$ we have $\widehat{\text{conv}}(\text{vert}(\sigma)) \subseteq \sigma$. To see $\sigma \subseteq \widehat{\text{conv}}(\text{vert}(\sigma))$, let $t \in \sigma \backslash \text{vert}(\sigma)$. Then there is no $X \in \mathcal{V}^*(\mathcal{M})$ such that $X(t) = 0$ and $X(\sigma \backslash \{t\}) = +$, and so by the Farkas Property there is a $Y \in \mathcal{V}(\mathcal{M})$ such that $\text{supp}(Y) \subseteq \sigma$ and $+ \in Y(\sigma \backslash \{t\}) \subseteq \{0, +\}$. Since $\mathcal{M}$ is acyclic, we conclude $Y(t) = -$, and so $t \in \widehat{\text{conv}}(\text{vert}(\sigma))$.

It's straightforward to check that for each $\tau \in \mathcal{F}$ we have $\text{vert}(\widehat{\text{conv}}(\tau)) = \tau$. Since the map $\tau \to \widehat{\text{conv}}(\tau)$ is rank-preserving, we have (1).

Because $\mathcal{M}$ itself is acyclic, each element of $\mathcal{F}$ is acyclic. If $\tau \in \mathcal{F}$ and ω is a face of τ then $(\tau \backslash \omega)^+ \in \mathcal{V}^*(\mathcal{M}(\tau))$, and so there is a $Y \in \mathcal{V}^*(\mathcal{M})$ whose restriction to τ is $(\tau \backslash \omega)^+$. For each $w \in \widehat{\text{conv}}(\omega) \backslash \omega$, there is a $\hat{\omega}^+ \{w\}^- \in \mathcal{V}(\mathcal{M})$ with $\hat{\omega} \subseteq \omega$, and since $Y \perp \hat{\omega}^+ \{w\}^-$, we have $Y(w) = 0$. Similarly, for every $t \in \widehat{\text{conv}}(\tau) \backslash (\tau \cup \widehat{\text{conv}}(\omega))$, we have $Y(t) = +$. Thus $Y(\widehat{\text{conv}}(\omega)) = \{0\}$ and $Y(\widehat{\text{conv}}(\tau) \backslash \widehat{\text{conv}}(\omega)) = +$. Since $(E \backslash \widehat{\text{conv}}(\tau))^+ \in \mathcal{V}^*(\mathcal{M})$, we have $(E \backslash \widehat{\text{conv}}(\tau))^+ \circ Y = (E \backslash \widehat{\text{conv}}(\omega) \in \mathcal{V}^*(\mathcal{M})$, and so $\widehat{\text{conv}}(\omega) \in \mathcal{D})$. Thus $\text{vert}(\widehat{\text{conv}}(\omega)) \in \mathcal{F}$. But we saw in (1) that $\text{vert}(\widehat{\text{conv}}(\omega)) = \omega$. Thus every face of τ is in $\mathcal{F}$.

By definition $\widehat{\text{conv}}(\sigma \cap \tau) \subseteq \widehat{\text{conv}}(\sigma) \cap \widehat{\text{conv}}(\tau)$. The inverse map in the proof of (1) showed us that for each $\sigma, \tau \in \mathcal{F}$ there are positive covectors $(E \backslash \widehat{\text{conv}}(\sigma))^+$ and $(E \backslash \widehat{\text{conv}}(\tau))^+$ of $\mathcal{M}$. Their composition is $(E \backslash (\widehat{\text{conv}}(\sigma) \cap \widehat{\text{conv}}(\tau))^+$. Each element of $\widehat{\text{conv}}(\sigma) \cap \widehat{\text{conv}}(\tau)$ is in the 0 set of each factor of this composition, hence is in $\widehat{\text{conv}}(\sigma) \cap \widehat{\text{conv}}(\tau)$. □

Proof of Proposition 9.54: Since $\hat{\mathcal{M}}$ is acyclic, $\hat{\mathcal{M}}$ has at least one face not containing f.

We first check that $\mathcal{F}$ is a strong fan. The first three conditions for a strong fan follow from Lemma 9.56. To see circuit compatibility, let $\alpha^+\beta^- \in \mathcal{V}(\mathcal{M})$ with $\alpha \subseteq \sigma \in \mathcal{F}$ and $\beta \subseteq \tau \in \mathcal{F}$. Then $\alpha^+\beta^- f^\epsilon \in \mathcal{V}(\hat{\mathcal{M}})$ for some $\epsilon \in \{0, +, -\}$. Thus $\alpha^+\beta^- f^\epsilon$ is orthogonal to the covectors $((E\cup\{f\})\backslash\widehat{\mathrm{conv}}(\sigma))^+$ and $((E\cup\{f\})\backslash\widehat{\mathrm{conv}}(\tau))^+$.

If $\alpha^+\beta^- f^\epsilon \cdot ((E\cup\{f\})\backslash\widehat{\mathrm{conv}}(\sigma))^+ = \{0\}$ then $\beta \subseteq \widehat{\mathrm{conv}}(\sigma)$ and $\epsilon = 0$. Otherwise, $\epsilon = +$. Likewise, if $\alpha^+\beta^- f^\epsilon \cdot ((E\cup\{f\})\backslash\widehat{\mathrm{conv}}(\tau))^+ = \{0\}$ then $\alpha \subseteq \widehat{\mathrm{conv}}(\tau)$ and $\epsilon = 0$, and otherwise $\epsilon = -$. Since $\alpha^+\beta^- f^\epsilon$ must be orthogonal to both, we see $\epsilon = 0$ and $\alpha \cup \beta \subseteq \widehat{\mathrm{conv}}(\sigma) \cap \widehat{\mathrm{conv}}(\tau) = \widehat{\mathrm{conv}}(\sigma \cap \tau)$. Also, by Lemma 9.36, $\alpha \cup \beta \subseteq \mathrm{vert}(E)$, so $\alpha \cup \beta \subseteq \mathrm{vert}(\widehat{\mathrm{conv}}(\sigma \cap \tau)) = \sigma \cap \tau$.

Thus $\mathcal{F}$ is a strong fan. Every maximal element of $\mathcal{F}$ spans X^0 for some $X \in \mathcal{C}^*(\hat{\mathcal{M}})$, and so has rank r. Now we consider the final condition, on elements of $\mathcal{F}$ of rank $r - 1$. The set of proper faces of $\hat{\mathcal{M}}$ is the poset of cells in a PL cell decomposition of an r-sphere, so every element σ of $\mathcal{F}$ of rank $r - 1$ is in exactly two elements of this set. If neither of these elements contains f, then they are both in $\mathcal{F}$. Otherwise, one of these elements is the 0 set of a positive $Y \in \mathcal{V}^*(\hat{\mathcal{M}})$ with $\sigma \cup \{f\} \in Y^0$. Thus $Y\backslash\{f\}$ is a positive covector of $\mathcal{M}$, giving us a facet containing σ. □

To show that every oriented matroid has a triangulation, we could just choose an arbitrary acyclic lifting and take stellar subdivisions of the resulting fan decomposition until we have a triangulation. But we'll get one more directly by choosing our lifting so that the resulting fan decomposition is simplicial.

Proof of Proposition 9.42: Let $r = \mathrm{rank}(\mathcal{M})$. For convenience, we assume $[n]$ is the set of elements of $\mathcal{M}$. Let $\mathcal{M}'$ be the oriented matroid on $\{1', \ldots, n'\}$ that is isomorphic to $\mathcal{M}$ by the isomorphism $i \to i'$. Let $\mathcal{M}_0$ be the extension of $\mathcal{M}'$ by a coloop f, and for each $i \in [n]$ let $\mathcal{M}_i = \mathcal{M}_{i-1} \cup e_i$ be the lexicographic extension $\mathcal{M}_{i-1}[(e_i')^+ f^+]$. Let $\hat{\mathcal{M}} = \mathcal{M}_n([n] \cup \{f\})$.

We can see by induction on i that if $Y \in \mathcal{V}^*(\mathcal{M}_i)$ and $Y(f) \neq 0$ then $Y^0 \cap [i]$ is independent. When $i = 0$ this is vacuously true. For larger i, $Y\backslash i \in \mathcal{V}^*(\mathcal{M}_{i-1})$, so the induction hypothesis tells us that $Y^0 \cap [i-1]$ is independent. Since Y is a composition of signed circuits, there is a signed circuit $Z \leq Y$ of $\mathcal{M}_i$ with $Z(f) \neq 0$. Extend $Y^0 \cap [i-1]$ to a maximal independent subset $\{x_1, \ldots, x_r\}$ of Z^0. Since $Z(f) \neq 0$, we have $\chi_{\mathcal{M}_i}(f, x_1, \ldots, x_r) \neq 0$. Thus either $i \in \{x_2, \ldots, x_r\}$ or $\chi_{\mathcal{M}_i}(i, x_1, \ldots, x_r) \neq 0$. Either way, $(Y^0 \cap [i-1]) \cup \{i\}$ is independent.

Since every covector of $\hat{\mathcal{M}}$ is $Y\backslash\{1', \ldots, n'\}$ for some covector Y of $\mathcal{M}_n$, we have that every positive covector X of $\hat{\mathcal{M}}$ with f in its support has X^0

independent. But these sets X^0 are exactly the cells in the induced subdivision of $\mathcal{M}$. We conclude that this subdivision is a triangulation. □

9.6 Superposition

We'll prove most of the results discussed at the end of Section 9.3, as well as several other results, using *superposition*, introduced in Anderson (1999).

To see the motivation, let $\mathcal{G}$ be a fan in a real vector space and C a cone contained in the underlying space of $\mathcal{G}$. Then $\{F \cap C : F \in \mathcal{G}, F \cap C \neq \emptyset\}$ is a fan whose underlying space is C. Each element of this fan arises from two sets of vertices whose convex hulls intersect: Such a thing lends itself to oriented matroid generalization. More generally, if $\mathcal{G}$ and $\mathcal{G}'$ are fans, then $\{F \cap F' : F \in \mathcal{G}, F' \in \mathcal{G}'\}$ is a fan whose underlying space is the intersection of the underlying spaces of $\mathcal{G}$ and $\mathcal{G}'$, which we call the *superposition* of $\mathcal{G}$ and $\mathcal{G}'$.

Definition 9.57 Let $\mathcal{M}$ be an oriented matroid on elements E and $\mathcal{T}_1$ and $\mathcal{T}_2$ be abstract simplicial complexes whose elements are independent subsets of E.

Let $\tilde{E} = \{\tilde{e} : e \in E\}$ be a copy of E, and let the **double** of $\mathcal{M}$ be the extension $\mathcal{M}_D = \mathcal{M} \cup \tilde{E}$ of $\mathcal{M}$ with $\tilde{e}$ parallel to e for each e. For each $\tau \subseteq E$ let $\tilde{\tau} = \{\tilde{e} : e \in \tau\}$. We define the **superposition** $\Sigma_{\mathcal{M}}(\mathcal{T}_1, \mathcal{T}_2)$ to be $\{\sigma^+ \tilde{\tau}^- \in \mathcal{V}(\mathcal{M}_D) \setminus \{0\} : \sigma \in \mathcal{T}_1, \tau \in \mathcal{T}_2\}$.

When context makes the oriented matroid clear we write $\Sigma_{\mathcal{M}}(\mathcal{T}_1, \mathcal{T}_2)$ as $\Sigma(\mathcal{T}_1, \mathcal{T}_2)$.

In our typical examples $\mathcal{T}_2$ will be an oriented matroid triangulation, and so our intuition from the realizable case is that $\Sigma(\mathcal{T}_1, \mathcal{T}_2)$ should be like the poset of non-$\{0\}$ cones in a subdivision of "the underlying space of $\mathcal{T}_1$." In particular, it should be isomorphic to the poset of cells in a PL cell complex that is a subdivision of $\|\mathcal{T}_1\|$. The following two propositions shed some light on how well this intuition plays out.

Proposition 9.58 *Let $\mathcal{M}$ be an oriented matroid on elements E and $\mathcal{T}_1$ and $\mathcal{T}_2$ be abstract simplicial complexes whose elements are independent subsets of E. Then $\Sigma(\mathcal{T}_1, \mathcal{T}_2)$ is an order ideal in $\mathcal{V}(\mathcal{M}_D) \setminus \{0\}$. In particular, $\Sigma(\mathcal{T}_1, \mathcal{T}_2)$ is isomorphic to the poset of cells in a PL cell complex.*

Proof: That $\Sigma(\mathcal{T}_1, \mathcal{T}_2)$ is an order ideal in $\mathcal{V}(\mathcal{M}_D) \setminus \{0\}$ follows immediately from $\mathcal{T}_1$ and $\mathcal{T}_2$ being abstract simplicial complexes. A topological representation of $(\mathcal{M}_D)^*$ gives us a PL cell complex whose poset of cells is isomorphic to $\mathcal{V}(\mathcal{M}_D)\setminus\{0\}$, and this isomorphism identifies $\Sigma(\mathcal{T}_1, \mathcal{T}_2)$ with a subcomplex. □

Remark 9.59 Proposition 9.58 is why we have focused on simplicial objects – triangulations and simplicial fans – rather than more general convex objects. Without the assumption that $\mathcal{T}_1$ and $\mathcal{T}_2$ are abstract simplicial complexes, $\Sigma(\mathcal{T}_1, \mathcal{T}_2)$ need not be an order ideal in $\mathcal{V}(\mathcal{M}_D)\setminus\{0\}$.

Remark 9.60 Our use of $\mathcal{M}_D$ is reminiscent of the Lawrence construction (Section 8.4). The sequence of oriented matroids considered in the Lawrence construction was

1. the original oriented matroid $\mathcal{M}$,
2. its dual $\mathcal{M}^*$,
3. the oriented matroid obtained from $\mathcal{M}^*$ by adding an *antiparallel* copy of each element, and
4. the dual $\mathcal{M}_L$ of this oriented matroid.

We then worked with $\mathcal{V}^*(\mathcal{M}_L)$. In our current construction we begin at the second stage of this sequence and add parallel rather than antiparallel elements. As in the last stage of the Lawrence construction, we draw conclusions based on the covector set of the dual $\mathcal{M}_D^*$.

Notation 9.61 In our discussion of superposition we'll assume some topological realization of $(\mathcal{M}_D)^*$ has been fixed, so that $\Sigma(\mathcal{T}_1, \mathcal{T}_2)$ indexes the cells in a fixed cell complex given by that realization. We'll denote the cell indexed by $\sigma^+\tilde{\tau}^-$ by $\sigma|\cap|\tau$. We'll denote the underlying space of this cell complex by $\|\Sigma(\mathcal{T}_1, \mathcal{T}_2)\|$.

The notation $\sigma|\cap|\tau$ is intended to suggest the intuition from the realizable case, in which it represents the intersection of two geometric simplices, without asserting an actual intersection of geometric objects in our generalization.

The following tells us the dimension of a cell $\sigma|\cap|\tau$. This dimension is one less than the rank of $\sigma^+\tilde{\tau}^-$ in $\mathcal{V}^*(\mathcal{M}_D^*) = \mathcal{V}(\mathcal{M}_D)$.

Proposition 9.62 *If* $\sigma^+\tilde{\tau}^- \in \mathcal{V}(\mathcal{M}_D)$ *then its rank in* $\mathcal{V}(\mathcal{M}_D)$ *is* $|\sigma \cup \tilde{\tau}| - \mathrm{rank}_\mathcal{M}(\sigma \cup \tau)$.

Proof: First, note that $\mathrm{rank}_\mathcal{M}(\sigma \cup \tau) = \mathrm{rank}_\mathcal{M}(\sigma \cup \tilde{\tau})$, since replacing an element e of a set with an element parallel to e doesn't change the rank. So take a maximal independent subset I of $\sigma \cup \tilde{\tau}$, and let $(\sigma \cup \tilde{\tau})\setminus I = \{x_1, \ldots, x_k\}$. For each $j \in [k]$ let X_j be a signed circuit with support contained in $I \cup \{j\}$. Then $0, X_1, X_1 \circ X_2, \ldots, X_1 \circ \cdots \circ X_k$ is a maximal chain in $\mathcal{V}(\mathcal{M}_D)(\sigma \cup \tilde{\tau})$.

9.6 Superposition

Since $\sigma^+ \tilde{\tau}^-$ is a maximal element of $\mathcal{V}(\mathcal{M}_D)(\sigma \cup \tilde{\tau})$ and covector sets are pure posets, the result follows. □

Our first goal is to understand $\Sigma(\mathcal{S}, \mathcal{T})$ when $\mathcal{S}$ is a 1-simplex and $\mathcal{T}$ is a partial recursive triangulation.

Notation 9.63 If σ is a finite set then $\mathcal{S}_\sigma$ denotes the simplicial complex whose unique maximal element is σ.

Recall from Section 6.6.4 that each independent set $\{e_0, e_\infty\}$ of elements of an oriented matroid $\mathcal{N}$ defines an oriented matroid program $(\mathcal{N}, e_\infty, e_0)$, and such a program induces orientations on some of the 1-cells in the cell decomposition of a sphere given by a topological representation of $\mathcal{N}$. Now consider $\{a, b\}$ independent in $\mathcal{M}$, and $\mathcal{T}$ a partial recursive triangulation of $\mathcal{M}$. Our intuition from the realizable case suggests that $\|\Sigma(\mathcal{S}_{\{a,b\}}, \mathcal{T})\|$ should be a one-dimensional ball, which we could choose to orient from an end $a|\cap|\sigma_a$ to an end $b|\cap|\sigma_b$. Perhaps this orientation on 1-cells in $\|\Sigma(\mathcal{S}_{\{a,b\}}, \mathcal{T})\|$ coincides with the partial orientation coming from the program $(\mathcal{M}_D^*, a, b)$? The next proposition explores this intuition. Its proof is work, but with the result in hand we'll be able to easily prove some of the results we've hoped for.

Proposition 9.64 (Anderson (1999)) *Let $\mathcal{M}$ be an oriented matroid, $\mathcal{T}$ a partial recursive triangulation of $\mathcal{M}$, and $\{a, b\}$ independent in $\mathcal{M}$.*

$\|\Sigma(\mathcal{S}_{\{a,b\}}, \mathcal{T})\|$ *is a one-dimensional manifold with boundary. The oriented matroid program* $((\mathcal{M}_D)^*, a, b)$ *induces compatible orientations on all 1-cells in* $\|\Sigma(\mathcal{S}_{\{a,b\}}, \mathcal{T})\|$. *(That is, whenever a 0-cell is contained in two 1-cells, one of these 1-cells is directed toward the 0-cell while the other is directed away.) Exactly one connected component of* $\|\Sigma(\mathcal{S}_{\{a,b\}}, \mathcal{T})\|$ *is a one-dimensional ball directed from an endpoint $a|\cap|\sigma_a$ to an endpoint $b|\cap|\sigma_b$, for some $\sigma_a, \sigma_b \in \mathcal{T}$. The remaining connected components are circles in which each 0-cell has the form $\{a, b\}|\cap|\omega$.*

In particular, there is exactly one $\sigma_a \in \mathcal{T}$ such that $a|\cap|\sigma_a \in \|\Sigma(\mathcal{S}_{\{a,b\}}, \mathcal{T})\|$, and every circle in $\|\Sigma(\mathcal{S}_{\{a,b\}}, \mathcal{T})\|$ is the underlying space of a non-Euclidean cycle of $((\mathcal{M}_D)^*, a, b)$.

Proof: We induct on rank($\mathcal{M}$), with the rank 1 case being vacuous.

For future reference, observe that if $\omega \subset E$ then each element of ω is a loop in $(\mathcal{M}_D)/\tilde{\omega}$, and $(\mathcal{M}/\tilde{\omega})_D = ((\mathcal{M}_D)/\tilde{\omega})\backslash\omega$. So $\mathcal{M}_D/\tilde{\omega}$, $(\mathcal{M}/\tilde{\omega})_D$, and $(\mathcal{M}_D/\tilde{\omega})\backslash\omega$ coincide on nonloops, and we need not be careful to distinguish them.

Assuming the induction hypothesis, we first show that every element of $\|\Sigma(\mathcal{S}_{\{a,b\}}, \mathcal{T})\|$ of the form $a|\cap|\sigma$ is a minimal element that is less than exactly one maximal element. Assume by way of contradiction that $a|\cap|\sigma$ is not minimal. Every smaller element has the form $a|\cap|\tau$ with $\tau \subset \sigma$. Eliminating a between $a^+\tilde{\sigma}^-$ and $\tilde{\tau}^+a^-$ gives a signed circuit supported on σ, contradicting that σ is independent.

Now that we know $a|\cap|\sigma$ is minimal, we consider two cases.

- If σ is a basis, then there is a signed circuit Y with support contained in $\tilde{\sigma} \cup \{b\}$ with $Y(b) = +$, and composing $a^+\tilde{\sigma}^-$ with Y shows that $\{a,b\}^+\tilde{\sigma}^- \in \Sigma(\mathcal{S}_{\{a,b\}}, \mathcal{T})$. Since σ is a maximal element of $\mathcal{T}$, this is the unique maximal element greater than $a^+\tilde{\sigma}^-$.
- If σ is not a basis, then by the induction hypothesis $\Sigma_{\mathcal{M}/\sigma}(\mathcal{S}_{\{b\}}, \text{link}(\sigma))$ contains a single element $b^+\tilde{\tau}^-$. This is the restriction of a signed circuit Y of $\mathcal{M}_D$. Composing the signed circuit $a^+\tilde{\sigma}^-$ with Y, we get $\{a,b\}^+(\tilde{\sigma} \cup \tilde{\tau})^- \in \Sigma(\mathcal{S}_{\{a,b\}}, \mathcal{T})$. Conversely, for each independent $\sigma' \supseteq \sigma$, $\pm a^+\tilde{\sigma}^-$ are the only signed circuits with support contained in $\{a\} \cup \tilde{\sigma}'$. Thus every element of $\Sigma(\mathcal{S}_{\{a,b\}}, \mathcal{T})_{>a^+\tilde{\sigma}^-}$ must have the form $\{a,b\}^+(\tilde{\sigma} \cup \tilde{\omega})^- \in \Sigma(\mathcal{S}_{\{a,b\}}, \mathcal{T})$. To see that ω must be τ, note that $a^+ \in \mathcal{V}(\mathcal{M}/\sigma)$ and $\{a,b\}^+\tilde{\omega}^- \in \mathcal{V}(\mathcal{M}/\sigma)$. Thus $b^+\tilde{\omega}^- \in \mathcal{V}(\mathcal{M}/\sigma)$, and so by uniqueness of τ we have $\omega = \tau$.

Similarly, every element of $\|\Sigma(\mathcal{S}_{\{a,b\}}, \mathcal{T})\|$ of the form $b|\cap|\sigma$ is a minimal element that is less than exactly one maximal element.

Next we show that every minimal element of $\Sigma(\mathcal{S}_{\{a,b\}}, \mathcal{T})$ of the form $\{a,b\}^+\tilde{\omega}^-$ is in exactly two maximal elements. By the induction hypothesis there is a unique $\nu_a \in \text{link}(\omega)$ such that $a^+\tilde{\nu}_a^- \in \mathcal{V}((\mathcal{M}/\omega)_D)$, and $a^+\tilde{\nu}_a^-$ is the restriction of some $Y \in \mathcal{V}(\mathcal{M}_D)$ with support contained in $\{a\} \cup \tilde{\nu}_a \cup \tilde{\omega}$. The composition $\{a,b\}^+\tilde{\omega}^- \circ Y = \{a,b\}^+(\tilde{\omega} \cup \tilde{\nu}_a)^-$ is one element of $\Sigma(\mathcal{S}_{\{a,b\}}, \mathcal{T})$ that is greater than $\{a,b\}^+\tilde{\omega}^-$. Similarly we get an element $\{a,b\}^+(\tilde{\omega} \cup \tilde{\nu}_b)^-$. To see that $\omega \cup \nu_a$ and $\omega \cup \nu_b$ are distinct, assume by way of contradiction otherwise. In $\mathcal{M}_D/\tilde{\omega}$ we can eliminate a between $\tilde{\nu}_a^+a^-$ and $\{a,b\}^+$ to get $(\tilde{\nu}_a \cup \{b\})^+ = (\tilde{\nu}_b \cup \{b\})^+$. Eliminating b between $(\tilde{\nu}_b \cup \{b\})^+$ and $\tilde{\nu}_b^+b^-$ shows $\tilde{\nu}_b$ dependent in $\mathcal{M}_D/\tilde{\omega}$, and so $\nu_b \cup \omega$ is dependent, contradicting that $\nu_b \cup \omega$ is an element of $\mathcal{T}$.

Assume by way of contradiction that there is one more element $\{a,b\}^+(\tilde{\omega} \cup \tilde{\alpha})^-$ with $\alpha \neq \emptyset$. Then $\{a,b\}^+\tilde{\alpha}^- \in \mathcal{V}(\mathcal{M}_D/\tilde{\omega})$. Since this element is a composition of conformal signed circuits of $\mathcal{M}_D/\tilde{\omega}$, $\{a,b\}^+$ is a signed circuit of $\mathcal{M}_D/\tilde{\omega}$, and $\tilde{\alpha}$ is independent in $\mathcal{M}_D/\tilde{\omega}$, there must be a conformal signed circuit of $\mathcal{M}_D/\tilde{\omega}$ of the form $\{a\}^+\tilde{\alpha}_a^-$ or of the form $\{b\}^+\tilde{\alpha}_b^-$. Our induction hypothesis tells us that $\alpha_a = \nu_a$ and $\alpha_b = \nu_b$. Since $\alpha \notin \{\nu_a, \nu_b\}$, we must have both $\{a\}^+\tilde{\alpha}_a^-$ and $\{b\}^+\tilde{\alpha}_b^-$ signed circuits of $\mathcal{M}_D/\tilde{\omega}$, and $\alpha = \nu_a \cup \nu_b$. Eliminating a between $\{a,b\}^-$ and $a^+\tilde{\alpha}_a^-$, we see $(\tilde{\alpha}_a \cup \{b\})^- \in \mathcal{V}(\mathcal{M}_D/\tilde{\omega})$.

9.6 Superposition

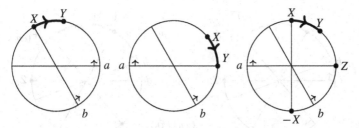

Figure 9.3 Orientations on 1-cell $\{a,b\}|\cap|\sigma$ induced by $((\mathcal{M}_D)^*, a, b)$.

Then eliminating b between $(\tilde{\alpha}_a \cup \{b\})^-$ and $b^+ \tilde{\alpha}_b^-$, we see $\alpha_a \cup \alpha_b$ is dependent in $\mathcal{M}_D/\tilde{\omega}$, and so $\alpha \cup \omega$ is dependent in $\mathcal{M}$. But $\alpha \cup \omega \in \mathcal{T}$, a contradiction.

We have now established that every minimal element of $\Sigma(\mathcal{S}_{\{a,b\}}, \mathcal{T})$ is contained in exactly one or two maximal elements, and thus that $\|\Sigma(\mathcal{S}_{\{a,b\}}, \mathcal{T})\|$ is a one-dimensional manifold with boundary. Further, we have shown that points on the boundary can only have the forms $a|\cap|\sigma$ or $b|\cap|\sigma$. Now we consider the orientations on 1-cells arising from the oriented matroid program $((\mathcal{M}_D)^*, a, b)$. Let $((S_e, S_e^+, S_e^-) : e \in E \cup \tilde{E})$ be our topological representation of $(\mathcal{M}_D)^*$. First let's review how a 1-cell $\{a,b\}|\cap|\sigma$ is oriented by this program. Let $P = \bigcap_{e \notin \{a,b\} \cup \sigma} S_e$ be the pseudocircle containing $\{a,b\}|\cap|\sigma$. Since σ is independent in $\mathcal{M}_D$, $(\mathcal{M}_D)^*$ has no signed cocircuit with support contained in σ, and so S_a and S_b intersect P in distinct points. Thus $((\mathcal{M}_D)^*, a, b)$ does indeed induce an orientation on $\{a,b\}|\cap|\sigma$. Let X and Y be the signed circuits of $\mathcal{M}_D$ corresponding to the endpoints of $\{a,b\}|\cap|\sigma$. Then $\{a,b\}|\cap|\sigma$ is oriented from the X endpoint to the Y endpoint if one of the following holds (see Figure 9.3):

- $X(b) = 0$
- $Y(a) = 0$
- $X(a) = Y(a) = +$ and the elimination Z of a between $-X$ and Y satisfies $Z(b) = +$.

In the first case, X is a boundary point of $\|\Sigma(\mathcal{S}_{\{a,b\}}, \mathcal{T})\|$ of the form $a|\cap|\sigma_a$, and in the second case Y is a boundary point of $\|\Sigma(\mathcal{S}_{\{a,b\}}, \mathcal{T})\|$ of the form $b|\cap|\sigma_b$. Thus we see the orientations on edges incident to boundary points of $\|\Sigma(\mathcal{S}_{\{a,b\}}, \mathcal{T})\|$. Now on to the third case.

Consider a 0-cell of the form $\{a,b\}|\cap|\omega$. As we have seen, the 1-cells in $\|\Sigma(\mathcal{S}_{\{a,b\}}, \mathcal{T})\|$ incident to $\{a,b\}|\cap|\omega$ are $\{a,b\}|\cap|(\omega \cup \tau_a)$ and $\{a,b\}|\cap|(\omega \cup \tau_b)$, where $a^+ \tilde{\tau}_a^-$ and $b^+ \tilde{\tau}_b^-$ are vectors of $(\mathcal{M}_D)/\omega$. We'll show that $\{a,b\}|\cap|(\omega \cup \tau_a)$ is oriented toward $\{a,b\}|\cap|\omega$ while the 1-cell $\{a,b\}|\cap|(\omega \cup \tau_b)$ is oriented away from $\{a,b\}|\cap|\omega$.

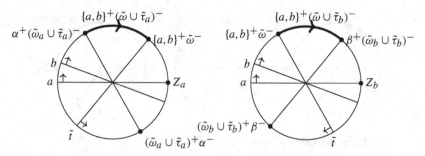

Figure 9.4 Orienting the 1-cells $\{a,b\}|\cap|(\omega \cup \tau_a)$ and $\{a,b\}|\cap|(\omega \cup \tau_b)$.

We first find the orientation on the cell $\{a,b\}|\cap|(\omega \cup \tau_a)$. The other endpoint of $\{a,b\}|\cap|(\omega \cup \tau_a)$ is $\alpha|\cap|(\omega_a \cup \tau_a)$, where $a \in \alpha \subseteq \{a,b\}$ and $\omega_a \subseteq \omega$. Eliminating a between the vectors $\tilde{\tau}_a^+ a^-$ and $\{a,b\}^+$ of $(\mathcal{M}_D)/\omega$, we see that $(\tilde{\tau}_a \cup \{b\})^+$ is a vector of $(\mathcal{M}_D)/\tilde{\omega}$, which is the restriction of a vector Z_a of $\mathcal{M}_D$ with support contained in $\tilde{\omega} \cup \tilde{\tau}_a \cup \{b\}$. Since $\omega \cup \tau_a$ is independent, this Z_a is the unique signed circuit with support contained in $\tilde{\omega} \cup \tilde{\tau}_a \cup \{b\}$ such that $Z_a(t) = +$ for some $t \in \tilde{\tau}_a$. Thus it's the elimination of a between $(\tilde{\omega}_a \cup \tilde{\tau}_a)^+ \alpha^-$ and $\{a,b\}^+ \tilde{\omega}^-$. This tells us that this cell is oriented from $\alpha_1 |\cap|(\omega_a \cup \tau_a)$ to $\{a,b\}|\cap|\omega$. (See the left-hand side of Figure 9.4.)

Now we find the orientation on the cell $\{a,b\}|\cap|(\omega \cup \tau_b)$. The other endpoint of $\{a,b\}|\cap|(\omega \cup \tau_a)$ is $\beta|\cap|(\omega_b \cup \tau_b)$, where $b \in \beta \subseteq \{a,b\}$ and $\omega_b \subseteq \omega$. From the signed circuit $\tilde{\tau}_b^- b^+$ of $(\mathcal{M}_D)/\omega$ we get a signed circuit Z_b of $\mathcal{M}_D$ with support contained in $\tilde{\omega} \cup \tilde{\tau}_b \cup \{b\}$ such that $Z_b(\tilde{\tau}_b) = -$ and $Z_b(b) = +$. This Z_b is the elimination of a between $(\tilde{\omega}_b \cup \tilde{\tau}_b)^- \{a,b\}^+$ and $\tilde{\omega}^+ \{a,b\}^-$. The right-hand side of Figure 9.4 shows the resulting picture in the pseudocircle containing the cell corresponding to $(\tilde{\omega} \cup \tilde{\tau}_b)^- \{a,b\}^+$, from which we see that this cell is oriented from $\{a,b\}|\cap|\omega$ to $\beta|\cap|(\omega_b \cup \tau_b)^-$.

It remains to show that $\|\Sigma(\mathcal{S}_{\{a,b\}}, \mathcal{T})\|$ has exactly one connected component that is a 1-ball. We've seen that every 1-ball component is oriented with a source of the form $a|\cap|\sigma_a$ and a sink of the form $b|\cap|\sigma_b$, and so these must be the endpoints of that component. We've also seen that no element of the form $a|\cap|\sigma_a$ is in a circle connected component. Thus it's enough to show that $\|\Sigma(\mathcal{S}_{\{a,b\}}, \mathcal{T})\|$ has exactly one element of the form $a|\cap|\sigma_a$. Let $t \in \mathcal{T}^0$ such that $\{a,t\}$ is independent and consider $\Sigma(\mathcal{S}_{\{a,t\}}, \mathcal{T})$. The definition of triangulation tells us that $t^+ \tilde{t}^-$ is the unique element of $\Sigma(\mathcal{S}_{\{a,t\}}, \mathcal{T})$ of the form $t^+ \tau^-$, and so $t|\cap|t$ is the endpoint of a 1-ball connected component of $\|\Sigma(\mathcal{S}_{\{a,t\}}, \mathcal{T})\|$. The other endpoint is $a|\cap|\sigma_a$ for some $\sigma_a \in \mathcal{T}$. Our orientation conclusion tells us that each 1-ball connected component of $\|\Sigma(\mathcal{S}_{\{a,t\}}, \mathcal{T})\|$ has one end of the form $t|\cap|\sigma_t$. Since t itself is the only σ_t possible, we see that

9.6 Superposition

$\|\Sigma(\mathcal{S}_{\{a,t\}}, \mathcal{T})\|$ has exactly one 1-ball connected component, and so σ_a is the unique element of $\mathcal{T}$ with $a^+\tilde{\sigma}_a^- \in \mathcal{V}(\mathcal{M}_D)$. □

Now we can quickly prove a number of facts about triangulations, beginning with Proposition 9.46.

Lemma 9.65 *1. If $\mathcal{T}$ is a triangulation of $\mathcal{M}$ and $\sigma \in \mathcal{T}$ then $\mathrm{link}(\sigma)$ is triangulation of $\mathcal{M}/\sigma$.*

2. Let $\mathcal{T}$ be a simplicial fan on $\mathcal{M}$ with the property that if $\mathcal{M} \cup f$ is an extension of $\mathcal{M}$ with $f \in \widehat{\mathrm{conv}}(E)$ then there is a unique $\tau \in \mathcal{T}$ such that $\tau^+ f^- \in \mathcal{V}(\mathcal{M} \cup f)$. Then for each $\sigma \in \mathcal{T}$, $\mathrm{link}(\sigma)$ has a similar property with respect to $\mathcal{M}/\sigma$.

Proof: 1. By induction it's enough to prove the result when σ is a singleton $\{s\}$. Every condition on a triangulation except extension compatibility is immediate.

To see extension compatibility, let $(\mathcal{M}/s) \cup f$ be an extension and $\omega \in \mathrm{link}(\{s\})$ such that $\omega^+ f^- \in \mathcal{V}((\mathcal{M}/s) \cup f)$. By the result of Exercise 6.4 there is an extension $\mathcal{M} \cup f$ of $\mathcal{M}$ such that $(\mathcal{M}/s) \cup f = (\mathcal{M} \cup f)/s$ and $\mathcal{M} \cup f$ weak maps to the extension $\mathcal{N}$ of $\mathcal{M}$ by an element f parallel to s. So $\mathcal{M} \cup f$ has a vector X with $\mathrm{supp}(X) \subseteq \omega \cup \{f, s\}$ and $X \backslash s = \omega^+ f^-$. The weak map condition tells us that $X \geq Y$ for some signed circuit Y of $\mathcal{N}$. Since both $\omega \cup \{s\}$ and $\omega \cup \{f\}$ are independent in $\mathcal{N}$, we see $Y = s^+ f^-$, and so $X = (\omega \cup \{s\})^+ f^-$. Extension compatibility of $\mathcal{T}$ then tells us that there can only be one such ω in $\mathrm{link}(\{s\})$.

The proof of (2) is similar. □

Proof of Proposition 9.46: Lemma 9.65 tells us that (1) implies (2) and (3) implies (2).

Let $\mathcal{T}$ be a partial recursive triangulation of $\mathcal{M}$. Let ω be a maximal simplex of $\mathcal{T}$. If the size of ω were less than $\mathrm{rank}(\mathcal{M})$ then $\emptyset = \mathrm{link}(\omega)$ would be a triangulation of $\mathcal{M}/\omega$, contradicting the definition. Additionally, consider an extension $\mathcal{M} \cup f$ of $\mathcal{M}$ with $f \in \widehat{\mathrm{conv}}(E)$. Then $\mathcal{T}$ is a partial recursive triangulation of $\mathcal{M} \cup f$, and Proposition 9.64 tells us that there is a unique σ in $\mathcal{T}$ such that $\sigma^+ f^-$ is a circuit of $\mathcal{M} \cup f$. This tells us that (2) implies (3) and that (2) implies that $\mathcal{T}$ is a simplicial fan. To complete the proof that (2) implies (1), consider $\sigma \in \mathcal{T}$ of rank $r - 1$. Then $\mathrm{link}(\sigma)$ is a partial recursive triangulation of the rank 1 oriented matroid $\mathcal{M}/\sigma$. If σ is not contained in a facet of $\mathcal{M}$ then $\mathcal{M}/\sigma$ is totally cyclic, and so $\mathrm{link}(\sigma)$ consists of two 0-simplices $\{a\}$ and $\{b\}$. Thus $\sigma \cup \{a\}$ and $\sigma \cup \{b\}$ are the two maximal simplices containing σ. □

Corollary 9.66 *Let $\mathcal{T}$ be a triangulation of $\mathcal{M}$. For every two maximal elements v, ω of $\mathcal{T}$ there is a sequence $v = \tau_1, \ldots, \tau_k = \omega$ of maximal elements such that τ_i and τ_{i+1} share a codimension 1 face for each i.*

Proof: Using lexicographic extensions we can get an oriented matroid $\mathcal{M} \cup \{a, b\}$ with $a \in \widehat{\text{conv}}(v)$, $b \in \widehat{\text{conv}}(\omega)$, and with both a and b in general position. Thus every maximal element of $\Sigma(\mathcal{S}_{\{a,b\}}, \mathcal{T})$ is $\{a,b\}^+\tilde{\tau}^-$ for some maximal element τ of $\mathcal{T}$, and every minimal element is $\{a,b\}^+\tilde{\sigma}^-$ for some codimension 1 element σ of $\mathcal{T}$. The 1-ball in $\|\Sigma(\mathcal{S}_{\{a,b\}}, \mathcal{T})\|$ consists of 1-cells giving our sequence $\tau_1, \ldots, \tau_k$. □

We now know that triangulations are the same thing as recursive triangulations: We use this to prove that strong triangulations are triangulations.

Proof of Proposition 9.47: Let $\mathcal{T}$ be a strong triangulation of $\mathcal{M}$, and assume by way of contradiction that $\mathcal{T}$ is not a recursive triangulation. Then there are elements $\omega \cup \sigma, \omega \cup \{t\}$ of $\mathcal{T}$ such that $\sigma^+t^- \in \mathcal{V}(\mathcal{M}/\omega)$. This σ^+t^- must be the restriction of a vector of $\mathcal{M}$, and so there are disjoint subsets ω_+, ω_- of ω such that $(\sigma \cup \omega_+)^+(\{t\} \cup \omega_-)^- \in \mathcal{V}(\mathcal{M})$. But $\sigma \cup \omega_+$ and $\{t\} \cup \omega_-$ are both elements of $\mathcal{T}$, contradicting the definition of strong triangulation. □

We can also use Proposition 9.64 to prove a generalization.

Proposition 9.67 *Let $\mathcal{M}$ be an oriented matroid, $\mathcal{T}$ a partial triangulation of $\mathcal{M}$, and σ independent in $\mathcal{M}$.*

1. *Every maximal element of $\|\Sigma(\mathcal{S}_\sigma, \mathcal{T})\|$ has the form $\sigma|\cap|\tau$.*
2. *Every codimension 1 element of $\|\Sigma(\mathcal{S}_\sigma, \mathcal{T})\|$ of the form $\sigma|\cap|\tau$ is in exactly two maximal elements.*
3. *Every remaining codimension 1 element of $\|\Sigma(\mathcal{S}_\sigma, \mathcal{T})\|$ has the form $(\sigma \setminus \{s\})|\cap|\tau$ for some $s \in \sigma$. Such an element is contained in only one maximal element.*

Proof: We induct on $\text{rank}(\mathcal{M})$, and within this on $|\sigma|$. When $\text{rank}(\mathcal{M}) \leq 2$ then the results follow from realizability of $\mathcal{M}$. When $|\sigma| = 1$ a maximal element of $\Sigma(\mathcal{S}_\sigma, \mathcal{T})$ is a signed circuit $\sigma^+\tau^-$ of $\mathcal{V}(\mathcal{M}_D)$ and there are no codimension 1 elements of $\|\Sigma(\mathcal{S}_\sigma, \mathcal{T})\|$, and when $|\sigma| = 2$ the result follows from Proposition 9.64.

Above this, to see the first statement we consider a maximal element $\hat{\sigma}|\cap|\tau$ of $\|\Sigma(\mathcal{S}_\sigma, \mathcal{T})\|$ and assume by way of contradiction that $\hat{\sigma} \subset \sigma$. Let $s \in \sigma \setminus \hat{\sigma}$. There is no $X \in \mathcal{V}(\mathcal{M}_D)$ with $\text{supp}(X) \subseteq \{s\} \cup \tilde{\tau}$, since otherwise we could assume $X(s) = +$ and find that $(\hat{\sigma}^+\tilde{\tau}^-) \circ X = (\hat{\sigma} \cup \{s\})^+\tilde{\tau}^- \in \mathcal{V}(\mathcal{M}_D)$,

9.6 Superposition

contradicting maximality of $\hat{\sigma}|\cap|\tau$. Hence s is a nonloop in $\mathcal{M}_D/\tilde{\tau}$. Since $\text{link}_\mathcal{T}(\tau)$ is a partial recursive triangulation of $\mathcal{M}_D/\tilde{\tau}$, by Proposition 9.64 there is an $\omega \in \text{link}_\mathcal{T}(\tau)$ such that $\{s\}^+\tilde{\omega}^- \in \mathcal{V}(\mathcal{M}_D/\tilde{\tau})$. This is the restriction of some $Y \in \mathcal{V}(\mathcal{M}_D)$ with support contained in $\{s\} \cup \tilde{\omega} \cup \tilde{\tau}$, and the composition $(\hat{\sigma}^+\tilde{\tau}^-) \circ Y = (\hat{\sigma} \cup \{s\})^+(\tilde{\tau} \cup \tilde{\omega})^-$ gives us an element $(\hat{\sigma} \cup \{s\})|\cap|(\tau \cup \omega)$ contradicting maximality of $\hat{\sigma}|\cap|\tau$. This proves the first statement.

To see the second statement, consider $\sigma|\cap|\tau$ of codimension 1. It's less than some $\sigma|\cap|(\tau \cup \omega) \in \|\Sigma(\mathcal{S}_\sigma, \mathcal{T})\|$. Thus in $\mathcal{V}(\mathcal{M}_D/\tilde{\tau})$ we have an element $\sigma^+\tilde{\omega}^-$. Since $\tilde{\tau} \cup \tilde{\omega}$ is independent in $\mathcal{M}_D$, we know that $\tilde{\omega}$ is independent in $\mathcal{M}_D/\tilde{\tau}$, and so at least one element of σ is a nonloop in $\mathcal{M}_D/\tilde{\tau}$.

We claim that every pair s_1, s_2 of elements of σ that are nonloops in $\mathcal{M}_D/\tilde{\tau}$ are either parallel or antiparallel in $\mathcal{M}_D/\tilde{\tau}$. To see this, assume by way of contradiction that $\{s_1, s_2\}$ is independent in $\mathcal{M}_D/\tilde{\tau}$ and consider the cell complex $\|\Sigma_{\mathcal{M}/\tau}(\mathcal{S}_{\{s_1, s_2\}}, \text{link}(\tau))\|$. Proposition 9.64 tells us that exactly one connected component of this cell complex is a 1-ball, and one endpoint of the 1-ball is $\{s_1\}|\cap|v_1$ for some $v_1 \in \text{link}(\tau)$. This endpoint is contained in a 1-cell $\{s_1, s_2\}|\cap|v_2$. Thus we have $X_1, X_2 \in \mathcal{V}(\mathcal{M}_D)$ with $\text{supp}(X_i) \subseteq \{s_1, s_2\} \cup \tilde{v}_i \cup \tilde{\tau}$, with $X_i(\{s_1, s_2\}) = +$, and with $X_i(\tilde{v}_i) = -$ for each i. The composition $\sigma^+\tilde{\tau}^- \circ X_i$ is $\sigma^+(v_i \cup \tilde{\tau})^- \in \Sigma(\mathcal{S}_\sigma, \mathcal{T})$. Thus we have a chain $\sigma^+\tilde{\tau}^- < \sigma^+(v_1 \cup \tilde{\tau})^- < \sigma^+(v_2 \cup \tilde{\tau})^-$ in $\Sigma(\mathcal{S}_\sigma, \mathcal{T})$, contradicting that $\sigma|\cap|\tau$ has codimension 1.

At the same time, the elements of σ that are nonloops in $\mathcal{M}_D/\tilde{\tau}$ can't all be parallel, since $\sigma^+ \in \mathcal{V}(\mathcal{M}_D/\tilde{\tau})$. Thus the set of elements of σ that are nonloops in $\mathcal{M}_D/\tilde{\tau}$ partitions into two nonempty sets σ_1, σ_2 with each element of σ_1 antiparallel to each element of σ_2. By Proposition 9.64, there is a unique $\omega_1 \in \text{link}(\tau)$ such that $\{s\}^+\tilde{\omega}_1^- \in \mathcal{V}(\mathcal{M}_D/\tilde{\tau})$ for each element s of σ_1, and likewise there is a unique $\omega_2 \in \text{link}(\tau)$ such that $\{s\}^+\tilde{\omega}_2^- \in \mathcal{V}(\mathcal{M}_D/\tilde{\tau})$ for each element s of σ_2. These two elements of $\mathcal{V}(\mathcal{M}_D/\tilde{\tau})$ are restrictions of elements Y_1, Y_2 of $\mathcal{V}(\mathcal{M}_D)$, and the compositions $\sigma^+\tilde{\tau}^- \circ Y_1 = \sigma^+(\tilde{\tau} \cup \tilde{\omega}_1)^-$, $\sigma^+\tilde{\tau}^- \circ Y_2 = \sigma^+(\tilde{\tau} \cup \tilde{\omega}_2)^-$ are elements of $\Sigma(\mathcal{S}_\sigma, \mathcal{T})$ that are greater than $\sigma^+\tilde{\tau}^-$. To see that they are distinct, assume by way of contradiction that $\omega_1 = \omega_2$. Then in $\mathcal{M}_D/\tilde{\tau}$, we can eliminate between the vectors $\{s_1\}^+\tilde{\omega}_1^-$, $\{s_2\}^+\tilde{\omega}_1^-$, and $\{s_1, s_2\}^+$ to see that ω_1 is dependent, a contradiction. We conclude that $\sigma|\cap|\tau$ is in exactly two maximal cells.

The third statement applies the same ideas in a shorter argument. If $\hat{\sigma} \subset \sigma$, $\hat{\sigma}^+\tilde{\tau}^- \in \Sigma(\mathcal{S}_\sigma, \mathcal{T})$, and $s \in \sigma \backslash \hat{\sigma}$, then either $(\hat{\sigma} \cup \{s\})^+\tilde{\tau}^- \in \Sigma(\mathcal{S}_\sigma, \mathcal{T})$ or s is a nonloop in $\mathcal{M}_D/\tilde{\tau}$, and applying Proposition 9.64 to s and $\text{link}(\tau)$ leads to a unique element of $\Sigma(\mathcal{S}_\sigma, \mathcal{T})$ of the form $(\hat{\sigma} \cup \{s\})^+(\tilde{\tau} \cup \tilde{\omega})^-$. If $\hat{\sigma}|\cap|\tau$ is codimension 1, this element must be $\sigma^+\tilde{\tau} \cup \tilde{\omega})^-$, and so $\hat{\sigma} = \sigma \backslash \{s\}$. □

9.7 Triangulations of Euclidean Oriented Matroids

The aim of this section is to prove the following.

Theorem 9.68 (Anderson (1999)) *Let $\mathcal{M}$ be Euclidean, $\mathcal{S}$ a simplicial fan on $\mathcal{M}$, and $\mathcal{T}$ a triangulation of $\mathcal{M}$. There is a PL homeomorphism $\|\mathcal{S}\| \to \|\Sigma(\mathcal{S},\mathcal{T})\|$, taking each simplex $\|\sigma\|$ to the union of all cells of the form $\sigma|\cap|\tau$.*

Theorem 9.44 is a corollary of this: For triangulations $\mathcal{T}_1$ and $\mathcal{T}_2$ of $\mathcal{M}$, $\Sigma(\mathcal{T}_1,\mathcal{T}_2) \cong \Sigma(\mathcal{T}_2,\mathcal{T}_1)$, so Theorem 9.68 tells us that $\|\mathcal{T}_1\|$ and $\|\mathcal{T}_2\|$ have isomorphic PL cell decompositions.

We use another idea from PL topology, *elementary collapse*. Consider a pair (α,β) of cells in a PL cell complex P such that α is the only element of P containing β. In particular, β is a codimension 1 face of P, and there is a PL deformation retraction from the underlying space of P to the underlying space of $P\setminus\{\alpha,\beta\}$. This retraction is called an *elementary collapse*. (See Section 9.7.1 for detailed definitions.) A cell complex is *collapsible* if there is a sequence of elementary collapses retracting the underlying space of P to a point. We will use the following theorem of PL topology.

Theorem 9.69 (Corollary 3.28 in Rourke and Sanderson 1982) *Every collapsible PL manifold is a PL ball.*

Our approach is to reduce the problem to fans $\mathcal{S}$ with a single maximal simplex such that $\|\Sigma(\mathcal{S},\mathcal{T})\|$ can be shown to have a PL subdivision that is a collapsible PL manifold. Our first step is to show that stellar subdivision and superposition work nicely together.

Lemma 9.70 *Let $\mathcal{S}$ and $\mathcal{T}$ be simplicial fans on $\mathcal{M}$ and $\mathcal{S}'$ an $\mathcal{M}$-stellar subdivision of $\mathcal{S}$. There is a PL homeomorphism $\|\Sigma(\mathcal{S},\mathcal{T})\| \to \|\Sigma(\mathcal{S}',\mathcal{T})\|$ taking each element $\sigma|\cap|\tau$ to the union of all elements of the form $\omega|\cap|\tau$ with ω in the subdivision of σ.*

Proof: $\mathcal{S}'$ is the stellar subdivision of $\mathcal{S}$ by an element x in a simplex σ_0 of $\mathcal{S}$ with σ_0.

For each $\sigma^+\tilde{\tau}^- \in \Sigma(\mathcal{S},\mathcal{T})$ let $P_{\sigma,\tau}$ be the union of all elements of $\|\Sigma(\mathcal{S}',\mathcal{T})\|$ of the form $\omega|\cap|\tau$ with $\omega \in \mathrm{sub}(\sigma)$. By Corollary 9.14, it's enough to show that the set of all $P_{\sigma,\tau}$ is a PL cell complex isomorphic to $\Sigma(\mathcal{S},\mathcal{T})$.

We first check that every element $\omega|\cap|\tau$ of $\|\Sigma(\mathcal{S}',\mathcal{T})\|$ is in at least one $P_{\sigma,\tau}$. We have $\omega \in \mathrm{sub}(\sigma)$ for some σ. If $\omega = \sigma$ then $\omega|\cap|\tau = P_{\sigma,\tau}$. Otherwise $\sigma \supseteq \sigma_0$ and $x \in \omega$. Eliminating x between $\omega^+\tilde{\tau}^-$ and $\sigma_0^+x^-$ gives us $\sigma^+\tilde{\tau}^- \in \Sigma(\mathcal{S},\mathcal{T})$, and $\omega|\cap|\tau \in P_{\sigma,\tau}$.

It is straightforward to check that $P_{\sigma,\tau} \cap P_{\sigma',\tau'} = P_{\sigma\cap\sigma',\tau\cap\tau'}$ for all $\sigma,\sigma',\tau,\tau'$, and that $P_{\sigma,\tau} \supseteq P_{\sigma',\tau'}$ if and only if $\sigma^+\tilde{\tau}^- > (\sigma')^+(\tilde{\tau}')^-$.

9.7 Triangulations of Euclidean Oriented Matroids

We next check that if $P_{\sigma',\tau'} \subset P_{\sigma,\tau}$ then $P_{\sigma',\tau'} \subseteq \partial P_{\sigma,\tau}$. It's enough to check this when $\sigma^+\tilde{\tau}^-$ covers $(\sigma')^+(\tilde{\tau}')^-$ in $\Sigma(\mathcal{S}',\mathcal{T})$: We need to check that every maximal cell $\omega|\cap|\tau'$ in $P_{\sigma',\tau'}$ is in exactly one maximal cell of $P_{\sigma,\tau}$. There are two cases. If $\sigma' = \sigma$ then ω is a maximal element of sub(σ) and the only maximal cell of $P_{\sigma,\tau}$ containing $\omega|\cap|\tau'$ is $\omega|\cap|\tau$. Otherwise $\sigma' \subset \sigma$. By Proposition 9.67, $\sigma' = \sigma \setminus \{s\}$. Certainly every maximal element of sub($\sigma \setminus \{s\}$) is in only one maximal element of sub(σ), and so $\omega|\cap|\tau'$ is in only one maximal element of $P_{\sigma,\tau}$.

It remains to see that each $P_{\sigma,\tau}$ is a PL ball. This is clear when sub(σ) = σ. Otherwise, the composition $(\sigma^+\tilde{\tau}^-) \circ \{x\}^+\sigma_0^- = (\sigma \cup \{x\})^+\tau^-$ is a tope of $(\mathcal{M}_D(\sigma \cup \tau \cup \{x\}))^*$ that has $\sigma^+\tilde{\tau}^-$ as a facet. Another tope is $(\{x\}^+\sigma_0^-) \circ (\sigma^+\tilde{\tau}^-) = ((\sigma \setminus \sigma_0) \cup \{x\})^+(\sigma_0 \cup \tilde{\tau})^-$. Choose a maximal chain in the tope poset $\mathcal{T}(\mathcal{M}_D(\sigma \cup \tau \cup \{x\}), (\sigma \cup \{x\})^+\tau^-)$ containing $((\sigma \setminus \sigma_0) \cup \{x\})^+(\sigma_0 \cup \tilde{\tau})^-$. As discussed in Section 4.5, the induced order on facets of $(\sigma \cup \{x\})^+\tau^-$ is a recursive coatom ordering in which facets of the form $(\sigma \setminus \{s\} \cup \{x\})^+\tilde{\tau}^-$ with $s \in \sigma_0$ come first. But these are exactly the maximal cells in $P_{\sigma,\tau}$. We conclude that $P_{\sigma,\tau}$ is a shellable (and hence PL) ball of the same dimension as $\sigma|\cap|\tau$. □

9.7.1 Collapsing

For a detailed discussion of collapsing, see chapter 3 of Rourke and Sanderson (1982).

Definition 9.71 Let $X \supset Y$ be polyhedra and B a PL n-ball such that $X = Y \cup B$ and $Y \cap B$ is an $(n-1)$-dimensional PL ball in the boundary of B. Then we say there is an **elementary collapse** from X to Y. We say X **collapses** to Y if there is a sequence $X = Y_1, \ldots, Y_k = Y$ such that for each i there is an elementary collapse from Y_i to Y_{i+1}. We say X is **collapsible** if X collapses to a point.

Let $P \supset Q$ be PL cell complexes. A **collapsing sequence** from P to Q is a sequence of pairs $(\omega_1, \nu_1), \ldots, (\omega_k, \nu_k)$ of elements of P such that each ω_i is the unique element of $P \setminus \bigcup_{j<i}\{\omega_j, \nu_j\}$ with ν_i as a proper subset and $P \setminus Q = \bigcup_{j=1}^k \{\omega_j, \nu_j\}$.

If there is an elementary collapse from X to Y then there is a particularly well-behaved deformation retraction from X to Y. Each term (ω_i, ν_i) demonstrates an elementary collapse from the underlying space of $P \setminus \bigcup_{j<i}\{\omega_j, \nu_j\}$ to the underlying space of $P \setminus \bigcup_{j\leq i}\{\omega_j, \nu_j\}$, so a collapsing sequence demonstrates that the underlying space of P collapses to the underlying space of Q. For brevity we will say this as "P collapses to Q."

Lemma 9.72 *Let $\mathcal{M}$ be Euclidean, $\mathcal{T}$ a partial triangulation of $\mathcal{M}$, and σ independent in $\mathcal{M}$. Then there is an extension $\mathcal{M}'$ of $\mathcal{M}$ and a simplicial complex $\mathcal{S}$ obtained from $\mathcal{S}_\sigma$ by a sequence of $\mathcal{M}'$-stellar subdivisions such that for each element ω of $\mathcal{S}'$, $\|\Sigma_{\mathcal{M}'}(\mathcal{S}_\omega, \mathcal{T})\|$ has a PL subdivision that is collapsible.*

Proof: We induct on $|\sigma|$. When $|\sigma| \in \{1, 2\}$ the result follows from Proposition 9.64.

Let $f, g \in \sigma$. Our argument will use stellar subdivisions in the face $\{f, g\}$ of σ, after taking a sequence of extensions in the convex hull of $\{f, g\}$. The idea of the proof is given in Figure 9.6. We take extensions $c_1, \ldots, c_k, d_1, \ldots, d_{k+1}$ and stellar subdivisions of σ by these new elements, indicated by the dashed lines, to get a simplicial complex with maximal simplices $\omega = (\sigma \backslash \{f, g\}) \cup \{c_i, d_j\}$ such that $\|\Sigma(\mathcal{S}_\omega, \mathcal{T})\|$ collapses to $\|\Sigma(\mathcal{S}_{\omega \backslash \{d_j\}}, \mathcal{T})\|$. The induction hypothesis then lets us collapse further.

Since $(\mathcal{M}, f, g)$ is Euclidean and extension of $\mathcal{M}$ by an element parallel to an element of $\mathcal{M}$ cannot introduce non-Euclidean cycles, $(\mathcal{M}_D, f, g)$ is Euclidean as well. Recall that this means that for each $X \in \mathcal{C}^*(\mathcal{M}_D)$ there's an extension $\mathcal{M} \cup c$ by a nonloop such that $\{f, g, c\}$ and $X^0 \cup \{c\}$ are both dependent.

Our first sequence of extensions and stellar subdivisions is illustrated in Figure 9.5. Let $\omega_1 | \cap | \tau_1, \ldots, \omega_k | \cap | \tau_k$ be all of the 0-cells of $\|\Sigma(\mathcal{S}_\sigma, \mathcal{T})\|$ of the form $\omega | \cap | \tau$ with $\{f, g\} \subseteq \omega$. For each i, $\omega_i \cup \tilde{\tau}_i$ is a minimal dependent set in $\mathcal{M}_D$, and so $(\omega_i \backslash \{f, g\}) \cup \tilde{\tau}_i$ is independent and has rank less than rank$(\mathcal{M})$. Thus there is an $X_i \in \mathcal{C}^*(\mathcal{M}_D)$ with $(\omega_i \backslash \{f, g\}) \cup \tilde{\tau}_i \subseteq X^0$ and $X_i(f) \neq 0$. By Proposition 6.68, we can take an extension $\mathcal{M}^{(1)} := \mathcal{M}_D \cup \{c_1, \ldots, c_k\}$ such that $\{f, g, c_i\}$ is rank 2 and $X_i^0 \cup \{c_i\}$ is rank $r - 1$ for each i. We orient the c_i so that one of the signed circuits Y_i with support $\{f, g, c_i\}$ has $Y_i(f) = +$ and $Y_i(c_i) = -$.

Let $X_i^{(i)}$ be the extension of X_i to a cocircuit of $\mathcal{M}^{(1)}$. Since $X_i^{(1)} \in \mathcal{C}^*(\mathcal{M}_D)$ and $\omega_i^+ \tilde{\tau}_i^- \in \mathcal{V}(\mathcal{M}^{(1)})$, these two vectors are orthogonal, so $0 \in X_i^{(1)} \cdot \omega_i^+ \tilde{\tau}_i^- = X_i^{(1)}(f) \boxplus X_i^{(1)}(g)$. Thus $X_i^{(1)}(f) = -X_i^{(1)}(g) \neq 0$. Likewise, $X_i^{(1)} \perp Y_i$ so $0 \in X_i^{(1)} \cdot Y_i = X_i^{(1)}(f) Y_i(f) \boxplus X_i^{(1)}(g) Y_i(g) = \pm (Y_i(f) \boxplus (-Y_i(g))$. We conclude $Y_i = \{f, g\}^+ \{c_i\}^-$.

Thus the rank 2 oriented matroid $\mathcal{M}^{(1)}(\{f, g, c_1, \ldots, c_k\})$ is realizable by an arrangement $(p_e : e \in \{f, g, c_1, \ldots, c_k\})$ on an affine line with p_f the leftmost point and p_g the rightmost. Rename elements as necessary so that the left-to-right order of points is $f = c_0, c_1, \ldots, c_k, c_{k+1} = g$.

Now take a sequence of stellar subdivisions: Let $D_0 = \mathcal{S}_\sigma$. For each $i \in [k]$ the ith subdivision is the stellar subdivision of D_{i-1} by c_i in the

9.7 Triangulations of Euclidean Oriented Matroids

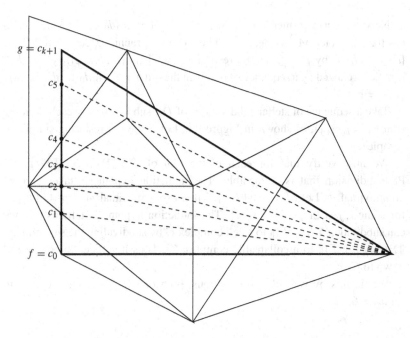

Figure 9.5 A partial triangulation T (fine lines), a simplex σ (bold lines), and extensions $c_1, \ldots, c_k$ that allow us to subdivide $\Sigma(\mathcal{S}_\sigma, \mathcal{T})$ into simplices that are easier to understand.

simplex $\{c_{i-1}, g\}$. (This is illustrated by the dashed lines in Figure 9.5.) Each of these is an $\mathcal{M}^{(1)}$-stellar subdivision, so by Proposition 9.49, $\|\Sigma(D_k, \mathcal{T})\|$ is PL homeomorphic to $\|\Sigma(\mathcal{S}_\sigma, \mathcal{T})\|$.

Let $\sigma' = \sigma \setminus \{f, g\}$. Each maximal simplex of D_k has the form $\sigma' \cup \{c_i, c_{i+1}\}$, which we denote σ_i. Figure 9.5 suggests that all 0-cells of $\|\Sigma(\mathcal{S}_{\sigma_i}, \mathcal{T})\|$ lie in $\|\Sigma((\mathcal{S}_{(\sigma_i \setminus \{c_i\})} \cup \mathcal{S}_{(\sigma_i) \setminus \{c_j\}}), \mathcal{T})\|$: We now verify this. Assume by way of contradiction there is a 0-cell $(\nu \cup \{c_i, c_{i+1}\}) | \cap | \tau$ of $\|\Sigma(\mathcal{S}_{\sigma_i}, \mathcal{T})\|$. Using elimination between the signed circuits $(\nu \cup \{c_i, c_{i+1}\})^+ \tilde{\tau}^-, \{f, g\}^+ \{c_i\}^-$, and $\{f, g\}^+ \{c_{i+1}\}^-$, we get a 0-cell $(\nu \cup \{f, g\}) | \cap | \tau$ of $\|\Sigma(\mathcal{S}_\sigma, \mathcal{T})\|$. This 0-cell gave rise to some c_j in our earlier sequence of extensions, with $\nu \cup \tilde{\tau} \cup \{c_j\} \subseteq Z_j^0$ for some $Z_j \in \mathcal{V}^*(\mathcal{M}^{(1)}(\{f, g, c_1, \ldots, c_k\}))$. Orthogonality of Z_j and $(\nu \cup \{c_i, c_{i+1}\})^+ \tilde{\tau}^-$ tells us that $Z_j(c_i) = -Z_j(c_{i+1})$, and so orthogonality of Z with the signed circuits $\pm Y$ with support contained in $\{c_i, c_{i+1}, c_j\}$ tells us that $Y(i) = Y(i+1)$. Since $\mathcal{M}^{(1)}(\{f, g, c_1, \ldots, c_k\})$ is acyclic (realizable by an affine point arrangement), we conclude that $Y = \pm \{c_i, c_{i+1}\}^+ \{c_j\}^-$. But this says that p_{c_j} lies between p_{c_i} and $p_{c_{i+1}}$ in our rank 2 affine point arrangement, a contradiction.

Now we get a further extension $\mathcal{M}' := \mathcal{M}^{(2)} \cup \{d_0,\ldots,d_k\}$: For each $i \in \{0,\ldots,k\}$, let $\mathcal{M}^{(2)} \cup \{d_0,\ldots,d_i\}$ be the lexicographic extension of $\mathcal{M}^{(2)} \cup \{d_0,\ldots,d_{i-1}\}$ by $[c_i^+ c_{i+1}^+]$. The restriction $\mathcal{M}'(\{c_0,c_1,\ldots,c_{k+1},d_0,\ldots,d_k\})$ is rank 2, realized by a sequence of points in the order $f = c_0, d_0, c_1, d_1, \ldots, d_k, c_{k+1} = g$.

Take a sequence of stellar subdivisions of D_k, subdividing by each d_i in the simplex $\{c_i, c_{i+1}\}$, as shown in Figure 9.6. Let $\mathcal{S}'$ be our resulting simplicial complex.

We now verify that for each element ω of $\mathcal{S}'$, $\|\Sigma_{\mathcal{M}'}(\mathcal{S}_\omega, \mathcal{T})\|$ has a PL subdivision that is collapsible. The induction hypothesis takes care of nonmaximal ω. Let $\sigma' = \sigma \setminus \{f,g\}$: Then every maximal ω is $\sigma' \cup \{c_i, d_j\}$ for some i, j with $j \in \{i, i+1\}$. The induction hypothesis tells us that we can subdivide $\|\Sigma(\mathcal{S}_\omega, \mathcal{T})\|$ to a cell complex C by subdividing the subcomplex $\|\Sigma(\mathcal{S}_{\sigma' \cup \{c_i\}}, \mathcal{T})\|$ to a collapsible complex C'. Thus it suffices to collapse C down to C'.

We do this by way of two bijections, both of which one can observe in Figure 9.6.

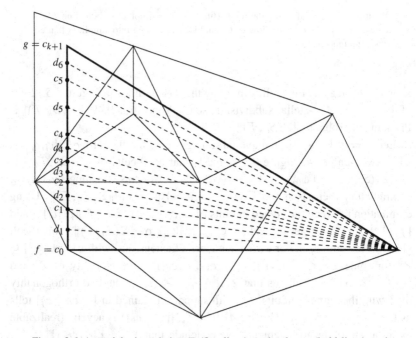

Figure 9.6 A partial triangulation T (fine lines), a simplex σ (bold lines), and extensions $c_1,\ldots,c_k,d_1,\ldots,d_{k+1}$ that allow us to subdivide $\Sigma(\mathcal{S}_\sigma, \mathcal{T})$ into a cell complex that is easy to understand.

9.7 Triangulations of Euclidean Oriented Matroids

The first is a bijection

$$\{\alpha^+\tilde{\tau}^- \in \Sigma(\mathcal{S}_\omega, \mathcal{T}) : d_j \in \alpha\} \to \{\alpha^+\tilde{\tau}^- \in \Sigma(\mathcal{S}_{\sigma' \cup \{c_i, c_j\}}, \mathcal{T}) : c_i, c_j \in \alpha\},$$

$$(\beta \cup \{d_j\})^+ \tilde{\tau}^- \to (\beta \cup \{c_i, c_j\})^+ \tilde{\tau}^-.$$

It's well defined because we can eliminate d_j between $(\beta \cup \{d_j\})^+ \tilde{\tau}^-$ and $\{c_i, c_j\}^+ d_j^-$ to get $(\beta \cup \{c_i, c_j\})^+ \tilde{\tau}^-$, and it's clearly injective. To see that it's surjective, recall that $\Sigma(\mathcal{S}_{\sigma' \cup \{c_i, c_j\}}, \mathcal{T})$ has no signed circuits of the form $(\gamma \cup \{c_i, c_j\})^+ \tilde{\tau}^-$, so every element $(\beta \cup \{c_i, c_j\})^+ \tilde{\tau}^-$ is a composition of signed circuits of the forms $(\gamma \cup \{c_i\})^+ \tilde{v}^-$ and $(\gamma \cup \{c_j\})^+ \tilde{v}^-$. Each covector Y is orthogonal to each of these and to $\{c_i, c_j\}^+ d_j^-$. By checking cases we verify that $Y \perp (\beta \cup \{d_j\})^+$.

The second bijection is

$$\{\alpha^+\tilde{\tau}^- \in \Sigma(\mathcal{S}_\omega, \mathcal{T}) : c_i, d_j \in \alpha\} \to \{\alpha^+\tilde{\tau}^- \in \Sigma(\mathcal{S}_{\sigma' \cup \{c_i, c_j\}}, \mathcal{T}) : c_i, c_j \in \alpha\},$$

$$(\beta \cup \{c_i, d_j\})^+ \tilde{\tau}^- \to (\beta \cup \{c_i, c_j\})^+ \tilde{\tau}^-.$$

The same elimination as before shows that this map is well defined, and it is certainly injective. To see that it's surjective, once again we can write each element $(\beta \cup \{c_i, c_j\})^+ \tilde{\tau}^-$ of the codomain as a composition of signed circuits of the forms $(\gamma_i \cup \{c_i\})^+ \tilde{v}_i^-$ and $(\gamma_j \cup \{c_j\})^+ \tilde{v}_j^-$. For each $(\gamma_j \cup \{c_j\})^+ \tilde{v}_j^-$ we can eliminate c_j between it and $d_j^+ \{c_i \cup c_j\}^-$ to get a Z such that $(\gamma_i \cup \{c_i\})^+ \tilde{v}_i^- \circ Z$ has the form $(\gamma \cup \{c_i, d_j\})^+ \tilde{v}^-$. The composition of all such vectors is $(\beta \cup \{c_i, d_j\})^+ \tilde{\tau}^-$.

Together these bijections give us a pairing $((\alpha \cup \{c_i, d_j\})^+ \tilde{\tau}^-, (\alpha \cup \{d_j\})^+ \tilde{\tau}^-)$ of the elements of $C \backslash C'$, and using these we collapse C to C'. □

9.7.2 Proof of Theorem 9.68

Lemma 9.73 *Let $\mathcal{N}$ be an oriented matroid and S_1, S_2 sets of elements such that $S_1^+ \in \mathcal{C}(\mathcal{N})$ and $(S_1 \cup S_2) \backslash \{s\}$ is independent for each $s \in S_1$. Let τ be independent in $\mathcal{N}$, and let $\mathcal{S}$ be the simplicial complex whose maximal elements are the sets $(S_1 \cup S_2) \backslash \{s\}$ with $s \in S_1$. Then $\|\Sigma(\mathcal{S}, \mathcal{S}_\tau)\|$ is either empty or a PL ball.*

Proof: If $\Sigma(\mathcal{S}, \mathcal{S}_\tau)$ is nonempty then the composition of all of its elements, together with S_1^+, is a vector $B = (S_1 \cup S_3)^+ \tilde{\omega}^-$ for some $S_3 \subseteq S_2$ and some $\omega \subseteq \tau$. Consider the corresponding cell in a topological representation of $\mathcal{N}_D(S_1 \cup S_3 \cup \tilde{\omega})^*$. One 0-cell in the boundary of this tope is S_1^+. We can use the methods of Chapter 4, with a maximal chain in the tope poset $\mathcal{T}(\mathcal{N}_D(S_1 \cup S_3 \cup \tilde{\omega})^*, B)$ containing $S_1^- \circ B$, to get a shelling of the boundary of the cell B

in which cells containing the 0-cell corresponding to S_1^+ come last. Thus the cells not containing this 0-cell constitute a PL cell decomposition of a ball. But this cell complex is $\|\Sigma(\mathcal{S}, \mathcal{S}_\tau)\|$. □

Proof of Theorem 9.68: We appeal to Corollary 9.14. We'll show that the set of all $\|\Sigma(\mathcal{S}_\sigma, \mathcal{T})\|$ is a PL cell complex that is isomorphic as a poset to $\|\mathcal{S}\|$.

The only work is in showing for each $\alpha \in \mathcal{S}$ that $\|\Sigma(\mathcal{S}_\alpha, \mathcal{T})\|$ is a PL ball with boundary $\bigcup_{\sigma \subset \alpha} \|\Sigma(\mathcal{S}_\sigma, \mathcal{T})\|$. We prove this by induction on $r = \text{rank}(\mathcal{M})$, and within this on $|\alpha|$. When $r = 1$, the result is clear. When $|\alpha| \in \{1, 2\}$ the result follows from Proposition 9.64. Above this, we first show that $\|\Sigma(\mathcal{S}_\alpha, \mathcal{T})\|$ is a PL manifold.

First consider a 0-cell of the form $\sigma|\cap|\tau$. Using the recursive definition of triangulation, we see $\text{link}(\tau)$ is a partial triangulation of $\mathcal{M}/\tau$, and $\sigma^+ \in \mathcal{C}(\mathcal{M}/\tau)$. By the induction hypothesis applied to each proper subset of σ we get a PL cell decomposition of $\|\Sigma(\partial \mathcal{S}_\sigma, \text{link}(\tau))\|$ with a cell for each proper subset of σ. Thus $\|\Sigma(\partial \mathcal{S}_\sigma, \text{link}(\tau))\|$ is a PL sphere of dimension $|\sigma| - 2$. But applying Lemma 9.73 with $S_1 = \sigma$, $S_2 = \emptyset$, and $\nu \in \mathcal{T}$ gives us another PL cell decomposition P of this sphere into cells $C_\nu := \|\Sigma_{\mathcal{M}/\tau}(\partial \mathcal{S}_\sigma, \nu)\|$ for some $\nu \in \mathcal{T}$. Each element $\alpha^+ \tilde{\beta}^-$ of $\Sigma(\partial \mathcal{S}_\sigma, \text{link}(\tau))$ is the restriction of some $Y \in \mathcal{V}(\mathcal{M}_D)$, and $\sigma^+ \tilde{\tau}^- \circ Y = \sigma^+(\tilde{\tau} \cup \tilde{\beta})^- \in \Sigma_\mathcal{M}(\mathcal{S}_\sigma, \mathcal{T})$. Thus we have a well-defined poset map $P \to \Sigma_\mathcal{M}(\mathcal{S}_\sigma, \mathcal{T})_{\geq \sigma^+ \tilde{\tau}^-}$, taking each C_ν to $\sigma^+(\tilde{\tau} \cup \tilde{\nu})^-$. Indeed, this map is a poset isomorphism, and so we get a simplicial isomorphism from $\Delta(P \cup \{\hat{0}\})$ to $\Sigma_\mathcal{M}(\mathcal{S}_\sigma, \mathcal{T})_{\geq \sigma^+ \tilde{\tau}^-}$. Thus the neighborhood $\|\Sigma_\mathcal{M}(\mathcal{S}_\sigma, \mathcal{T})_{\geq \sigma^+ \tilde{\tau}^-}\|$ of $\sigma|\cap|\tau$ is a PL $(|\sigma| - 1)$-ball with $\sigma|\cap|\tau$ in its interior.

By a similar argument, for a 0-cell of the form $\hat{\sigma}|\cap|\tau$ with $\hat{\sigma} \subset \sigma$, the induction hypothesis tells us that $\|\Sigma_{\mathcal{M}/\tau}(\partial \mathcal{S}_{\hat{\sigma}} * (\sigma \setminus \hat{\sigma}), \text{link}(\tau))\|$ is a PL ball, and Lemma 9.73 gives us a cell decomposition of this ball showing that $\hat{\sigma}|\cap|\tau$ has a neighborhood that's a PL ball with $\hat{\sigma}|\cap|\tau$ in its boundary.

We conclude that $\|\Sigma(\mathcal{S}_\sigma, \mathcal{T})\|$ is a PL $(|\sigma| - 1)$-manifold with boundary $\|\Sigma(\partial \mathcal{S}_\sigma, \mathcal{T})\|$.

Since a PL subdivision of a PL manifold is a PL manifold, the subdivision of each $\|\Sigma(\mathcal{S}_\sigma, \mathcal{T})\|$ that we found in Lemma 9.72 is a PL manifold, as well as being collapsible. Thus by Theorem 9.69 this subdivision, and thus $\|\Sigma(\mathcal{S}_\sigma, \mathcal{T})\|$ itself, is a PL ball.

We conclude that $\{\|\Sigma(\mathcal{S}_\sigma, \mathcal{T})\| : \sigma \in \mathcal{S}\}$ is a PL cell decomposition of $\|\Sigma(\mathcal{S}, \mathcal{T})\|$ that is isomorphic as a poset to $\mathcal{S}$. So the result follows from Proposition 9.14. □

9.8 Final Remarks

For much more on oriented matroid triangulations, see Santos (2002).

One motivation for considering oriented matroid triangulations arises from topology. The theory of *combinatorial differential manifolds* seeks to model a smooth manifold by a simplicial complex, which plays the role of the topological space, and a collection of oriented matroids, which plays the role of a coordinate atlas.

Unfortunately, to describe the idea we need to use the word *triangulation* in a different sense, by discussing *triangulations of manifolds*. Let M be a smooth manifold. A *triangulation* of M is a homeomorphism $\eta: X \to M$, where X is a geometric simplicial complex and η is smooth on every closed simplex. Let p be a vertex of X. Each 1-simplex with p as an endpoint maps under η to a smooth curve in M, and we can take a tangent at $\eta(p)$ to this curve to get a ray in the tangent space $T_{\eta(p)}M$. (This is illustrated on the left-hand side of Figure 9.7.) More generally, if $\sigma \in \text{star}(p)$ then the set of tangent vectors to $\eta(\sigma)$ at $\eta(p)$ is a cone in $T_{\eta(p)}M$, and the set of all such cones is a simplicial fan whose underlying space is the tangent space $T_{\eta(p)}M$. To combinatorialize this, we note that the poset of cones in our fan is isomorphic to the boundary of star(p), and so the boundary of star(p) is a triangulation of the oriented matroid $\mathcal{M}_p$ of our vector arrangement in $T_{\eta(p)}M$. If we think of star(p) as a crude analog to a coordinate chart at $\eta(p)$, then we can think of a realization of $\mathcal{M}_p$ as strengthening this to a piecewise-linear coordinate chart.

We can do something similar at every point in M, not just the images of 0-simplices. This is illustrated on the right-hand side of Figure 9.7. Each point

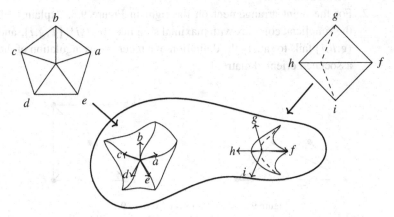

Figure 9.7 Two parts of a triangulation of a smooth manifold and the resulting vector arrangements at two points.

in M is $\eta(q)$ for some q in the interior of a simplex Δ_q. For each vertex v of star(Δ_q) there is a line seqment in X with endpoints q and v, and in M we can take a tangent at $\eta(q)$ to the image of this line segment. Again this gives us an oriented matroid $\mathcal{M}_q$ such that the boundary of the star of Δ_q is a triangulation of $\mathcal{M}_q$.

The success of this viewpoint in Gel'fand and MacPherson (1992) led MacPherson to propose a purely combinatorial definition, in which a combinatorial differential manifold is an abstract simplicial complex, together with a combinatorial "atlas" of oriented matroids. For a detailed definition, see MacPherson (1993) and Anderson (1999). The simplicial complex is not a priori assumed to be a topological manifold, nor are the oriented matroids assumed to be realizable, but the conditions on the oriented matroid "coordinate chart" $\mathcal{M}_p$ at a vertex p do their combinatorial best to make star(p) model the face poset of a triangulation of an arrangement of rays realizing $\mathcal{M}_p$. In particular, star(p) is a triangulation of $\mathcal{M}_p$. MacPherson asked whether every combinatorial differential manifold is a topological manifold: This amounts to asking whether every triangulation of a totally cyclic oriented matroid is a topological sphere. See Anderson (1999) for details.

Exercises

9.1   1. For the point arrangement on the left in Figure 9.8, explain how the simplicial complex with maximal simplices $\{a,b,c\}$ and $\{a,b,d\}$ fails to satisfy the definition of a recursive triangulation of the associated oriented matroid.
2. For the point arrangement on the right in Figure 9.8, explain how the simplicial complex with maximal simplices $\{e, f, i\}$, $\{f, g, i\}$, and $\{g, h, i\}$ fails to satisfy the definition of a recursive triangulation of the associated oriented matroid.

Figure 9.8 Illustrations for Exercise 9.1.

9.2 Prove that if $\mathcal{M}$ is an oriented matroid on elements E, $\mathcal{F}$ is a decomposition of $\mathcal{M}$, and $\mathcal{M} \cup f$ is an extension of $\mathcal{M}$ with $f \in \widehat{\mathrm{conv}}(E)$ then there is a unique $\sigma \in \mathcal{F}$ with $\sigma^+ f^- \in \mathcal{V}(\mathcal{M} \cup f)$.

9.3 This exercise will prove that Proposition 9.64 can be strengthened if we assume $\mathcal{T}$ to be a strong triangulation.

Theorem 9.74 (Rambau 2002) *Let $\mathcal{T}$ be a strong partial triangulation of $\mathcal{M}$ and $\{a,b\}$ independent in $\mathcal{M}$. Then $\|\Sigma(\mathcal{S}_{\{a,b\}}, \mathcal{T})\|$ consists of a single path.*

The method is to find a total order on the minimal elements of $\Sigma(\mathcal{S}_{\{a,b\}}, \mathcal{T})$ and then show that this order reflects the order in which the corresponding 0-cells must appear in the path contained in $\|\Sigma(\mathcal{S}_{\{a,b\}}, \mathcal{T})\|$.

1. Let $\{a\}^+ \tilde{\tau}_a^-$ and $\{b\}^+ \tilde{\tau}_b^-$ be the unique elements of $\Sigma(\mathcal{S}_{\{a,b\}}, \mathcal{T})$ of these forms. Prove that if $\tau_a = \tau_b$ then these are the only two 0-cells of $\Sigma(\mathcal{S}_{\{a,b\}}, \mathcal{T})$.
This settles a special case of the theorem. For the remainder of the problem, we assume $\tau_a \neq \tau_b$.

2. For $c = \sigma_1^+ \tilde{\tau}_1^-, d = \sigma_2^+ \tilde{\tau}_2^- \in \Sigma(\mathcal{S}_{\{a,b\}}, \mathcal{T})$, we say $c \prec d$ if there is an $X \in \mathcal{C}(\mathcal{M})$ such that $a \in X^+ \subseteq \{a\} \cup \tilde{\tau}_2$ and $X^- \subseteq \tilde{\tau}_1$.
Prove that $c \prec d$ if and only if there is a $Y \in \mathcal{C}(\mathcal{M})$ such that $X^+ \subseteq \tilde{\tau}_2$ and $b \in X^- \subseteq \{b\} \cup \tilde{\tau}_1$.

3. Prove that for each pair c, d of distinct minimal elements of $\Sigma(\mathcal{S}_{\{a,b\}}, \mathcal{T})$ either $c \prec d$ or $d \prec c$, but not both.

4. Let $\{a\}^+ \tilde{\tau}_a^- = c_0, c_1, \ldots, c_k = \{b\}^+ \tilde{\tau}_b^-$ be the sequence of minimal elements of $\Sigma(\mathcal{S}_{\{a,b\}}, \mathcal{T})$ corresponding to 0-cells on the path in $\|\Sigma(\mathcal{S}_{\{a,b\}}, \mathcal{T})\|$, listed in the order they appear on this path. Prove that $c_i \prec c_{i+1}$ for each i.

5. Now assume by way of contradiction that there is a minimal element $d = \{a,b\}^+ \tilde{\tau}^-$ of $\Sigma(\mathcal{S}_{\{a,b\}}, \mathcal{T})$ that is not an element of the sequence $c_0, c_1, \ldots, c_k$. Prove that $d \not\prec c_0$, $c_k \not\prec d$, and there is no i such that $c_i \prec d \prec c_{i+1}$. Conclude a contradiction.

10

Spaces of Oriented Matroids

10.1 Comparing Oriented Matroids and Subspaces of $\mathbb{R}^n$

Thinking of oriented matroids as combinatorial analogs to subspaces of $\mathbb{R}^n$ leads to topological questions concerning the strength of this analogy. This chapter will survey results and open questions in this area. Most proofs are omitted.

In Chapter 4 we got a solid answer to such a question, in the form of the Topological Representation Theorem. Let V be a rank r subset of $\mathbb{R}^n$ and $\mu(V)$ its oriented matroid. One topological object associated to V is its unit sphere $S(V)$, and a combinatorial analog to $S(V)$ is $\mathcal{V}^*(\mu(V))\backslash\{0\}$. This analogy is as strong as we could hope for: $\mathcal{V}^*(\mu(V))\backslash\{0\}$ is canonically isomorphic to the poset of cells in a regular cell decomposition of $S(V)$. In particular, we get a homeomorphism $S(V) \to \|\Delta(\mathcal{V}^*(\mu(V))\backslash\{0\})\|$, and so $\mu(V)$ has its own combinatorially defined "unit sphere." This suggested a question: If $\mathcal{M}'$ is a nonrealizable oriented matroid, is $\mathcal{V}^*(\mathcal{M}')\backslash\{0\}$ also isomorphic to the poset of cells in a regular cell decomposition of a sphere? Of course, in Chapter 4 we found the answer to be "yes."

In Section 10.2 we'll follow a similar line of reasoning on *extension spaces*. Recall that $\mathcal{E}(\mathcal{M})$ is the set of proper single-element extensions of $\mathcal{M}$ (extensions by a nonloop, non-coloop f), ordered by weak maps. Returning to V a rank r subset of $\mathbb{R}^n$ and its oriented matroid $\mu(V)$, we have another analog to $S(V)$. The vector arrangement $\mathcal{A}$ of orthogonal projections of the unit coordinate vectors onto V is a realization of $\mu(V)$, and each $\mathbf{v} \in S(V)$ defines an extension of $\mathcal{A}$, realizing an element $\epsilon_\mathcal{A}(\mathbf{v})$ of $\mathcal{E}(\mathcal{M})$. In Section 6.5 we saw that this analogy between points in the sphere and elements of $\mathcal{E}(\mathcal{M})$ is again quite nice: There is a regular cell decomposition of $S(V)$ with cells indexed by elements of $\mathcal{E}(\mathcal{M})$ so that the map from cells to $\mathcal{E}(\mathcal{M})$ is a poset map. Unlike the situation with the Topological Representation Theorem, this map may not

296

10.1 Comparing Oriented Matroids and Subspaces of $\mathbb{R}^n$

be surjective, so we only get an embedding of $S(V)$ into $\|\Delta\mathcal{E}(\mu(V))\|$. Faith in the philosophy that "$\mu(V)$ is a strong combinatorial analog to V," as well as experimentation with small examples might lead one to conjecture that this embedding is a homotopy equivalence, and that $\|\Delta\mathcal{E}(\mathcal{M})\|$ has the homotopy type of a sphere even when $\mathcal{M}$ is not realizable. We'll examine this conjecture in Section 10.2.

Going beyond individual oriented matroids, we next examine the assertion: *If rank r oriented matroids on $[n]$ are a lot like rank r subspaces of $\mathbb{R}^n$, then the poset of all rank r oriented matroids on $[n]$ should topologically be a lot like the space of all rank r subspaces of $\mathbb{R}^n$.* The poset in question is the MacPhersonian MacP(r, n), partially ordered by weak maps, and the space in question is the Grassmannian $G(r, \mathbb{R}^n)$. As in our $\mathcal{V}^*(\mathcal{M})$ and $\mathcal{E}(\mathcal{M})$ discussions, we have a canonical map $\mu \colon G(r, \mathbb{R}^n) \to \text{MacP}(r, n)$ taking our objects in $\mathbb{R}^n$ to their combinatorial analogs. But unlike those previous discussions, we no longer have a regular cell decomposition of our topological space isomorphic to a subposet of our poset of interest. Indeed, this line of thought breaks down on several levels. The Universality Theorem tells us that preimages $\mu^{-1}(\mathcal{M})$ need not be open balls – in fact, they can have homotopy type as complicated as they like. Additionally, any hope that a relation $\mathcal{M} \geq \mathcal{N}$ corresponds to a relation $\mu^{-1}(\mathcal{M}) \supseteq \mu^{-1}(\mathcal{N})$ was quashed in Section 5.2.1, where we saw realizable $\mathcal{M}$ and $\mathcal{N}$ such that $\mathcal{M} \rightsquigarrow \mathcal{N}$ but $\mu^{-1}(\mathcal{M}) \cap \mu^{-1}(\mathcal{N}) = \emptyset$, as well as $\mathcal{M}$ and $\mathcal{N}$ such that $\mathcal{M} \rightsquigarrow \mathcal{N}$ and $\mu^{-1}(\mathcal{M}) \cap \mu^{-1}(\mathcal{N}) \neq \emptyset$ but $\mu^{-1}(\mathcal{M}) \not\supseteq \mu^{-1}(\mathcal{N})$.

Without the regular cell decompositions of our earlier discussions, it's not obvious how μ should lead us to any topological map from $G(r, \mathbb{R}^n)$ to $\|\Delta\text{MacP}(r, n)\|$, or any other nice topological comparisons between Grassmannians and MacPhersonians. Section 10.4 will address this in two ways. The elegant way is a canonical map from the Grassmannian to a quite weird space associated to the MacPhersonian. The less elegant way is a result on the existence of triangulations of the Grassmannian that each induce a map $\tilde{\mu} \colon G(r, \mathcal{R}^n) \to \|\text{MacP}(r, n)\|$ that is simplicial with respect to the triangulation, and a second result that the homotopy type of $\tilde{\mu}$ is independent of the choice of triangulation. A theorem of McCord gives us a kind of equivalence between the elegant weird route and the murky simplicial one.

There are plenty of reasons not to expect $\tilde{\mu}$ to preserve much interesting topological structure. Besides the points we have already raised, μ is far from being surjective: For a fixed $r \geq 3$, as n goes to infinity the proportion of MacP(r, n) that is realizable goes to 0. (See Section 2.7.4 for remarks on this.) Surprisingly, all of the results we have on $\tilde{\mu}$ and homotopy are positive. In particular, the conjecture that $\tilde{\mu}$ is always a homotopy equivalence remains open.[1]

We'll survey results in this area in a broader context. We can generalize $\mu\colon G(r,\mathbb{R}^n) \to \operatorname{MacP}(r,n)$ in two different ways. First of all, let $\mathcal{M}$ be a rank n oriented matroid with a realization $\mathcal{A} = (\mathcal{H}_e : e \in E)$ in $\mathbb{R}^n$ and define the **combinatorial Grassmannian** $G(r,\mathcal{M})$ to be the set of all rank r strong map images of $\mathcal{M}$. Again, we partially order $G(r,\mathcal{M})$ by weak maps. We get a map $\mu_{\mathcal{A}}\colon G(r,\mathbb{R}^n) \to G(r,\mathcal{M})$ sending each V to the oriented matroid given by the signed hyperplane arrangement $(V \cap \mathcal{H}_e : e \in E)$.

Problem 10.1 Verify that if $\mathcal{M}$ is the coordinate oriented matroid on $[n]$ and $\mathcal{A}$ is the arrangement of coordinate hyperplanes then $G(r,\mathcal{M}) = \operatorname{MacP}(r,n)$ and $\mu_{\mathcal{A}} = \mu$.

As another level of generalization, why stick to a single subspace of $\mathbb{R}^n$?

Definition 10.2 Let $1 \leq r_1 < \cdots < r_k \leq n$ be integers.

Let V be an n-dimensional vector space over $\mathbb{R}$. A $(r_1,\ldots,r_k)$-**flag** in V is a sequence $V_1 \subset \cdots \subset V_k \subseteq V$ of linear subspaces with $\dim(V_i) = r_i$ for each i. The $(r_1,\ldots,r_k)$-**flag space** $G(r_1,\ldots,r_k,V)$ is the set of all $(r_1,\ldots,r_k)$-flags in V, viewed as a subspace of $G(r_1,V) \times \cdots \times G(r_k,V)$.

Let $\mathcal{M}$ be a rank n oriented matroid. A $(r_1,\ldots,r_k)$-**flag** in $\mathcal{M}$ is a sequence $\mathcal{M}_1 \leftarrow \cdots \leftarrow \mathcal{M}_k \leftarrow \mathcal{M}$ of oriented matroids, related by strong maps, with $\operatorname{rank}(\mathcal{M}_i) = r_i$ for each i. The $(r_1,\ldots,r_k)$-**flag poset** $G(r_1,\ldots,r_k,\mathcal{M})$ is the set of all $(r_1,\ldots,r_k)$-flags in $\mathcal{M}$, viewed as a subposet of $G(r_1,\mathcal{M}) \times \cdots \times G(r_k,\mathcal{M})$.

As in our preceding discussion, a realization $\mathcal{A}$ of $\mathcal{M}$ in V determines a map $\mu_{\mathcal{A}}\colon G(r_1,\ldots,r_k,V) \to G(r_1,\ldots,r_k,\mathcal{M})$.

There are both positive and negative results on these variations on μ, which we'll discuss in Sections 10.3 and 10.5.

10.2 Extension Spaces

In Section 6.5 we saw how a realization $\tilde{\mathcal{A}}$ of $\mathcal{M}$ has an adjoint $\mathcal{A}_{\text{adj}}$ with oriented matroid $\mathcal{M}_{\text{adj}}$, and that the partition of S^{r-1} given by the hyperplanes in $\mathcal{A}_{\text{adj}}$ coincides with the partition into preimages $\epsilon_{\tilde{\mathcal{A}}}^{-1}(\mathcal{N})$. Proposition 6.42 took this further and showed that for every $\mathcal{M}$ with an adjoint $\mathcal{M}_{\text{adj}}$, we get an injection

$$S^{r-1} \cong \|\Delta \mathcal{V}^*(\mathcal{M}_{\text{adj}})\backslash\{0\}\| \hookrightarrow \|\Delta \mathcal{E}(\mathcal{M})\|.$$

---
[1] A claim of a proof (Biss 2003) was retracted (Biss 2009).

As we saw in Section 6.5, this need not be surjective, even for realizable $\mathcal{M}$: Different realizations of $\mathcal{M}$ may admit different realizable extensions. Nonetheless, in small examples one can check that $\|\Delta\mathcal{E}(\mathcal{M})\|$ retracts to the order complex of the image of $\epsilon_{\mathcal{A}}$. This brings us to progressively more optimistic questions:

1. If $\mathcal{M}$ is realized by an arrangement $\mathcal{A}$ then is the resulting map $S^{r-1} \to \|\Delta\mathcal{E}(\mathcal{M})\|$ a homotopy equivalence?
2. If $\mathcal{M}_{\text{adj}}$ is an adjoint for $\mathcal{M}$ then is the map $\|\Delta\mathcal{V}^*(\mathcal{M}_{\text{adj}})\backslash\{0\}\| \hookrightarrow \|\Delta\mathcal{E}(\mathcal{M})\|$ a homotopy equivalence?
3. If $\mathcal{M}$ is a rank r oriented matroid, is $\|\Delta\mathcal{E}(\mathcal{M})\| \simeq S^{r-1}$?

The last question was answered in the negative in Mnëv and Richter-Gebert (1993), which gave two examples of oriented matroids with disconnected extension space. In the same year it was shown (Sturmfels and Ziegler 1993) that there is indeed a homotopy equivalence when $\mathcal{M}$ is *strongly Euclidean*. Strong equivalence is a highly restrictive condition – stronger than being realizable – but, for instance, all rank 3 oriented matroids are strongly Euclidean. The conjecture for realizable oriented matroids, known as the *Extension Space Conjecture*, was only disproved in 2016 by Liu, by a surprising approach.

Theorem 10.3 (Liu 2020) *For each N let $\mathcal{A}_N$ be the vector arrangement comprising N copies of each element of $\{e_i - e_j : 1 \leq i \leq j \leq 4\}$. Let $\tilde{\mathcal{A}}_N$ be a configuration obtained by perturbing each vector in $\mathcal{A}_N$ by a small random displacement in $\mathbb{R}^4$. For large enough N, with probability greater than 0, $\tilde{\mathcal{A}}_N$ contains a subconfiguration $\mathcal{A}$ such that the oriented matroid dual to the oriented matroid of $\mathcal{A}$ has disconnected extension space.*

Liu notes that the value of N required for these arguments to work is "roughly 10^5." This leaves open the question for realizable oriented matroids with ranks between 3 and roughly 10^5.

There is a beautiful relationship between extension spaces and zonotopal tilings, via the *Bohne–Dress Theorem*: See Richter-Gebert and Ziegler (1994) for details.

10.3 $G(1, \mathcal{M})$ and $G(r-1, \mathcal{M})$

We can say a little more about combinatorial Grassmannians before leaving the realm of regular cell decompositions.

For each $S \subseteq \{0, +, -\}^E$ such that $S = -S$, let $S/\pm$ be the quotient identifying each X with $-X$. Note that $S/\pm$ is still a poset.

Proposition 10.4 $G(1,\mathcal{M}) \cong (\mathcal{V}^*(\mathcal{M})\backslash\{0\})/\pm$.

In particular, if $\mathcal{M}$ is rank r then $\|\Delta G(1,\mathcal{M})\|$ is homeomorphic to $G(1,\mathbb{R}^n)$, and if $\mathcal{A}$ is a signed hyperplane arrangement realizing $\mathcal{M}$ then the partition of $G(1,\mathbb{R}^n)$ by preimages of $\mu_\mathcal{A}: G(1,\mathbb{R}^n) \to G(1,\mathcal{M})$ gives a regular cell decomposition with cells indexed by $G(1,\mathcal{M})$, and hence a homeomorphism $G(1,\mathbb{R}^n) \to \|\Delta G(1,\mathcal{M})\|$.

Proof: We can specify an element of $G(1,\mathcal{M})$ by its set of nonzero covectors: This can be any pair $\{\pm X\}$ from $\mathcal{V}^*(\mathcal{M})\backslash\{0\})$. This verifies the first sentence. The Topological Representation Theorem tells us that $\mathcal{V}^*(\mathcal{M})\backslash\{0\})$ indexes the cells in an antipodally symmetric regular cell decomposition of S^{r-1}, and identifying antipodal points in S^{r-1} gives us a regular cell decomposition of $G(1,\mathcal{M})$ with cells indexed by $(\mathcal{V}^*(\mathcal{M})\backslash\{0\})/\pm$, giving us the second assertion. A realization of $\mathcal{M}$ by a signed hyperplane arrangement $\mathcal{A}$ gives us a particular topological representation of $\mathcal{M}$ that partitions $G(1,\mathbb{R}^n)$ into the preimages of $\mu_\mathcal{A}$, giving us the final assertion. □

In particular, $\|\Delta\mathrm{MacP}(1,n)\|$ is homeomorphic to $G(1,\mathbb{R}^n)$, and since duality gives us an isomorphism $\mathrm{MacP}(r,n) \cong \mathrm{MacP}(n-r,n)$ we have that $\|\Delta\mathrm{MacP}(n-1,n)\| \cong G(n-1,\mathbb{R}^n) \cong G(1,\mathbb{R}^n)$. But more complicated oriented matroid Grassmannians have no such duality. Indeed, the news is bad.

Proposition 10.5 *For a rank r oriented matroid $\mathcal{M}$, both the function*

$$\pi : \mathcal{E}(\mathcal{M}) \to G(r-1,\mathcal{M}),$$
$$\pi(M \cup f) = (\mathcal{M} \cup f)/f$$

and the induced function

$$\Delta\pi : \Delta\mathcal{E}(\mathcal{M}) \to \Delta G(r-1,\mathcal{M})$$

are surjective. For each $\mathcal{N} \in G(1,\mathcal{M})$, the preimage $\pi^{-1}(\mathcal{N})$ consists of two extensions with opposite localizations.

The first statement is almost obvious from Proposition 5.29 on topological representations of strong maps: Given a centrally symmetric topological representation $(\mathcal{S}_e : e \in E)$ of $\mathcal{M}$ in S^{r-1} and $\mathcal{N} \in G(r-1,\mathcal{M})$, we get a centrally symmetric embedding $S^{r-2} \hookrightarrow S^{r-1}$ so that $S_f := \nu(S^{r-2})$ intersects our arrangement in a topological representation of $\mathcal{N}$. A choice of sides for S_f gives us a signed pseudosphere $\mathcal{S}_f$, and we claim that $(\mathcal{S}_e : e \in E \cup \{f\})$ is an arrangement of signed pseudospheres. To see this, we use the Las Vergnas characterization of extensions (Theorem 6.7). Let $\sigma : \mathcal{C}^*(\mathcal{M}) \to \{0,+,-\}$ be the candidate for a localization that we get from $\mathcal{S}_f$.

Thus $\sigma(X)$ indicates which of S_f, S_f^+, S_f^- contains the 0-cell corresponding to X. Because our topological representation is centrally symmetric we have that $\sigma(-X) = -\sigma(X)$ for all X. So we need only check that each restriction of σ to a rank 2 minor $\mathcal{M}/A$ is a localization. For such a minor, the intersection $S_A = \bigcap_{e \in A} S_e$ is a pseudocircle intersecting S_f in a topological representation of $\mathcal{N}/A$. Thus this intersection is either the entire circle S_A or two antipodal points. In the latter case, central symmetry tells us that the sides of $S_A \cap S_f$ in S_A are $S_f^+ \cap S_A$ and $S_f^- \cap S_A$. So indeed, the restriction of σ to $\mathcal{M}/A$, given by the restriction of σ to signed cocircuits in S_A, is the localization of an extension $\mathcal{M} \cup f$ of $\mathcal{M}$. Since $\mathcal{V}^*(\mathcal{M} \cup f)$ coincides with the set of sign vectors we get from the arrangement $(\mathcal{S}_e : e \in E \cup \{f\})$, we conclude that $(\mathcal{S}_e : e \in E \cup \{f\})$ is a topological representation of $\mathcal{M} \cup f$. To see the second statement, note that for a chain $\gamma = \{\mathcal{N}_1 < \cdots \mathcal{N}_k\}$ in $G(r-1, \mathcal{M})$, in our construction of an extension corresponding to each $\mathcal{N}_i$ we could choose our sides for S_f consistently. For instance, we could let X be a signed cocircuit of $\mathcal{M}$ that's not a signed cocircuit of any $\mathcal{N}_i$ and choose sides so that $X \in S_f^+$. Such a choice defines a chain in $\mathcal{E}(\mathcal{M})$ that maps to γ. The resulting set of extensions of $\mathcal{M}$ is a chain that maps to γ.

Corollary 10.6 *For a rank r oriented matroid $\mathcal{M}$, the map $\|\Delta \pi\| : \|\Delta \mathcal{E}(\mathcal{M})\| \to \|\Delta G(r-1, \mathcal{M})\|$ is a double cover.*

While we've mentioned the existence of $\mathcal{M}$ for which $\|\Delta \mathcal{E}(\mathcal{M})\|$ is disconnected, we have not delved into the details enough to assert that $\|\Delta G(r-1, \mathcal{M})\|$ is disconnected in these cases as well. But here is a smoking gun: One of the examples in Mnëv and Richter-Gebert (1993) is a rank 4 $\mathcal{M}$ for which $\|\Delta \mathcal{E}(\mathcal{M})\|$ has a component consisting of a single point $\{\mathcal{N}\}$. Corollary 10.6 tells us that $\pi(\mathcal{N})$ is also its own component of $G(r-1, \mathcal{M})$. So in contrast to Proposition 10.4, we have the following.

Proposition 10.7 *There are rank 4 oriented matroids $\mathcal{M}$ such that $\|\Delta G(3, \mathcal{M})\|$ is disconnected.*

10.4 McCord's Theorem and Triangulations

Of our various examples of maps $m: T \to P$ between spaces and posets outlined in Section 10.1, we have exhausted those for which P is easily seen to be isomorphic to the poset of cells in a regular cell decomposition of T. Now we need to develop a framework for studying the maps $\mu_A: G(r_1, \ldots, r_k, \mathbb{R}^n) \to G(r_1, \ldots, r_k, \mathcal{M})$, including the special case $\mu: G(r, \mathbb{R}^n) \to \text{MacP}(r, n)$.

Definition 10.8 Let P be a poset. The **up topology** on P is the topology in which the open sets are the upper order ideals of P.

It is easy to verify that the up topology is indeed a topology. A map from a topological space to a poset P that is continuous with respect to the up topology on P is sometimes called *upper semi-continuous*, but in this chapter we will simply call it continuous.

Example 10.9 Let P be the poset of cells in a regular cell decomposition of a space T, and let $\tau: T \to P$ be the map sending each $t \in T$ to the smallest cell containing t. Then τ is continuous.

Example 10.10 Let P be a poset of finite rank, and let $\|\Delta P\|$ be a realization of its order complex. For each chain γ in ΔP and each x in the interior of $\|\gamma\|$, let $M(x) = \max(\gamma)$. Then $M: \|\Delta P\| \to P$ is continuous.

Proposition 10.11 *For each rank n oriented matroid $\mathcal{M}$ realized by a signed hyperplane arrangement $\mathcal{A}$ in $\mathbb{R}^n$, the map*

$$\mu_{\mathcal{A}}: G(r_1,\ldots,r_k,\mathbb{R}^n) \to G(r_1,\ldots,r_k,\mathcal{M})$$

is continuous.

Proof: This map is the restriction of a product of maps $\mu_{\mathcal{A}}: G(r_i,\mathbb{R}^n) \to G(r_i,\mathcal{M})$, so it suffices to show that each of these maps is continuous.

For each $\mathcal{N} \in G(r_i,\mathcal{M})$, the complement of $G(r_i,\mathcal{M})_{\geq \mathcal{N}}$ is an order ideal, so by Exercise 7.1 the preimage of this complement is closed. Thus $\mu_{\mathcal{A}}^{-1}(G(r_i,\mathcal{M})_{\geq \mathcal{N}})$ is open. Since every open set in $G(r_i,\mathcal{M})$ is a union of sets of the form $G(r_i,\mathcal{M})_{\geq \mathcal{N}}$, the result follows. □

The map $\mu_{\mathcal{A}}$, viewed as a continuous map between spaces, is the canonical, elegant map promised in Section 10.1. As promised, its codomain is strange. For instance, the up topology on a poset is typically not Hausdorff: If $p < q$ then every open set containing p also contains q. But consider the following remarkable consequence of a theorem of McCord.

Theorem 10.12 (McCord 1967) *Let P be a poset of finite rank, and let $M: \|\Delta P\| \to P$ be the map described in Example 10.10. Then M is a weak homotopy equivalence.*

(A continuous function $f: X \to Y$ is a *weak homotopy equivalence* if each induced map $f_*: \pi_i(X) \to \pi_i(Y)$ of homotopy groups is an isomorphism.)

Example 10.13 Let's return to Example 10.9, in which P is the poset of cells in a regular cell complex with underlying space T. Proposition 4.18 gave

10.4 McCord's Theorem and Triangulations

us a homeomorphism $h\colon T \to \|\Delta P\|$ such that $M \circ h = \tau$. Thus the map $\tau\colon T \to P$ is a weak homotopy equivalence.

Recall that a **triangulation** of a space T is a homeomorphism $\eta\colon \|X\| \to T$ from a geometric simplicial complex to T.

Definition 10.14 Let T be a topological space, $f\colon T \to S$ be a function, and $\eta\colon \|X\| \to T$ be a triangulation. We say that η **respects** f if, for each $\omega \in X$, the composition $f \circ \eta$ is constant on the interior of $\|\omega\|$.

Corollary 10.15 *Let T be a topological space, P a poset, and $f\colon T \to P$ continuous. Let $\eta\colon \|X\| \to T$ be a triangulation that respects f.*

1. *The function $F\colon X \to P$ taking each $\omega \in X$ to the value of $f \circ \eta$ on the interior of ω is a poset map.*
2. *Let $\tilde{f}\colon T \to \|\Delta P\|$ be the composition $T \xrightarrow{\eta^{-1}} \|\Delta X\| \xrightarrow{\|\Delta F\|} \|\Delta P\|$. Then we have a commutative diagram*

$$\begin{array}{ccc} T & \xrightarrow{\tilde{f}} & \|P\| \\ & \searrow f & \downarrow M \\ & & P, \end{array}$$

where M is the map of Example 10.10, and $\tilde{f}$ is a homotopy equivalence if and only if f is a weak homotopy equivalence.

Given one of our maps $\mu_{\mathcal{A}}\colon G(r_1,\ldots,r_k,\mathbb{R}^n) \to G(r_1,\ldots,r_k,\mathcal{M})$, if we can find a triangulation of each $G(r_1,\ldots,r_k,\mathbb{R}^n)$ that respects $\mu_{\mathcal{A}}$ then, via Corollary 10.15, we get both a topological map $\tilde{\mu}_{\mathcal{A}}\colon G(r_1,\ldots,r_k,\mathbb{R}^n) \to \|\Delta G(r_1,\ldots,r_k,\mathcal{M})\|$ and a nice relationship between this map and $\mu_{\mathcal{A}}$.

Finding an explicit such triangulation for each $\mu_{\mathcal{A}}\colon G(r_1,\ldots,r_k,\mathbb{R}^n) \to G(r_1,\ldots,r_k,\mathcal{M})$ would be a horrendous task, but we can appeal to theorems on the existence of triangulations with good properties. The following is a consequence of the *Semi-algebraic Triangulation Theorem* (Hironaka 1975).

Theorem 10.16 *If T is an algebraic variety in $\mathbb{R}^N$ and $\{\Gamma_\alpha : \alpha \in I\}$ a partition of T into finitely many semi-algebraic sets then there is a triangulation $\eta\colon \|X\| \to T$ such that for each $\omega \in X$, the interior of $\|\omega\|$ is mapped into a single Γ_α, and $\eta(\|\omega\|)$ is a semialgebraic set.*

Further, if $\phi\colon \|Y\| \to T$ is another such triangulation then there is a PL homeomorphism $h\colon \|X\| \to \|Y\|$ so that $\eta = \phi \circ h$.

Applying this to the double cover of the Plücker embedding of $G(r_1,\ldots,r_k,\mathbb{R}^n)$ gives us a triangulation $\eta\colon |X| \to G(r_1,\ldots,r_k,\mathbb{R}^n)$

respecting $\mu_{\mathcal{A}}$. Further, if $\nu\colon |Y| \to G(r_1,\ldots,r_k,\mathbb{R}^n)$ is another such triangulation, then the PL homeomorphism promised in the theorem tells us that there is a simplicial complex Z and subdivision maps $\alpha\colon |Z| \to |X|$ and $\beta\colon |Z| \to |Y|$ such that $h = \beta \circ \alpha^{-1}$. It is not hard to check that the maps $\tilde{\mu}_{\mathcal{A}}$ arising from η and from ν are each homotopic to the map $\tilde{\mu}_{\mathcal{A}}$ arising from $\eta \circ \alpha = \nu \circ \beta$. Thus we have the following.

Corollary 10.17 *Let $\mathcal{M}$ be a rank n oriented matroid, $\mathcal{A}$ a realization of $\mathcal{M}$ in $\mathbb{R}^n$, and $M\colon \|\Delta G(r_1,\ldots,r_k,\mathcal{M})\| \to G(r_1,\ldots,r_k,\mathcal{M})$ the map of Example 10.10.*

There is a map $\tilde{\mu}_{\mathcal{A}}\colon G(r_1,\ldots,r_k,\mathbb{R}^n) \to \|\Delta G(r_1,\ldots,r_k,\mathcal{M})\|$, canonical up to homotopy, such that $\tilde{\mu}_{\mathcal{A}} \circ M = \mu_{\mathcal{A}}$. The map $\tilde{\mu}_{\mathcal{A}}$ is a homotopy equivalence if and only if $\mu_{\mathcal{A}}$ is a weak homotopy equivalence.

10.5 Positive Results on Combinatorial Grassmannians

As one might hope, there are strong results on posets of rank 2 oriented matroids, but the proofs are surprisingly tricky. We have the following results from the PhD theses of Abawonse and Babson.

Theorem 10.18 (Abawonse 2023) *$\|\Delta\mathrm{MacP}(2,n)\|$ is homeomorphic to $G(2,\mathbb{R}^n)$, and $\|\Delta\mathrm{MacP}(1,2,n)\|$ is homeomorphic to $G(1,2,\mathbb{R}^n)$.*

Duality gives us similar homeomorphisms $\|\Delta\mathrm{MacP}(n-2,n)\| \cong G(n-2,\mathbb{R}^n)$ and $\|\Delta\mathrm{MacP}(n-2,n-1,n)\| \cong G(n-2,n-1,\mathbb{R}^n)$.

Theorem 10.19 (Babson 1994) *For every $\mathcal{M}$ of rank $n \geq 2$ we have that $\|\Delta G(2,\mathcal{M})\|$ is homotopy equivalent to $G(2,\mathbb{R}^n)$, and $\|\Delta G(1,2,\mathcal{M})\|$ is homotopy equivalent to $G(1,2,\mathbb{R}^n)$.*

The news on homotopy type of MacPhersonians is much better than that which we saw for general combinatorial Grassmannians in Proposition 10.7.

Theorem 10.20 (Anderson 1998) *For all r and n the induced map of homotopy groups $\tilde{\mu}_*\colon \pi_i(G(r,\mathbb{R}^n)) \to \pi_i(\|\Delta\mathrm{MacP}(r,n)\|)$ is an isomorphism if $i \in \{0,1\}$ and a surjection if $i = 2$.*

The proof of the $i = 0$ case is easy: We only need to see that $\|\Delta\mathrm{MacP}(r,n)\|$ is connected. For each r-subset B of $[n]$ let $\mathcal{M}_B$ be the unique element of $\mathrm{MacP}(r,n)$ whose set of nonloops is B. These oriented matroids $\mathcal{M}_B$ are exactly the minimal elements of $\mathrm{MacP}(r,n)$, and certainly for each pair B,C of r-sets there is an $\mathcal{M}_{B,C} \in \mathrm{MacP}(r,n)$ that has both B and C as bases.

10.5 Positive Results on Combinatorial Grassmannians

For each pair $\mathcal{M}, \mathcal{N} \in \mathrm{MacP}(r,n)$, by choosing a basis B of $\mathcal{M}$ and C of $\mathcal{N}$ we get relations $\mathcal{M} \geq \mathcal{M}_B \leq \mathcal{M}_{B,C} \geq \mathcal{M}_C \leq \mathcal{N}$. Thus every two vertices of $\|\Delta \mathrm{MacP}(r,n)\|$ are connected by a path in the 1-skeleton of length at most 4.

The proof that $\tilde{\mu}_*: \pi_i(G(r, \mathbb{R}^n)) \to \pi_i(\|\Delta \mathrm{MacP}(r,n)\|)$ is surjective for $i \in \{1,2\}$ is more work, but injectivity when $i = 1$ comes from elementary algebraic topology. Back in Section 1.5 we defined the *oriented Grassmannian* $OG(r, \mathbb{R}^n)$, whose elements are subspaces of $\mathbb{R}^n$ equipped with an orientation. Define the **oriented MacPhersonian** $O\mathrm{MacP}(r,n)$ to be the poset of all rank r chirotopes on $[n]$. An orientation of V determines a preferred chirotope for $\mu(V)$, so we have $O\mu: OG(r, \mathbb{R}^n) \to O\mathrm{MacP}(r,n)$, leading to a commutative diagram

$$\begin{array}{ccc} OG(r, \mathbb{R}^n) & \xrightarrow{\widetilde{O\mu}} & \|\Delta O\mathrm{MacP}(r,n)\| \\ \downarrow & & \downarrow \\ G(r, \mathbb{R}^n) & \xrightarrow{\tilde{\mu}} & \|\Delta \mathrm{MacP}(r,n)\|, \end{array}$$

where the vertical maps are double covers. $OG(r, \mathbb{R}^n)$ is simply connected, and so $\pi_1(G(r, \mathbb{R}^n)) = \mathbb{Z}_2$. Since the two orientations of an element of $G(r, \mathbb{R}^n)$ map under $O\mu$ to distinct elements of $O\mathrm{MacP}(r,n)$, from the Homotopy Lifting Property of double covers we see that nonzero elements of $\pi_1(G(r, \mathbb{R}^n))$ must map under $\tilde{\mu}$ to nonzero elements of $\pi_1(\|\Delta \mathrm{MacP}(r,n)\|)$.

From the double cover and the result on surjectivity in π_1, we see one more homotopy result.

Corollary 10.21 $\|\Delta O\mathrm{MacP}(r,n)\|$ *is simply connected.*

This offers a first glimpse into the application of algebraic topology to oriented matroids. It fits into a broader approach that's important in the proof of Theorem 10.19 and the following.

Theorem 10.22 (Anderson and Davis 2002) *For all r and n, the map $\tilde{\mu}^*: H^*(\|\Delta \mathrm{MacP}(r,n)\|; \mathbb{Z}_2) \to H^*(G(r,n); \mathbb{Z}_2)$ is a split surjection.*

A **fiber bundle** with **fiber** F is a topological map $f: E \to B$ that locally behaves like $B \times F \to B$: that is, each $b \in B$ has a neighborhood U and a homeomorphism $h_U: U \times F \to f^{-1}(U)$ such that the diagram

$$\begin{array}{ccc} U \times F & \xrightarrow{h_U} & E \\ \downarrow{\pi_1} & & \downarrow{f} \\ U & \hookrightarrow & B \end{array}$$

commutes. We will write $F \hookrightarrow E \xrightarrow{f} B$ for a fiber bundle $f : E \to B$ with fiber F. A fiber bundle has associated to it long exact sequences in homotopy groups and in cohomology that allow us to relate the topologies of F, E, and B. This is often useful as a tool to study a complicated space E in terms of simpler B and F.

Example 10.23 For $r_1 < \cdots < r_k < n$, we have fiber bundles

$$G(r_1, \mathbb{R}^{r_2}) \hookrightarrow G(r_1, \ldots, r_k, \mathbb{R}^n) \to G(r_2, \ldots, r_k, \mathbb{R}^n)$$

and

$$G(r_1, \ldots, r_{k-1}, \mathbb{R}^{r_k}) \hookrightarrow G(r_1, \ldots, r_k, \mathbb{R}^n) \to G(r_k, \mathbb{R}^n).$$

Example 10.24 Associated to each Grassmannian $G(r, \mathbb{R}^n)$, we have the **tautological bundle**

$$\mathbb{R}^r \hookrightarrow \{(V, \mathbf{x}) : V \in G(r, \mathbb{R}^n), \mathbf{x} \in V\} \to G(r, \mathbb{R}^n)$$

and the associated sphere bundle

$$S^{r-1} \hookrightarrow \{(V, \mathbf{x}) : V \in G(r, \mathbb{R}^n), \mathbf{x} \in S(V)\} \to G(r, \mathbb{R}^n).$$

A map of fiber bundles from $f : E \to B$ to $f' : E' \to B'$ is a commutative diagram of continuous maps

$$\begin{array}{ccc} E & \longrightarrow & E' \\ \downarrow f & & \downarrow f' \\ B & \longrightarrow & B'. \end{array}$$

A map of fiber bundles induces maps between their associated long exact sequences, allowing us to compare the topologies of the various spaces.

It is usually too optimistic to hope for a poset map $P \to Q$ to induce a fiber bundle $\|\Delta P\| \to \|\Delta Q\|$. But there is a weaker notion that gets us to many of the same tools.

Definition 10.25 A **quasifibration** is a continuous surjective map $f : E \to B$ such that for each $b \in B$ and $e \in f^{-1}(b)$, each induced map $f_* : \pi_i(E, f^{-1}(b), e) \to \pi_i(B, b)$ is an isomorphism.

Every fiber bundle is a quasifibration, and the same algebraic topology tools we've mentioned hold for quasifibrations and maps of quasifibrations. A special case of *Quillen's Theorem B* gives us a criterion for a map of posets to induce a quasifibration.

Theorem 10.26 (Quillen 1973, Babson 1994) *If $f : P \to Q$ is a poset map such that*

10.5 Positive Results on Combinatorial Grassmannians

1. *for every $p \in P$ and $q \in Q_{\leq f(p)}$, the set $f^{-1}(q) \cap P_{\leq p}$ is contractible, and*
2. *for every $p \in P$ and $q \in Q_{\geq f(p)}$, the set $f^{-1}(q) \cap P_{\geq p}$ is contractible,*

then $\|\Delta f\| \colon \Delta P\| \to \|\Delta Q\|$ is a quasifibration.

In proving Theorem 10.19, Babson showed that the projections $G(1,2,\mathcal{M}) \to G(2,\mathcal{M})$ and $G(1,2,\mathcal{M}) \to G(1,\mathcal{M})$ induce quasifibrations $\|\Delta G(1,2,\mathcal{M})\| \to \|\Delta G(2,\mathcal{M})\|$ and $\|\Delta G(1,2,\mathcal{M})\| \to \|\Delta G(1,\mathcal{M})\|$, and that there were maps of quasifibrations between the corresponding maps of flag spaces of $\mathbb{R}$ to these. (This did not solve the problem by itself; it just narrowed it to a manageable level.) The proof of Theorem 10.22 used the Topological Representation Theorem to show that the combinatorial analog to Example 10.24,

$$\{(\mathcal{M}, X) \colon \mathcal{M} \in \text{MacP}(r,n), X \in \mathcal{V}^*(\mathcal{M}) \setminus \{\mathbf{0}\}\} \to \text{MacP}(r,n),$$

induces a map of order complexes that is a quasifibration, and that there is a map of quasifibrations from the sphere bundle of Example 10.24 to this. This led to a construction of cohomology classes of $\|\Delta \text{MacP}(r,n)\|$ analogous to construction for $G(r, \mathbb{R}^n)$.

11

Hints on Selected Exercises

- Problem 1.37: See chapter 6 in Ziegler (1995) for a thorough development of Gale diagrams.
- Exercise 1.6: In some convex hexagons the diagonals $\overline{14}$, $\overline{25}$, and $\overline{36}$ intersect in a point, while in other convex hexagons these three diagonals have no common intersection point. Show that this distinction leads to a distinction in the oriented matroids of rank 2 subspaces of the corresponding elements of $G(3, \mathbb{R}^6)$.
- Exercise 1.7: Induct on D.
- Exercise 1.8: If you've never met the Vandermonde determinant, now is a good time to look it up.
- Exercise 5.1: Every strong map is a weak map. So, for instance, find a strong map image of $\mathcal{M}$ whose covector set contains no signed cocircuits of $\mathcal{M}$.
- Exercise 5.2: Add one more element to the oriented matroid $\mathcal{M}_1$ in Figure 5.6 that is realized by a point at the intersection of the lines $\overline{12}$ and $\overline{34}$.

  More inspiration might be found in the constructions in Anderson and Davis (2001).
- Exercise 6.4: By inducting on rank(A), reduce to the case $A = \{a\}$. For this restricted case, imagine we had a realization $(\mathcal{H}_e : e \in E \cup f)$ of $\mathcal{N}$ by signed hyperplanes together with a realization $(\mathcal{G}_e : e \in (E \backslash a) \cup f)$ of $(\mathcal{M}/a) \cup f$ in H_a with $G_e = H_e \cap H_a$ for all $e \neq f$. We could get a realization of the desired $\mathcal{M} \cup f$ by rotating H_a slightly around the axis G_f.

  Use this idea to define, for arbitrary $\mathcal{M}$, a localization for $\mathcal{M} \cup f$ in terms of a localization of the extension $(\mathcal{M}/a) \cup f$.
- Exercise 6.11: Apply the result of Exercise 6.10.

- Problem 7.18: The complete matrix is

$$\begin{pmatrix} 0 & 1 & t & t & 0 & t^2 & t & t^2 & 2t-t^2 & 2t-t^2 & t^2+2t & 1 \\ 0 & 1 & t & 1 & t & t & 2t-t^2 & t^2 & 2t-t^2 & 4t-t^2-2 & t^2 & 0 \\ 1 & 1 & 1 & 1 & 1 & 1 & 1 & 1 & 1 & 1 & 1 & 1 \end{pmatrix}.$$

- Exercise 7.1: This can be done either as a consequence of Proposition 5.21 or by a similar proof to that proposition, using the Plücker embedding.
- Exercise 7.20: $\Omega[14^+3^+]$.
- Exercise 7.3: The vertices of a regular octagon, together with its center e.
- Exercise 7.4: Add an element to Ω at the intersection of lines $\{1,5\}$ and $\{2,14\}$.
- Exercise 8.2: One can make a rank 5 example by tweaking $EFM[8]$: extending by a coloop, replacing some elements with lexicographic extensions involving that coloop, then deleting the coloop. Another approach: By Steinitz's Theorem it's enough to find a nonrealizable polytopal rank 3 or 4 oriented matroid.
- Exercise 9.3: For the fourth part, use realizability of rank r oriented matroids with at most $r+2$ elements. For the last part, be careful not to assume that $\prec$ is transitive unless you have a proof. (You don't need transitivity.)

References

Abawonse, Olakunle S. 2023. Triangulations of Grassmannians and flag manifolds. *Geom. Dedicata*, **217**(5), Paper No. 88, 25.

Anderson, Laura. 1998. Homotopy groups of the combinatorial Grassmannian. *Discrete Comput. Geom.*, **20**(4), 549–560.

Anderson, Laura. 1999. Topology of combinatorial differential manifolds. *Topology*, **38**(1), 197–221.

Anderson, Laura. 2001. Representing weak maps of oriented matroids. *European J. Combin.*, **22**(5), 579–586. Combinatorial geometries (Luminy, 1999).

Anderson, Laura. 2019. Vectors of matroids over tracts. *J. Combin. Theory Ser. A*, **161**, 236–270.

Anderson, Laura, and Davis, James F. 2001. There is no tame triangulation of the infinite real Grassmannian. *Adv. Appl. Math.*, **26**(3), 226–236.

Anderson, Laura, and Davis, James F. 2002. Mod 2 cohomology of combinatorial Grassmannians. *Selecta Math. (N.S.)*, **8**(2), 161–200.

Anderson, Laura, and Wenger, Rephael. 1996. Oriented matroids and hyperplane transversals. *Adv. Math.*, **119**(1), 117–125.

Ardila, Federico, Rincón, Felipe, and Williams, Lauren. 2017. Positively oriented matroids are realizable. *J. Eur. Math. Soc.*, **19**(3), 815–833.

Arkani-Hamed, Nima, Bourjaily, Jacob L., Cachazo, Freddy, Goncharov, Alexander B., Postnikov, Alexander, and Trnka, Jaroslav. 2014. *Scattering amplitudes and the positive Grassmannian*, https://arxiv.org/abs/1212.5605.

Arkani-Hamed, Nima, Bourjaily, Jacob, Cachazo, Freddy, Goncharov, Alexander B., Postnikov, Alexander, and Trnka, Jaroslav. 2016. *Grassmannian geometry of scattering amplitudes*. Cambridge: Cambridge University Press.

Babson, Eric. 1994. A combinatorial flag space. PhD thesis, MIT.

Bachem, Achim, and Kern, Walter. 1986. Adjoints of oriented matroids. *Combinatorica*, **6**(4), 299–308.

Bachem, Achim, and Kern, Walter. 1992. *Linear programming duality*. Berlin: Springer-Verlag.

Bachem, Achim, and Wanka, Alfred. 1988. Separation theorems for oriented matroids. *Discrete Math.*, **70**(3), 303–310.

Baker, Matthew, and Bowler, Nathan. 2019. Matroids over partial hyperstructures. *Adv. Math.*, **343**, 821–863.

Barmak, Jonathan Ariel. 2011. On Quillen's Theorem A for posets. *J. Combin. Theory Ser. A*, **118**(8), 2445–2453.
Bayer, Margaret, and Sturmfels, Bernd. 1990. Lawrence polytopes. *Canad. J. Math.*, **42**(1), 62–79.
Belkale, Prakash, and Brosnan, Patrick. 2003. Matroids, motives, and a conjecture of Kontsevich. *Duke Math. J.*, **116**(1), 147–188.
Billera, Louis J., and Munson, Beth Spellman. 1984a. Oriented matroids and triangulations of convex polytopes. In *Progress in combinatorial optimization (Waterloo, Ont., 1982)*. Toronto, ON: Academic Press, pp. 27–37.
Billera, Louis J., and Munson, Beth Spellman. 1984b. Polarity and inner products in oriented matroids. *European J. Combin.*, **5**(4), 293–308.
Billera, Louis J., and Munson, Beth Spellman. 1984c. Triangulations of oriented matroids and convex polytopes. *SIAM J. Algebraic Discrete Meth.*, **5**(4), 515–525.
Biss, Daniel K. 2003. The homotopy type of the matroid Grassmannian. *Ann. Math. (2)*, **158**(3), 929–952.
Biss, Daniel K. 2009. Erratum to "The homotopy type of the matroid Grassmannian" [MR2031856]. *Ann. Math. (2)*, **170**(1), 493.
Björner, Anders. 1995. Topological methods. In *Handbook of combinatorics, Vol. 1, 2*, (eds.) R. L. Graham, M. Grötschel, and L. Lovász. Amsterdam: Elsevier, pp. 1819–1872.
Björner, Anders, Las Vergnas, Michel, Sturmfels, Bernd, White, Neil, and Ziegler, Günter M. 1999. *Oriented matroids*, 2nd ed. Encyclopedia of Mathematics and Its Applications, vol. 46. Cambridge: Cambridge University Press.
Bland, Robert G. 1977. A combinatorial abstraction of linear programming. *J. Combin. Theory Ser. B*, **23**(1), 33–57.
Bland, Robert G., and Dietrich, Brenda L. 1988. An abstract duality. *Discrete Math.*, **70**(2), 203–208.
Bland, Robert G., and Las Vergnas, Michel. 1978. Orientability of matroids. *J. Combin. Theory Ser. B*, **24**(1), 94–123.
Bland, Robert G., and Las Vergnas, Michel. 1979. Minty colorings and orientations of matroids. In *Second International Conference on Combinatorial Mathematics (New York, 1978)*. Annals of the New York Academy of Sciences, vol. 319. New York: New York Academy of Sciences, pp. 86–92.
Boege, T. n.d. *On stable equivalence of semialgebraic sets*. https://taboege.de/blog/2022/02/On-stable-equivalence-of-semialgebraic-sets, accessed: 2023-02-08.
Bokowski, Jürgen G. 2006. *Computational oriented matroids*. Cambridge: Cambridge University Press.
Bokowski, Jürgen, and Sturmfels, Bernd. 1989. *Computational synthetic geometry*. Lecture Notes in Mathematics, vol. 1355. Berlin: Springer-Verlag.
Bruggesser, H., and Mani, P. 1971. Shellable decompositions of cells and spheres. *Math. Scand.*, **29**, 197–205.
Büchi, J. Richard, and Fenton, William E. 1988. Large convex sets in oriented matroids. *J. Combin. Theory Ser. B*, **45**(3), 293–304.
Buchstaber, Victor M., and Panov, Taras E. 2002. *Torus actions and their applications in topology and combinatorics*. University Lecture Series, vol. 24. Providence, RI: American Mathematical Society.

Cordovil, Raul. 1987. Polarity and point extensions in oriented matroids. *Linear Algebra Appl.*, **90**, 15–31.
da Silva, Ilda. 1987. Quelques propriétés des matroïdes orientés. PhD thesis, Université de Paris VI.
da Silva, Ilda. 1995. Axioms for maximal vectors of an oriented matroid; a combinatorial characterization of the regions determined by an arrangement of pseudohyperplanes. *European J. Combin.*, **16**(2), 125–145.
Dobbins, Michael Gene. 2011. Representations of polytopes. PhD thesis, Temple University.
Edelman, Paul H. 1984. The acyclic sets of an oriented matroid. *J. Combin. Theory Ser. B*, **36**(1), 26–31.
Folkman, Jon, and Lawrence, Jim. 1978. Oriented matroids. *J. Combin. Theory Ser. B*, **25**(2), 199–236.
Fukuda, Komei. 1982. Oriented matroid programming. PhD thesis, University of Waterloo (Canada).
Fukuda, Komei, and Tamura, Akihisa. 1990. Dualities in signed vector systems. *Portugal. Math.*, **47**(2), 151–165.
Galashin, Pavel, Karp, Steven N., and Lam, Thomas. 2022. The totally nonnegative Grassmannian is a ball. *Adv. Math.*, **397**, Paper No. 108123, 23.
Geelen, Jim, Gerards, Bert, and Whittle, Geoff. 2014. Solving Rota's conjecture. *Notices Amer. Math. Soc.*, **61**(7), 736–743.
Gel'fand, Israel M., and MacPherson, Robert D. 1982. Geometry in Grassmannians and a generalization of the dilogarithm. *Adv. Math.*, **44**(3), 279–312.
Gel'fand, Israel M., and MacPherson, Robert D. 1992. A combinatorial formula for the Pontrjagin classes. *Bull. Amer. Math. Soc. (N.S.)*, **26**(2), 304–309.
Gel'fand, Israel M., and Serganova, Vera V. 1987. Combinatorial geometries and the strata of a torus on homogeneous compact manifolds. *Uspekhi Mat. Nauk*, **42**(2), 107–134.
Gel'fand, Israel M., Goresky, R. Mark, MacPherson, Robert D., and Serganova, Vera V. 1987. Combinatorial geometries, convex polyhedra, and Schubert cells. *Adv. Math.*, **63**(3), 301–316.
Goodman, Jacob E., and Pollack, Richard. 1980a. On the combinatorial classification of nondegenerate configurations in the plane. *J. Combin. Theory Ser. A*, **29**(2), 220–235.
Goodman, Jacob E., and Pollack, Richard. 1980b. Proof of Grünbaum's conjecture on the stretchability of certain arrangements of pseudolines. *J. Combin. Theory Ser. A*, **29**(3), 385–390.
Hatcher, Allen. 2002. *Algebraic topology*. Cambridge: Cambridge University Press.
Higgs, Denis A. 1968. Strong maps of geometries. *J. Combin. Theory*, **5**, 185–191.
Hironaka, Heisuke. 1975. Triangulations of algebraic sets. In *Algebraic geometry (Proc. Sympos. Pure Math., Vol. 29, Humboldt State Univ., Arcata, Calif., 1974)*. Providence, RI: American Mathematical Society, pp. 165–185.
Hudson, John F. P. 1969. *Piecewise linear topology*. University of Chicago Lecture Notes prepared with the assistance of J. L. Shaneson and J. Lees. New York/Amsterdam: W. A. Benjamin.
Jaggi, Beat, and Mani-Levitska, Peter. 1988. *A simple arrangement of lines without the isotopy property*. preprint, Univ. Bern. 28 pages.

Jaggi, Beat, Mani-Levitska, Peter, Sturmfels, Bernd, and White, Neil. 1989. Uniform oriented matroids without the isotopy property. *Discrete Comput. Geom.*, **4**(2), 97–100.

Jarra, Manoel, and Lorscheid, Oliver. 2024. Flag matroids with coefficients. Adv. Math. **436**, Paper No. 109396, 46, DOI: 10.1016/j.aim.2023.109396.

Kapovich, Michael, and Millson, John J. 1998. On representation varieties of Artin groups, projective arrangements and the fundamental groups of smooth complex algebraic varieties. *Inst. Hautes Études Sci. Publ. Math.*, **88**, 5–95.

Kung, Joseph P. S. 1983. A characterization of orthogonal duality in matroid theory. *Geom. Dedicata*, **15**(1), 69–72.

Kung, Joseph P. S., Rota, Gian-Carlo, and Yan, Catherine H. 2009. *Combinatorics: the Rota way*. Cambridge Mathematical Library. Cambridge: Cambridge University Press.

Lafforgue, L. 2003. *Chirurgie des grassmanniennes*. CRM Monograph Series, vol. 19. Providence, RI: American Mathematical Society.

Las Vergnas, Michel. 1975a. *Coordinatizable strong maps of matroids*. preprint, 123 pages.

Las Vergnas, Michel. 1975b. Matroïdes orientables. *C. R. Acad. Sci. Paris Sér. A-B*, **280**, Ai, A61–A64.

Las Vergnas, Michel. 1978. Extensions ponctuelles d'une géométrie combinatoire orientée. In *Problèmes combinatoires et théorie des graphes (Colloq. Internat. CNRS, Univ. Orsay, Orsay, 1976)*. Colloq. Internat. CNRS, vol. 260. Paris: CNRS, pp. 265–270.

Las Vergnas, Michel. 1980. Convexity in oriented matroids. *J. Combin. Theory Ser. B*, **29**(2), 231–243.

Las Vergnas, Michel. 1984. Oriented matroids as signed geometries real in corank 2. In *Finite and infinite sets, Vol. I, II (Eger, 1981)*, (eds.) András Hajnal, László Lovász, and Vera T. Sós. Colloq. Math. Soc. János Bolyai, vol. 37. North-Holland, Amsterdam, pp. 555–565.

Lawrence, Jim. 1983. Lopsided sets and orthant-intersection by convex sets. *Pacific J. Math.*, **104**(1), 155–173.

Lawrence, Jim. 1984. *Shellability of oriented matroid complexes*. preprint, 8 pages.

Lee, Seok Hyeong, and Vakil, Ravi. 2013. Mnëv-Sturmfels universality for schemes. In *A celebration of algebraic geometry*. Clay Math. Proc., vol. 18. Providence, RI: American Mathematical Society, pp. 457–468.

Levi, Friedrich. 1926. Die Teilung der projektiven Ebene durch Gerade oder Pseudogerade. *Ber. Math.-Phys. Kl. Sachs. Akad. Wiss.*, **78**, 256–267.

Liu, Gaku. 2020. A counterexample to the extension space conjecture for realizable oriented matroids. *J. Lond. Math. Soc. (2)*, **101**(1), 175–193.

MacPherson, Robert. 1993. Combinatorial differential manifolds. In *Topological methods in modern mathematics (Stony Brook, NY, 1991)*. Houston, TX: Publish or Perish, pp. 203–221.

Mandel, Arnaldo. 1982. *Topology of oriented matroids*. PhD thesis, University of Waterloo.

Mayhew, Dillon, Whittle, Geoff, and Newman, Mike. 2014. Is the missing axiom of matroid theory lost forever? *Q. J. Math.*, **65**(4), 1397–1415.

Mayhew, Dillon, Newman, Mike, and Whittle, Geoff. 2018. Yes, the 'missing axiom' of matroid theory is lost forever. *Trans. Amer. Math. Soc.*, **370**(8), 5907–5929.

McCord, Michael C. 1967. Homotopy type comparison of a space with complexes associated with its open covers. *Proc. Amer. Math. Soc.*, **18**, 705–708.

Milnor, John W., and Stasheff, James D. 1974. *Characteristic classes*. Annals of Mathematics Studies, No. 76. Princeton, NJ: Princeton University Press; Tokyo: University of Tokyo Press.

Minty, George J. 1966. On the axiomatic foundations of the theories of directed linear graphs, electrical networks and network-programming. *J. Math. Mech.*, **15**, 485–520.

Mnëv, Nikolai E. Brief history of Universality theorem for moduli spaces of line arrangements. www.pdmi.ras.ru/ mnev/bhu.html, accessed: March 30, 2022.

Mnëv, Nikolai E. 1986. The topology of configuration varieties and convex polytopes varieties. PhD thesis, Leningrad State University. [In Russian]

Mnëv, Nikolai E. 1988. The universality theorems on the classification problem of configuration varieties and convex polytopes varieties. In *Topology and geometry— Rohlin Seminar*, (eds.) Oleg Yanovich Viro and Anatoly Moiseevich Vershik. Lecture Notes in Math., vol. 1346. Berlin: Springer, pp. 527–543.

Mnëv, Nicolai E., and Richter-Gebert, Jürgen. 1993. Two constructions of oriented matroids with disconnected extension space. *Discrete Comput. Geom.*, **10**(3), 271–285.

Munkres, James R. 1984. *Elements of algebraic topology*. Menlo Park, CA: Addison-Wesley.

Munson, Beth Spellman. 1981. Face lattices of oriented matroids. PhD thesis, Cornell University.

Oxley, James G. 1992. *Matroid theory*. Oxford Science Publications. New York: The Clarendon Press; Oxford University Press.

Postnikov, Alexander. 2006. *Total positivity, Grassmannians, and networks*. https://arxiv.org/abs/math/0609764

Postnikov, Alexander, Speyer, David, and Williams, Lauren. 2009. Matching polytopes, toric geometry, and the totally non-negative Grassmannian. *J. Algebraic Combin.*, **30**(2), 173–191.

Quillen, Daniel. 1973. Higher algebraic K-theory. I. In *Algebraic K-theory, I: Higher K-theories (Proc. Conf., Battelle Memorial Inst., Seattle, Wash., 1972)*. Lecture Notes in Math., vol. 341. Berlin: Springer, pp. 85–147.

Rambau, Jörg. 2002. Circuit admissible triangulations of oriented matroids. Geometric combinatorics (San Francisco, CA/Davis, CA, 2000). *Discrete Comput. Geom.*, **27**, 155–161.

Richter-Gebert, Jürgen. 1988. Kombinatorische Realisierbarkeitskriterien für orientierte Matroide. MPhil thesis, TU Darmstadt.

Richter-Gebert, Jürgen. 1989. Kombinatorische Realisierbarkeitskriterien für orientierte Matroide. (Criteria for the realizability of oriented matroids). *Mitt. Math. Semin. Gießen*, **194**, 1–112.

Richter-Gebert, Jürgen. 1993. Oriented matroids with few mutations. *Discrete Comput. Geom.*, **10**(3), 251–269.

Richter-Gebert, Jürgen. 1996a. *Realization spaces of polytopes*. Lecture Notes in Mathematics, vol. 1643. Berlin: Springer-Verlag.

Richter-Gebert, Jürgen. 1996b. Two interesting oriented matroids. *Doc. Math.*, **1**, No. 07, 137–148.

Richter-Gebert, Jürgen and Ziegler, Günter M. 1994. Zonotopal tilings and the Bohne-Dress theorem. In *Jerusalem combinatorics '93*. Contemp. Math., vol. 178. Providence, RI: American Mathematical Society, pp. 211–232.

Ringel, Gerhard. 1956. Teilungen der Ebene durch Geraden oder topologische Geraden. *Math. Z.*, **64**, 79–102.

Rockafellar, R. Tyrell. 1969. The elementary vectors of a subspace of R^N. In *Combinatorial mathematics and its applications (Proc. Conf., Univ. North Carolina, Chapel Hill, N.C., 1967)*. Chapel Hill, NC: University of North Carolina Press, pp. 104–127.

Rourke, Colin Patrick, and Sanderson, Brian Joseph. 1982. *Introduction to piecewise-linear topology*. Springer Study Edition. Berlin/New York: Springer-Verlag.

Santos, Francisco. 2002. Triangulations of oriented matroids. *Mem. Amer. Math. Soc.*, **156**(741), viii+80.

Scott, Joshua S. 2006. Grassmannians and cluster algebras. *Proc. London Math. Soc. (3)*, **92**(2), 345–380.

Seymour, Paul D. 1979. Matroid representation over GF(3). *J. Combin. Theory Ser. B*, **26**(2), 159–173.

Shor, Peter W. 1991. Stretchability of pseudolines is NP-hard. In *Applied geometry and discrete mathematics*, (eds.) Bernd Sturmfels and Peter Gritzman. DIMACS Ser. Discrete Math. Theoret. Comput. Sci., vol. 4. Providence, RI: American Mathematical Society. pp. 531–554.

Stanley, Richard P. 1997. *Enumerative combinatorics. Vol. 1*. Cambridge Studies in Advanced Mathematics, vol. 49. Cambridge: Cambridge University Press. With a foreword by Gian-Carlo Rota, corrected reprint of the 1986 original.

Sturmfels, Bernd. 1988. Some applications of affine Gale diagrams to polytopes with few vertices. *SIAM J. Discrete Math.*, **1**(1), 121–133.

Sturmfels, Bernd. 1989. On the matroid stratification of Grassmann varieties, specialization of coordinates, and a problem of N. White. *Adv. Math.*, **75**(2), 202–211.

Sturmfels, Bernd, and Ziegler, Günter M. 1993. Extension spaces of oriented matroids. *Discrete Comput. Geom.*, **10**(1), 23–45.

Suvorov, P. 1988. Isotopic but not rigidly isotopic plane systems of straight lines. In *Topology and geometry—Rohlin Seminar*, (eds.) Oleg Yanovich Viro. Lecture Notes in Math., vol. 1346. Berlin: Springer. pp. 545–556.

Tsukamoto, Yasuyuki. 2013. New examples of oriented matroids with disconnected realization spaces. *Discrete Comput. Geom.*, **49**(2), 287–295.

Tutte, W. T. 1958. A homotopy theorem for matroids. I, II. *Trans. Amer. Math. Soc.*, **88**, 144–174.

Vakil, Ravi. 2006. Murphy's law in algebraic geometry: badly-behaved deformation spaces. *Invent. Math.*, **164**(3), 569–590.

Vámos, Peter. 1978. The missing axiom of matroid theory is lost forever. *J. London Math. Soc. (2)*, **18**(3), 403–408.

Verkama, Emil. 2023. Repairing the Universality Theorem for 4-polytopes. MPhil thesis, Aalto University.

Vershik, Anatoliĭ M. 1988. Topology of the convex polytopes' manifolds, the manifold of the projective configurations of a given combinatorial type and representations of lattices. In *Topology and geometry—Rohlin Seminar*. Lecture Notes in Math., vol. 1346, (eds.) Oleg Yanovich Viro. Berlin: Springer. pp. 557–581.

Walker, James W. 1981. Homotopy type and Euler characteristic of partially ordered sets. *European J. Combin.*, **2**(4), 373–384.

White, Neil L. 1989. A nonuniform matroid which violates the isotopy conjecture. *Discrete Comput. Geom.*, **4**(1), 1–2.

Whitney, Hassler. 1935. On the abstract properties of linear dependence. *Amer. J. Math*, **57**(3), 509–533.

Winder, Robert O. 1966. Partitions of N-space by hyperplanes. *SIAM J. Appl. Math.*, **14**, 811–818.

Wu, Pei. 2021. A counterexample to Las Vergnas' strong map conjecture on realizable oriented matroids. *Discrete Comput. Geom.*, **65**(1), 143–149.

Zaslavsky, Thomas. 1975. Facing up to arrangements: face-count formulas for partitions of space by hyperplanes. *Mem. Amer. Math. Soc.*, **1**(1), vii+102.

Zaslavsky, Thomas. 1977. A combinatorial analysis of topological dissections. *Adv. Math.*, **25**(3), 267–285.

Živaljević, Rade T. 2016. On the continuous combinatorics of posets, polytopes and matroids. *Fundam. Prikl. Mat.*, **21**(6), 143–164.

Ziegler, Günter M. 1991. Some minimal nonorientable matroids of rank three. *Geom. Dedicata*, **38**(3), 365–371.

Ziegler, Günter M. 1995. *Lectures on polytopes*. Graduate Texts in Mathematics, vol. 152. New York: Springer-Verlag.

Index

acyclic, 93
acyclic cone, 264
acyclic reorientations, 94
adjoint, 178, 248
affine hyperplane, 9
affine oriented matroid, 189
affine polar, 244
affine representation of a vector
 arrangement, 11
affine space, 9
 of an oriented matroid, 189
affine span, 9
affine subspace, 9
Alexander Horned Sphere, 106
alternating oriented matroid, 41
antiparallel, 94, 126
 in an affine oriented matroid, 190
arrangement, 10
atom, 105
augmented face poset, 107, 124, 261

basis axioms for matroids, 79
basis of a chirotope, 68
Basis Exchange Axiom, Signed, 65
Basis Exchange Principle, 31, 33, 65
basis of an oriented matroid, 66
basis orientation, 15
Betsy Ross arrangement, 36
Bohne–Dress Theorem, 299
boundary, 163
bounded poset, 98, 105

$\mathcal{C}$, 4, 34, 45, 50
$\mathcal{C}^*$, 4, 34
Carathéodory's Theorem, 42, 91

$C(X)$, 113
chain, 100, 101
characteristic polynomial, 131
χ, 4, 34, 64, 69
chirotope, 4, 60, 64, 82, 130
 of a contraction, 90
Circuit Elimination Axiom, 45
circuit set of a matroid, 79
circuit-compatible, 267
circuits, signed, 4
closed orthant, 3
closed walk, 192
closure operator, 81
coatom, 105, 108, 110
cocircuit graph, 159, 191
cocircuits, signed, 4
collapse, 286, 287
collapsible, 286
collapsing sequence, 287
coloop, 78, 127
combinatorial Grassmannian, 298, 299, 304
combinatorial type of a convex polytope, 239
comodular, 79, 161
Composition Axiom, 44
composition of sign vectors, 25
conformal, 26, 51
connected, 96
contractible carrier, 148
Contractible Carrier Lemma, 148
contraction, 54, 129
 in realizable oriented matroids, 38
convex hull of a set of elements, 81, 129, 246, 267
convex hull of a set of covectors, 159
convex polytope, 10, 243

317

convex set of covectors, 104, 113, 242
convex set of elements, 134
coordinate oriented matroid, 77, 298
coproduct, 157
corank of a poset element, 101
corank of an oriented matroid, 67
covectors, 4
   of an oriented matroid, 59
covering relation, 51, 105
cryptomorphic, 24

decomposition of an arrangement of rays, 264
decreasing in an affine oriented matroid, 191
defining signed hyperplane, 243
degenerate hyperplane, 8
deletion, 54, 129
   in realizable oriented matroids, 38
dependence in a chirotope, 68
direct sum, 96, 128
direction in an affine oriented matroid, 191
dot product of sign vectors, 27
double of an oriented matroid, 277
dual, 18, 59
   of a realizable oriented matroid, 34
dual pairs, 28
Dual Pivoting Property, 31, 33, 68, 69

$\mathcal{E}(\mathcal{M})$, 182
edge of $G(\mathcal{M})$, 159
Edmonds–Fukuda–Mandel oriented matroid, 205
Edmonds–Mandel face lattice, 246, 263
elementary collapse, 287
elements of a realizable oriented matroid, 34
elements of an oriented matroid, 59
elimination of an element, 45
EM($\mathcal{M}$), 246
Euclidean, 155
Euclidean oriented matroid, 188
Euclidean oriented matroid program, 192
extension, 96
extension poset, 182
extension space, 182, 296
Extension Space Conjecture, 183, 299
extension-compatible, 267
exterior algebra, 19
exterior product, 21

face lattice of a polytope, 243
face lattice of an oriented matroid, 98
face of a cone, 264

face of a polytope, 10, 243
face of a set of elements, 266
face of an oriented matroid, 245, 266
facet, 243
   of a convex polytope, 118
   of a covector, 118
fan, 259, 264
   decomposition, 269
   on an oriented matroid, 267
Fano plane, 135
Farkas Lemma in $\mathbb{R}^r$, 28
Farkas Property, 55
feasible region, 197
fiber bundle, 305
filter, 105
fixed element, 219
fixing sequence, 219
flag poset, 298
flag space, 298
flags in $\mathbb{R}^n$, 298
flags, OM, 298
flat, 99, 126
framing of a strong map, 154, 212
framing of a subspace, 153

Gale diagram, 35, 242, 252
general position for an element, 177
generalized Euclidean property, 186
good cover, 131
good path, 169
good path pair, 169
$G(r, \mathbb{R}^n)$, 17
graded, 105
graph of an affine oriented matroid, 191
Grassmann–Plücker relations, 63
   3-term, 80
Grassmannian, 17, 23, 24, 143, 239, 297
Grassmannian poset, 154
Grassmannian realization space, 90, 215, 239

Hahn–Banach separation theorem, 212
Helly's theorem, 213
homothety, 217
hyperoperation, 27
hyperplane, 8
   arrangements, 8
   at infinity, 190
   of a chirotope, 68
   of an oriented matroid, 67
   realization of a strong map, 139
hypersum of signs, 27

Index

Incomparability Axiom, 45
increasing in an affine oriented matroid, 191
independence in a chirotope, 68
independence in oriented matroids, 66
inner product of sign vectors, 28
interior of a convex set of covectors, 164
intersection property, 185, 188
isomorphism of matroids, 92

$\Lambda(\mathcal{M})$, 251
Las Vergnas condition, 161, 300
Las Vergnas face lattice, 245, 263
lattice, 107
Lawrence construction, 37, 245, 278
length of a chain, 101
Levi Enlargement Lemma, 155, 186, 212
Levi's intersection property, 186
lexicographic extension, 176
lifting, 97, 205, 255, 273
linear extension, 105
linear independence of a vector arrangement, 31
linear programming, 158, 197
linked spheres, 128
localization, 160
loop, 78, 126
$LV(\mathcal{M})$, 245

MacPhersonian, 2, 70, 297
$\mathcal{M}_{adj}$, 178
$Mat(r, n)$, 3
matrix realization of a strong map, 139
$\mathcal{M}_{coord}$, 77
$\mathcal{M}[e_1^{s_1}, \ldots, e_k^{s_k}]$, 176
$\mathcal{M}_D$, 277
minor, 54
$\mathcal{M}^\square$, 247

negative half-space, 9
nerve, 131
non-Euclidean cycle, 193, 279
non-Pappus oriented matroid, 139

objective function, 197
OM of an arrangement of rays, 263
Ω_{14}^+, 221
order complex, 105, 148
order ideal, 105
orientable, 135
orientable matroid, 39
oriented Grassmannian, 24, 305

oriented MacPhersonian, 305
oriented matroid polytope, 242, 247
oriented matroid program, 191
orthant, 3
  in projective space, 238
orthogonality of sign vectors, 28

$P^\square$, 244
Pappus's Theorem, 139
paralellism class, 94
parallel, 94, 126
  in an affine oriented matroid, 190
  pseudohyperplanes, 190
partial recursive triangulation, 270
partial triangulation of a point set, 258
partially directed walk, 192
path in $G(\mathcal{M})$, 159
piecewise-linear, 108
Pivoting Property, 31, 32, 68, 69
PL ball, 260
PL cell complex, 260
PL cell decomposition, 261
PL homeomorphic, 260
PL manifold, 262
  with boundary, 262
PL sphere, 260
PL subdivision, 261
Plücker embedding, 22
polar, 242–244
  to an oriented matroid, 247
polarity, 243
polyhedron, 260
polytopal complex, 261
polytopal decomposition, 258
polytopal set of covectors, 166, 242
positive half-space, 9
positive sign vector, 93
positively oriented matroid, 239
positroid, 42, 239
preliminary path, 172
preliminary path pair, 172
product, 157
projective basis, 216
projective realization space, 216
proper lifting, 97
proper single-element extension, 96
pseudo-radial projection, 265
pseudocircle, 129, 160
pseudoconfiguration of points, 84
pseudohyperplane, 190
pseudoline, 190

Index

pseudosphere, 10, 86, 106
pseudosubspace, 190
pulling a vertex, 273
pure cell complex, 107
pure poset, 101, 107

quasifibration, 306
Quillen's Theorem B, 306

$\mathcal{R}(\mathcal{M})$, 216
Radon's theorem, 213
rank
　of a matroid, 79
　of an oriented matroid, 65, 67
　of a poset, 101
　of a poset element, 101
　of a realizable oriented matroid, 13
rank-preserving weak map, 142
rationally equivalent, 225
realizability
　in small rank, 90
　of strong maps, 139
　of strong maps, weak, 139
　of a weak map, strong, 145
　of a weak map, weak, 145
realizable, 2
realizable oriented matroid, 34
realization, 2
　of an oriented matroid fan, 269
　of a strong map, 139
　space, 23, 146, 216
　space of a convex polytope, 240
recursive triangulation, 271
recursive coatom ordering, 104, 110, 261, 287
reduced row-echelon form, 6, 62, 180
reducible element, 219
reduction sequence, 219
regular cell, 107
　complex, 107
　decomposition, 107
relative interior, 9
reorientation of a sign vector, 93
reorientation of an oriented matroid, 93, 126
$\mathcal{RP}(\mathcal{M})$, 216

$S(X, Y)$, 6
saturated chain, 101
Semi-algebraic Triangulation Theorem, 303
separation set, 6
sequence induced by a chain, 118
shellable, 109

shelling, 104, 108
Shor semialgebraic set, 227
$\Sigma_\mathcal{M}(\mathcal{T}_1, \mathcal{T}_2)$, 277
signed affine hyperplane, 9
signed affine point arrangement, 35
signed circuits of an oriented matroid, 45
signed cocircuits of an oriented matroid, 59
signed hyperplane, 9
　realization space, 214
simple, 94
Simplex Algorithm, 197, 198
simplicial complex, abstract, 105
simplicial complex, geometric, 105
simplicial cone, 265
simplicial fan, 265
　on an oriented matroid, 267
single-element extension, 96
skeleton, 108
sphere at infinity, 190
stable equivalence, 226
stable projection, 226
standard presentation of a PL cell complex, 262
stellar subdivision, 263
　in an oriented matroid fan, 271
strong fan, 267
Strong Circuit Elimination, 46
strong fan decomposition, 269
strong map, 136, 212, 298
strong map image, 136
strong triangulation, 269
strongly Euclidean, 299
subdivision of an oriented matroid, 275
submatroid, induced, 54
subspace realization of a strong map, 139
subthin, 104, 110
superposition, 261, 277
Sylvester–Gallai Theorem, 134
Symmetry Axiom, 44, 45

$\mathcal{T}(\mathcal{M})$, 114
$\mathcal{T}(\mathcal{M}, B)$, 116
tensor product, 20
thin, 104, 110
thin Schubert cells, 237
tope, 80, 93, 104, 114
tope poset, 116, 155
Topological Representation Theorem, 259, 261, 300, 307
total order, 105
totally cyclic, 93

totally nonnegative part of the Grassmannian, 239
triangulation
　of an arrangement of rays, 266
　of a manifold, 293
　of an oriented matroid, 269
　of a point set, 258
　of a space, 303

underlying oriented matroid of a fan, 264
underlying space of a PL cell complex, 261
underlying space of a regular cell complex, 107
uniform, 95, 212
universal property of coproducts, 157
universal property of products, 157
Universality Theorem, 214, 297
unsigned circuits of a chirotope, 68
unsigned cocircuits of a chirotope, 68
upper semi-continuous, 302

$\mathcal{V}$, 4, 34, 44, 50, 58
Vámos matroid, 199
$\mathcal{V}^*$, 4, 34, 58
vector arrangement, 4, 8
Vector Elimination Axiom, 44
vector of an oriented matroid, 4, 44, 129
vector realization of a strong map, 139
vector realization space, 147
vertex of a cone, 264
vertex of a PL cell complex, 261
vertex of a polytope, 10, 243
vertex set, 266
von Staudt constructions, 237

walk, 192
weak homotopy equivalence, 302
weak map, 142
weird, 185
wholly undirected walk, 192

Printed in the United States
by Baker & Taylor Publisher Services